T0183265

Lecture Notes in Artificial Intelligence 13467

Subseries of Lecture Notes in Computer Science

Series Editors

Randy Goebel
University of Alberta, Edmonton, Canada

Wolfgang Wahlster
DFKI, Berlin, Germany

Zhi-Hua Zhou
Nanjing University, Nanjing, China

Founding Editor

Jörg Siekmann
DFKI and Saarland University, Saarbrücken, Germany

More information about this subseries at https://link.springer.com/bookseries/1244

Kevin Buzzard · Temur Kutsia (Eds.)

Intelligent Computer Mathematics

15th International Conference, CICM 2022
Tbilisi, Georgia, September 19–23, 2022
Proceedings

Springer

Editors
Kevin Buzzard (ID)
Imperial College London
London, UK

Temur Kutsia (ID)
Johannes Kepler University Linz
Linz, Austria

ISSN 0302-9743 ISSN 1611-3349 (electronic)
Lecture Notes in Artificial Intelligence
ISBN 978-3-031-16680-8 ISBN 978-3-031-16681-5 (eBook)
https://doi.org/10.1007/978-3-031-16681-5

LNCS Sublibrary: SL7 – Artificial Intelligence

This Springer imprint is published by the registered company Springer Nature Switzerland AG
The registered company address is: Gewerbestrasse 11, 6330 Cham, Switzerland

Preface

This volume contains the papers presented at the 15th Conference on Intelligent Computer Mathematics (CICM 2022), held during September 19–23, 2022, in Tbilisi, Georgia. CICM was part of the Computational Logic Autumn Summit (CLAS 2022), which took place during September 19–30, 2022, and was hosted by Ivane Javakhishvili Tbilisi State University.

With the continuing, rapid progress of digital methods in communications, knowledge representation, processing, and discovery, the unique character and needs of mathematical information require unique approaches. Separate communities have developed theoretical and practical solutions for these challenges including computation, deduction, narration, and data management. CICM brings these communities together, offering a venue for discussing and developing solutions to problems posed by the integration of these diverse areas.

CICM was initially formed in 2008 as a joint meeting of communities involved in computer algebra systems, theorem provers, and mathematical knowledge management, as well as those involved in a variety of aspects of scientific document archives. Since then, the conference has been held annually: Birmingham (UK, 2008), Grand Bend (Canada, 2009), Paris (France, 2010), Bertinoro (Italy, 2011), Bremen (Germany, 2012), Bath (UK, 2013), Coimbra (Portugal, 2014), Washington D.C. (USA, 2015), Białystok (Poland, 2016), Edinburgh (UK, 2017), Hagenberg (Austria, 2018), Prague (Czech Republic, 2019), Bertinoro (Italy, 2020, virtual), and Timişoara (Romania, 2021, virtual). CICM 2022 was organized in hybrid mode.

CICM 2022 solicited 37 formal submissions. Each submission was assigned to at least three Program Committee (PC) members and was reviewed in single-blind mode. The committee decided to accept 24 papers including 17 regular ones, one project/survey paper, four short papers describing software systems and datasets, and two papers highlighting the development of systems and tools in the last year. Besides the formal submissions, eight papers describing work in progress were also presented at the conference.

The reviewing process included a response period in which the authors could respond and clarify points raised by the reviewers. In addition to the main sessions, a special doctoral program was organized, providing a forum for PhD students to present their research and get advice from senior members of the community. The EuroProofNet workshop on the development, maintenance, refactoring, and search of large libraries of proofs was also a part of CICM 2022.

The conference featured three invited talks:

- Erika Ábrahám (RWTH Aachen University): "SMT solving for Arithmetic Theories".
- Deyan Ginev (FAU Erlangen-Nürnberg and NIST): "Welcome to ar5iv! Wrestling with the open problems of scholarly writing".
- Sébastien Gouëzel (IRMAR, Université de Rennes 1): "Formalizing the change of variables formula for integrals in mathlib".

The PC work was managed using the EasyChair system, which was very helpful and convenient. We thank Besik Dundua (conference chair), Mikheil Rukhaia, Ana Idadze, and their colleagues in Tbilisi for the successful organization of the conference, which was a difficult task especially because of the challenges posed by the hybrid mode. The Kurt Gödel Society and Matthias Baaz provided invaluable support in the organization. We are grateful to Serge Autexier for his publicity work, Mădălina Eraşcu for serving as the doctoral program chair, and Florian Rabe for serving as the workshop chair. We also thank the authors of the submitted papers, the PC members and external reviewers, the workshop organizers, and the invited speakers and the participants of the conference.

August 2022   Kevin Buzzard
                                                               Temur Kutsia

Organization

Program Committee Chairs

Kevin Buzzard	Imperial College London, UK
Temur Kutsia	Johannes Kepler University Linz, Austria

Program Committee

Jesús Aransay	Universidad de La Rioja, Spain
David Aspinall	University of Edinburgh, UK
Serge Autexier	DFKI, Germany
Alexander Bentkamp	Vrije Universiteit Amsterdam, The Netherlands
Alex J. Best	Vrije Universiteit Amsterdam, The Netherlands
David Cerna	Czech Academy of Sciences, Institute of Computer Science, Czech Republic
Shaoshi Chen	KLMM, AMSS, Chinese Academy of Sciences, China
Sander Dahmen	Vrije Universiteit Amsterdam, The Netherlands
Mădălina Eraşcu	West University of Timisoara, Romania
William Farmer	McMaster University, Canada
Cezary Kaliszyk	University of Innsbruck, Austria
Fairouz Kamareddine	Heriot-Watt University, UK
Marie Kerjean	CNRS, LIPN, Université Sorbonne Paris Nord, France
Michael Kohlhase	FAU Erlangen-Nürnberg, Germany
Olexandr Konovalov	University of St Andrews, UK
Angeliki Koutsoukou-Argyraki	University of Cambridge, UK
Robert Y. Lewis	Brown University, USA
Heather Macbeth	Fordham University, USA
Barbara Morawska	Ahmedabad University, India
Daniele Nantes-Sobrinho	Universidade de Brasília, Brazil
Pedro Quaresma	University of Coimbra, Portugal
Florian Rabe	FAU Erlangen-Nürnberg, Germany
Claudio Sacerdoti Coen	University of Bologna, Italy
Christoph Schwarzweller	Gdansk University, Poland
Sofiène Tahar	Concordia University, Canada
Olaf Teschke	FIZ Karlsruhe, Germany
Wolfgang Windsteiger	Johannes Kepler University Linz, Austria

Additional Reviewers

Aksoy, Kubra
Asakura, Takuto
Baanen, Anne
Barhoumi, Oumaima
Buran, Michal
Cohen, Cyril
de Lima, Thaynara Arielly
Deniz, Elif
Eberl, Manuel
From, Asta Halkjær
Jakubův, Jan
Li, Haokun
Miquey, Étienne

Müller, Dennis
Nagashima, Yutaka
Naumowicz, Adam
O'Connor, Liam
Olarte, Carlos
Rashid, Adnan
Starosta, Štěpán
Stuckey, Peter J.
Thiemann, René
Vélez, M. Pilar
Wong, Thomas
Yu, Wensheng
Zhan, Bohua

Abstracts of Invited Talks

SMT Solving for Arithmetic Theories

Erika Ábrahám ⓘ

RWTH Aachen University, Germany
abraham@informatik.rwth-aachen.de

Propositional satisfiability is the problem of determining, for a formula of propositional logic, whether there is an assignment of truth values to its variables for which that formula evaluates to true. In the 90's, a technology called SAT solving has become impressively powerful, being able to check the satisfiability of huge real-world propositional logic problems with millions of clauses.

Based on this success, the question raised whether these technologies could be somehow extended to check also quantifier-free first-order logic formulas over different theories for satisfiability. This would be very helpful as a lot of real-world problems cannot be conveniently encoded in propositional logic. This was the motivation for the development of so-called Satisfiability Modulo Theories (SMT) solvers.

In this talk we review the algorithmic background of these developments for arithmetic theories, present our own SMT solver SMT-RAT and discuss interesting future research directions.

Keywords: Satisfiability checking · SMT solving · Real algebra

Welcome to ar5iv! Wrestling with the Open Problems of Scholarly Writing

Deyan Ginev[1,2]

[1] FAU Erlangen-Nürnberg, Germany
[2] NIST, USA
deyan.ginev@gmail.com

The ar5iv Lab was introduced in February 2022. It is an official, web-native preview of arXiv's corpus of scholarly preprints. In 2022, we can showcase a level of coverage and fidelity in converting LaTeX to HTML5 so as to receive wide community approval and initiate a new partnership under the arXiv Labs umbrella. This feat took us 15 years since the project was first presented to the CICM community.

In this talk, I will outline the history and strategic decisions that helped us get here. I will also explore some of the open problems that keep us at 75% success rate in conversion. The core task at hand continues to be one of rescuing documents written for the requirements of a printed page into a web-native future.

Keywords: ar5iv · arXiv · Scholarly HTML5 · Mathematical documents

Contents

Theorem Proving and Expression Transformation

Satisfiability, QBF, and SMT Solving

Computer-Aided Teaching

Datasets and System Entries

Invited Talk

A Formalization of the Change of Variables Formula for Integrals in mathlib

Sébastien Gouëzel[⊠] [iD]

IRMAR, CNRS UMR 6625, Université de Rennes 1, 35042 Rennes, France
`sebastien.gouezel@univ-rennes1.fr`

Abstract. We report on a formalization of the change of variables formula in integrals, in the mathlib library for Lean. Our version of this theorem is extremely general, and builds on developments in linear algebra, analysis, measure theory and descriptive set theory. The interplay between these domains is transparent thanks to the highly integrated development model of mathlib.

Keywords: Change of variables · Integral · Formalization · mathlib

1 Introduction

The change of variables formula in integrals is a basic tool in mathematics, playing an important role both in concrete computations of integrals and in more theoretical domains, notably Poincaré duality for de Rham cohomology. Its most basic formulation is the following:

Theorem 1.1. *Consider the vector space $\mathbb{R}^n$ with it standard Lebesgue measure, and $f : \mathbb{R}^n \to \mathbb{R}^n$ a C^1-diffeomorphism (i.e., f is a bijection, it is continuously differentiable, and so is its inverse). Then, for any integrable function $g : \mathbb{R}^n \to \mathbb{R}$, the function $x \mapsto |\det Df(x)| \cdot g(f(x))$ is also integrable and*

$$\int g(y) \, \mathrm{dLeb}(y) = \int |\det Df(x)| \cdot g(f(x)) \, \mathrm{dLeb}(x).$$

This paper is devoted to the description of a formalization of (a more sophisticated version of) this theorem, in the Lean proof assistant, developed at Microsoft Research by Leonardo de Moura [9], within the library `mathlib` [11]. Apart from its mathematical relevance, an interest of this theorem from the formalization point of view is that it mixes several domains of mathematics that are typically taught in different courses, notably linear algebra, calculus, measure theory (and descriptive set theory for the aforementioned more sophisticated version). Therefore, it can only be formalized in a library which is developed enough in all these directions, and in which all these areas can interact in a coherent way. This is the case of `mathlib`, but also of the main library of Isabelle/HOL (which

K. Buzzard and T. Kutsia (Eds.): CICM 2022, LNAI 13467, pp. 3–18, 2022.
https://doi.org/10.1007/978-3-031-16681-5_1

already contains a version of the above theorem) or of `mathcomp-analysis` in Coq (which does not contain a version of the above theorem at the time of this writing, but might in the near future). The need for such coherence in advanced mathematics libraries will be a guiding theme in this paper.

This paper is written both for mathematicians who want to learn more on advanced versions of the change of variables formula or on theorem provers, and for formalizers: we will explain design issues that show up at different places, and justify the specific choices that have been made in the `mathlib` formalization to solve these issues.

2   Sketch of Proof of Theorem 1.1

Let us sketch a proof of Theorem 1.1 as may be found in standard textbooks, to highlight the tools that are needed. Approximating the function g by characteristic functions of measurable sets, and then the measurable set by a compact set, it is sufficient to prove the following statement: if K is a compact set, then

$$\text{Leb}(f(K)) = \int_K |\det Df(x)| \, \mathrm{dLeb}(x).$$

We will check that each of these quantities is bounded above by the other one.

Fix $\delta > 0$. Let $\varepsilon > 0$. Cover K by boxes made from a grid of mesh ε, and denote the center of such a box B_i by c_i. By uniform continuity of Df on compact sets, if ε is small enough, the differential Df is arbitrarily close to $Df(c_i)$ on B_i. It follows that $f(B_i)$ is included in the image of B_i under the linear map $(1 + \delta)Df(c_i)$, uniformly in i. Then

$$\text{Leb}(f(K)) \leqslant \sum_i \text{Leb}(f(B_i)) \leqslant \sum_i \text{Leb}((1 + \delta)Df(c_i)(B_i))$$

$$= (1 + \delta)^n \sum_i |\det(Df(c_i))| \, \text{Leb}(B_i),$$

where the last equality follows from the fact that a matrix A rescales the volume according to $|\det A|$. If δ is small enough, then $|\det(Df(c_i))| \leqslant (1+\delta)|\det Df(x)|$ for any $x \in B_i$ by uniform continuity. Then the above sum can be bounded by

$$(1 + \delta)^n \sum_i \int_{B_i} (1 + \delta)|\det Df(x)| \, \mathrm{dLeb}(x).$$

Finally, denoting by K_ε the neighborhood of K given by the union of the B_i, we have proved that

$$\text{Leb}(f(K)) \leqslant (1 + \delta)^{n+1} \int_{K_\varepsilon} |\det Df(x)| \, \mathrm{dLeb}(x).$$

When ε tends to 0, then K_ε tends to K. The dominated convergence theorem shows that the integral over K_ε converges to the integral over K. Finally, as δ

is arbitrary, we have proved the inequality

$$\text{Leb}(f(K)) \leqslant \int_K |\det Df(x)| \, d\text{Leb}(x).$$

The converse inequality can be proved along the same lines, but one step is more delicate: one should show that $\text{Leb}(f(B_i)) \geqslant (1+\delta)^{-n} \text{Leb}(Df(c_i)(B_i))$, i.e., one should show that $f(B_i)$ is comparatively large. A new ingredient is needed there, to prove that f is locally surjective (in a quantitative way). This follows from the inverse function theorem. Equivalently, one can mimic the above computation but for the map f^{-1} (which is also a diffeomorphism). This concludes the proof.   □

3   A More Sophisticated Version of the Theorem

From the proof sketch above, it is obvious that some assumptions in Theorem 1.1 may be relaxed. For instance, it is not necessary that f is defined on the whole space: if it is a diffeomorphism between two open sets of $\mathbb{R}^n$, then the same proof will go through. It is also not necessary that the vector space is $\mathbb{R}^n$: any finite-dimensional real vector space with a Lebesgue measure (i.e., a translation invariant sigma-finite nonzero measure) will do, as such a measure has the same rescaling properties under linear maps.

On the other hand, since the proof relies on the inverse function theorem, it looks as though assuming that the map is defined on an open set and its differential is continuous and invertible can not be avoided. It turns out that this is not the case (and this may come as a surprise even to mathematicians who are very familiar with this theorem and its applications): all these assumptions can be dispensed with.

The most general version of the theorem is expressed in terms of a (slightly non-standard) notion of differentiability along a set:

Definition 3.1. *Consider a normed real vector space E, a map $f : E \to E$ and a continuous linear map $A : E \to E$. We say that A is a derivative of f at a point $x \in E$ along a set $s \subseteq E$ if, when y tends to x inside s, then $f(y) = f(x) + A(y - x) + o(y - x)$.*

When s is the whole space (or a neighborhood of x), this coincides with the usual notion of differentiability at x.

We say that a measure on a finite-dimensional real vector space is a *Lebesgue measure* if it is nonzero, sigma-finite, and invariant under left-translation. Such a measure always exists, and it is unique up to scalar multiplication. Such measures are also known as Haar measures in the more general context of locally compact topological groups.

Here is the general version of the change of variables formula.

Theorem 3.2. *Let E be a finite-dimensional normed real vector space endowed with a Lebesgue measure, $f : E \to E$ a map, s a Borel-measurable subset of E*

and $f'(x)$ a linear map on E for each x. Assume that f is injective on s, and that at every $x \in s$ the linear map $f'(x)$ is a derivative of f at x along s. Then $f(s)$ is also Borel-measurable, and any function $g : E \to \mathbb{R}$ satisfies the equality

$$\int_{f(s)} g(y) \, \mathrm{dLeb}(y) = \int_s |\det f'(x)| \cdot g(f(x)) \, \mathrm{dLeb}(x).$$

As in the proof sketch of Paragraph 2, this follows from a result on the measure of the image set:

$$\mathrm{Leb}(f(s)) = \int_s |\det f'(x)| \, \mathrm{dLeb}(x). \tag{1}$$

This theorem is proved in [4, Theorem 263D], with the difference that the emphasis there is on Lebesgue-measurable sets more than Borel-measurable ones. To obtain the fact that $f(s)$ is Borel-measurable if s is, one needs additionally the Lusin-Souslin theorem [5, Theorem 423I], an important and nontrivial result in descriptive set theory.

We will not give a full proof of Theorem 3.2, and refer the interested reader to [4] instead. Let us only stress where one can follow the sketch given in Sect. 2, and where one should depart from it. Let us focus on the proof of (1). Fix a small $\varepsilon > 0$. With the definition of differentiability along s, one may split s into countably many disjoint small sets s_i on which f is well approximated by a linear map A_i, up to ε. Then one would like to say that $\mathrm{Leb}(f(s_i))$ is close to $|\det(A_i)| \cdot \mathrm{Leb}(s_i)$, but one can not resort to the inverse function theorem.

For the direct inequality

$$\mathrm{Leb}(f(s_i)) \leq (|\det(A_i)| + \varepsilon) \cdot \mathrm{Leb}(s_i), \tag{2}$$

we fix $\delta > 0$ and we use a covering lemma (such as the Vitali or the Besicovitch covering theorems – see Sect. 6 for more on these theorems) ensuring that one can cover s_i with countably many balls $(B_{ij})_{j \in \mathbb{N}}$ whose measures add up to at most $\mathrm{Leb}(s_i) + \delta$. For each such ball B_{ij}, its image has measure bounded by $(|\det(A_i)| + \varepsilon)\mathrm{Leb}(B_{ij})$. Adding these estimates and letting δ tend to 0, we get (2).

For the converse inequality

$$(|\det(A_i)| - \varepsilon) \cdot \mathrm{Leb}(s_i) \leq \mathrm{Leb}(f(s_i)),$$

there is nothing to prove if A_i is not invertible. If it is invertible, one argues that f is invertible on s_i and that its inverse is close to A_i^{-1} (this is a non-standard version of the inverse function theorem). Then, one repeats the above computation for f^{-1} to get the desired estimate.

Adding all these inequalities over i one gets that $\mathrm{Leb}(f(s)) = \sum_i \mathrm{Leb}(f(s_i))$ is comparable to $\sum |\det(A_i)| \, \mathrm{Leb}(s_i)$, which is comparable to $\int_s |\det f'(x)| \, \mathrm{dLeb}(x)$ as $f'(x)$ is close to A_i on s_i. This concludes the proof. $\square$

A difficulty that we have ignored in this proof sketch is that the derivative along a set is in general not unique at points where the set is not fat enough. This has the unpleasant consequence that, in the statement of Theorem 3.2, the function $|\det f'(x)|$ is in general not Borel-measurable along s, as illustrated by the following example.

Example 3.3. Take $s = \mathbb{R} \times \{0\} \subseteq \mathbb{R}^2$, and $f(x,y) = (x,0)$. Let also t be any (possibly non-measurable) subset of $\mathbb{R}$, and set

$$f'(x,y) = \begin{pmatrix} 1 & 0 \\ 0 & 1_t(x) \end{pmatrix}.$$

Along horizontal directions, $f'(x,y)$ acts as the identity, so it is indeed a derivative of f along s. But $\det f'(x,y) = 1_t(x)$ is not a measurable function.

Nevertheless, Theorem 3.2 is still true in this example, since s and $f(s)$ have zero measure, so both the left and the right hand side in the statement of the theorem vanish. In general, $f'(x)$ is uniquely defined on a full measure subset of s (its *Lebesgue density points*, which are again studied using covering lemmas), and is measurable there. In particular, even though $|\det f'(x)|$ is not always Borel-measurable, it coincides almost everywhere with a Borel-measurable function.

For Theorem 3.2 to be true, it means that the theory of integration one uses should work smoothly with functions which are not Borel-measurable, but coincide almost everywhere with a Borel-measurable function. While this was not the case of the first definition of the integral in `mathlib`, it had already been refactored (for different reasons) before the start of this project to allow non-Borel measurable functions, so no modification of the library was needed on this side.

Let us explain why the definition of integral had been refactored prior to this work. `mathlib` initially contained a definition of the integral for which the integral of non-measurable functions was zero by convention. This seemed quite satisfactory and made it possible to prove many theorems, until the formalization of the fundamental theorem of calculus. This theorem reads as follows: if two functions $f, g : \mathbb{R} \to \mathbb{R}$ are continuous on an interval $[a,b]$ and g is the derivative of f there, then $\int g(x)\, d\mathrm{Leb}_{|[a,b]}(x) = f(b) - f(a)$. It turns out that this theorem as stated was not true in `mathlib`: we are not making any assumption on g outside $[a,b]$, so there is no reason why g should be measurable globally, which means that $\int g(x)\, d\mathrm{Leb}_{|[a,b]}(x)$ could be equal to 0 for no good reason. The first version of this theorem in `mathlib` therefore needed an additional assumption that g was measurable globally.

Without this assumption, it is true that g is null-measurable (i.e. almost everywhere measurable) with respect to the restricted measure $d\mathrm{Leb}_{|[a,b]}$: it coincides almost everywhere with a measurable function, namely the function equal to g on $[a,b]$ and to 0 elsewhere (which is indeed measurable by continuity of g on $[a,b]$).

To get a more satisfactory statement for the fundamental theorem of calculus, the definition of the integral in `mathlib` was therefore refactored to allow for almost everywhere measurable functions: if a function is almost everywhere measurable, then its integral is defined to be the integral of a measurable function which coincides with it almost everywhere, and otherwise the integral is defined to be 0.

This definition has several advantages. For instance, if two functions coincide almost everywhere then they have the same integral regardless of measurability issues. Moreover, it is exactly the kind of integration theory which is needed for Theorem 3.2 to hold! That this change was needed and fruitful would not have been noticed in a pure measure-theory library, and was really a consequence of the interaction of different domains of mathematics in `mathlib`.

Getting definitions right the first time is hard. Definitions should be driven by the theorems they enable, even in different domains, and one should not be afraid to refactor a core definition.

As an aside, let us note that the definition of integrals was refactored a third time in `mathlib`, to allow for functions that take values in spaces which are not second-countable. While the standard definition of integration (writing a function as a pointwise limit of simple functions) is easier to work out when the target space is second-countable, this restriction prevents some applications to complex analysis and spectral theory. When these limitations were noticed, the definition was changed again, for the better. Now, the functions that can be integrated in `mathlib` are the almost everywhere strongly measurable ones, i.e., the functions that coincide almost everywhere with a pointwise limit of simple functions. And there are several results ensuring that most concrete functions are almost everywhere strongly measurable functions – for instance measurable functions into second-countable spaces, or continuous functions from second-countable spaces.

4   The Formalized Version of the Theorem

Here is the full statement of the formalized version of Theorem 3.2.

```
theorem integral_image_eq_integral_abs_det_fderiv_smul
[normed_group E] [normed_space ℝ E] [finite_dimensional ℝ E]
[measurable_space E] [borel_space E]
(μ : measure E) [is_add_haar_measure μ]
[normed_group F] [normed_space ℝ F] [complete_space F]
{s : set E} {f : E → E} {f' : E → (E →L[ℝ] E)} (hs : measurable_set s)
(hf' : ∀ x ∈ s, has_fderiv_within_at f (f' x) s x)
(hf : set.inj_on f s) (g : E → F) :
∫ y in f '' s, g y ∂μ = ∫ x in s, |(f' x).det| · g (f x) ∂μ
```

Here is a rephrasing of the theorem for readers who are not familiar with Lean's syntax. We start with a finite-dimensional real normed vector space E, a measure μ on E which is assumed to be a Lebesgue measure (in more formal terms, an additive Haar measure), a subset s of E which is assumed to be measurable (assumption `hs`), and a function $f : E \to E$ which is injective on s (assumption `hf`). Consider also, for each $x \in E$, a continuous linear map $f'(x)$ on E. Assume that, for each $x \in s$, then $f'(x)$ is a derivative of f at x along s (assumption `hf'`). Then the change of variables formula holds: for any function

$g : E \to F$ (where F is any complete real vector space), then

$$\int_{y \in f(s)} g(y) \, \mathrm{d}\mu(y) = \int_{x \in s} |\det(f'(x))| g(x) \, \mathrm{d}\mu(x).$$

This corresponds perfectly to Theorem 3.2.

Let us list the different domains of mathematics that are involved in the statement of Theorem 3.2:

1. Analysis and topology: to talk about normed spaces and continuity.
2. Calculus: to make sense of derivatives.
3. Measure theory: to talk about integrals and measures of sets (including the definition of additive Haar measures).
4. Linear algebra: to talk about finite dimensional spaces, and also about determinants of linear maps.

All these domains should be formalized before the above formalized statement `integral_image_eq_integral_abs_det_fderiv_smul` can be merely written down and understood by the system.

There are also tools that show up in the proof of the theorem, but not in its statement:

5. Ordinals and transfinite induction (these show up in the proof of the covering theorems).
6. Linear maps rescale Lebesgue measures according to the absolute value of their determinants.
7. Covering theorems, like the Besicovitch and Vitali covering theorems.
8. Descriptive set theory, notably the theory of Polish spaces and analytic sets in them (they are instrumental in the proof of the Lusin-Souslin theorem).

All these should be formalized before the proof of the change of variables theorem.

This project resulted in 80 pull requests to `mathlib`, adding roughly 15,000 lines of code. Among these, most are devoted to the prerequisites presented above: the file on the change of variables formula itself has only 1259 lines, less than 10% of the total. In the topics above, Items 1–5 were already mature enough that they needed few additions. Items 6–8 form the bulk of the formalization of this project, with roughly 20% for 6, the remaining 70% being split evenly between 7 and 8.

`mathlib` is an open source project: everyone can submit pull requests, which are then submitted to a thorough review process. There are 25 maintainers of the project. When a maintainer is happy with a pull request (and several sanity checks have been automatically performed, as explained in [2]), then he can merge it to the main branch (and of course no maintainer can merge his own pull requests). Given the width of `mathlib`, no maintainer is expert in all areas: the pull requests in this project were therefore refereed by different maintainers depending on their domains. An important point is that the maintainers coordinate to ensure the unity of the whole library. For instance, the linear maps that are used in linear algebra are the same as those that are used in algebraic applications such as Galois theory, or in analytic applications such as derivatives of maps.

This inter-operability is extremely useful for a project such as the change of variables formula, that involves many different areas of mathematics: in a less coherent project, one would likely need to add glue to make sure that different modules can work together, and this would become quickly unwieldy at this level of complexity. In this respect, in the language of [10], the `mathlib` library is cathedral-like as it is complex and coherent, but its open-source development process also has some bazaar characteristics. The delicate balance between these two models is only possible thanks to the hard work of the maintainers, who should be thanked for their dedication.

The next three sections will be devoted to more in-depth discussion of the three main ingredients 6–8. Before that, let us make a few remarks on the formalized statement of the theorem.

Remark 4.1. There are several assumptions in this theorem that appear between brackets, like `[normed_group E]`. These are *typeclass assumptions*, that should be filled automatically by the system when the theorem is used. The only assumptions that should be checked by the user are those between parentheses, like `(hs : measurable_set s)`. The typeclass assumptions are checked by the system using special lemmas that are tagged as *instances*. For instance, the fact that the volume on $\mathbb{R}$ is a Lebesgue measure is an instance, as well as the fact that the product of two Lebesgue measures is a Lebesgue measure, so the theorem will automatically apply to the standard Lebesgue measure on $E = \mathbb{R} \times \mathbb{R}$ (and the fact that $\mathbb{R} \times \mathbb{R}$ is a finite-dimensional real normed vector space is also checked automatically by typeclass inference).

Remark 4.2. In this version of the theorem, E is a general finite-dimensional real vector space, and μ is a general Lebesgue measure on E (this is the content of the typeclass assumption `[is_add_haar_measure μ]`). One may wonder if one really needs this generality, and if it would not be more natural to have the theorem only on $\mathbb{R}^n$ with its canonical (product) Lebesgue measure. For instance, this is the way the analogue of this theorem is formulated in Isabelle/HOL. However, we found that this more restricted version is too limited.

As an illustration, let us recall a classical proof of the value of the Gaussian integral $\int_{\mathbb{R}} e^{-x^2/2} \, \mathrm{d}x = \sqrt{2\pi}$. Denoting by I the value of the integral, one can compute using a polar change of coordinates in $\mathbb{R} \times \mathbb{R}$, writing $(x, y) = (r\cos\theta, r\sin\theta)$ as follows:

$$I^2 = \int_{\mathbb{R} \times \mathbb{R}} e^{-(x^2 + y^2)/2} \, \mathrm{d}x \, \mathrm{d}y = \int_{r=0}^{+\infty} \int_{\theta=-\pi}^{\pi} e^{-r^2/2} \cdot r \, \mathrm{d}r \, \mathrm{d}\theta$$

$$= 2\pi \int_{r=0}^{\infty} r e^{-r^2/2} \, \mathrm{d}r = 2\pi,$$

where the computation $\int_{r=0}^{\infty} r e^{-r^2/2} \, \mathrm{d}r = 1$ follows from the fact that $r e^{-r^2/2}$ is the derivative of $-e^{-r^2/2}$.

This proof (which has been formalized in `mathlib`) uses the change of variables theorem in the space $E = \mathbb{R} \times \mathbb{R}$, with the product Lebesgue measure. But this space is *not* one of the spaces $\mathbb{R}^n$, defined as the space of functions from $\{0, \ldots, n-1\}$ to $\mathbb{R}$: the product space $\mathbb{R} \times \mathbb{R}$ and the function space $\{0, 1\} \to \mathbb{R}$ are obviously the same for a mathematician, but in all rigor they are different (albeit canonically isomorphic), and in particular they are definitely different from Lean's point of view. One could try to reformulate the above proof using $\mathbb{R}^2$ instead of $\mathbb{R} \times \mathbb{R}$, but then Fubini theorem (which has been used in the first step of the above proof) would not apply directly as it only makes sense for product spaces.

This is an important lesson: *more general theorems apply in more situations, so one should aim for generality in formalized mathematics library to improve their usability.*

Remark 4.3. The differentiability assumption inside a set is not completely standard in mathematics, but it shows up often in particular cases. Notably, in one dimension, one often talks about left derivatives and right derivatives, which are particular cases of Definition 3.1 where s is respectively $(-\infty, x]$ or $[x, +\infty)$. In differential geometry, one also often encounters functions which are differentiable on half-spaces or on submanifolds. With these examples in mind, derivatives in `mathlib` were defined from the start using Definition 3.1, so they were already general enough for Theorem 3.2. As most of the basics of analysis and topology in `mathlib`, this is strongly inspired from the Isabelle/HOL formalization of analysis described in [7].

Remark 4.4. There is a difference between the statement of Theorem 3.2 and its formalized version: in Theorem 3.2, we require the function g to be integrable on $f(s)$, but this assumption is nowhere to be seen in the formalized version (not even measurability or almost everywhere measurability of g). This makes the formalized version easier to use: the user does not need to worry about proving the integrability of the function.

The formalized version comes with a companion theorem, saying that g is integrable on $f(s)$ if and only if $x \mapsto |\det f'(x)| \cdot g(x)$ is integrable on s. Since the integral of non-integrable functions is defined to be 0 by convention, the theorem is then tautologically true for non-integrable functions as both sides of the statement vanish.

The companion theorem is not trivial, especially regarding measurability issues. Ultimately, it relies on the fact that the restriction of f to s is a measurable embedding, i.e., it maps measurable sets to measurable sets. This follows from the deep Lusin-Souslin theorem in descriptive set theory. A first version of the formalization had the additional assumption that g was integrable and avoided the Lusin-Souslin theorem. It was less satisfactory since it required more work from the end user of the theorem when applying it.

This rule is followed throughout `mathlib`: *One should try to minimize the assumptions of theorems to make them easier to apply for the users, even if this comes with a higher proof burden for the formalizer of the theorem.* (In our specific case, the Lusin-Souslin theorem indeed required several thousands additional lines of formalization.)

Let us give another silly example of this rule: the formula $\int f + g = \int f + \int g$ requires the functions f and g to be integrable. On the other hand, the formula $\int cf = c \int f$ is true whether f is integrable or not, so the latter formula should be given without integrability assumptions, even though the proof becomes more complicated as it requires several case distinctions. Indeed, if f is nonintegrable and c is nonzero, then cf is also nonintegrable so both sides vanish; if f is nonintegrable and c is zero, then cf becomes integrable as it is the zero function, but both sides vanish again; if f is integrable then cf is also integrable and one is back to the usual situation.

5   Linear Maps Rescale Lebesgue Measure According to Their Determinants

A basic ingredient of Theorem 3.2 is that it holds at an infinitesimal level, i.e., for the linearized map: a linear map should act on Lebesgue measure by multiplying it according to the absolute value of its determinant.

Here is the formalized statement:

```
lemma add_haar_image_linear_map
[normed_group E] [normed_space ℝ E] [measurable_space E] [borel_space E]
[finite_dimensional ℝ E] (μ : measure E) [is_add_haar_measure μ]
(f : E →ₗ[ℝ] E) (s : set E) :
μ (f " s) = ennreal.of_real (abs f.det) * μ s
```

The proof of this theorem splits into two steps.

Let us first prove it on $E = \mathbb{R}^n$, with its standard product measure. A linear map f on this space corresponds in a canonical way to a matrix M. Gaussian elimination shows that M can be written as the product of transvection matrices (i.e., matrices with ones on the diagonal and at most one non-zero off-diagonal coefficient) and a diagonal matrix. As the statement to be proved is clearly stable under multiplication (as the determinant of a product is the product of the determinants), it therefore suffices to check it for transvections and for diagonal matrices. For diagonal matrices, it follows readily from the fact that the measure is a product measure and from the elementary 1-dimensional situation. For transvections (whose determinant is 1), we should show that a transvection preserves Lebesgue measure. This follows from a straightforward computation using Fubini to single out the coordinate at which there is a nonzero entry in the matrix.

The key argument of this step is thus Gaussian elimination, which was not yet proved in `mathlib` before this project and was formalized with this goal in mind. This is probably the most unexpected outcome of this project!

Let us now turn to the case of a general finite-dimensional vector space E, with a general Lebesgue measure μ. There is no canonical basis, and therefore no way to identify a linear map with a matrix, and moreover a general Lebesgue measure has no product structure a priori, so that the above argument does not make sense. However, one can choose a basis, which yields an isomorphism A between $\mathbb{R}^n$ and E for some n. Let $g := A^{-1} \circ f \circ A$ be the conjugate of f under this isomorphism. Its determinant coincides with that of f, and moreover the first step applies to g. To conclude, it suffices to show that the image under A of the standard Lebesgue measure on $\mathbb{R}^n$ coincides with μ (or a scalar multiple of μ). We deduce this from uniqueness of Lebesgue measure up to scalar multiplication, which was already available in `mathlib` in the right generality: two Haar measures on a locally compact group are multiples of each other. This is a nontrivial fact, which had fortunately already been formalized by van Doorn with a totally unrelated application in mind, see [1].

Remark 5.1. Once the above theorem is available, one deduces that the volume of balls behaves like $\mu(B(x, r)) = r^d \mu(B(0, 1))$ where d is the dimension of the space. This fact makes it possible to apply the Vitali covering theorem (Theorem 6.1 below) to μ, and is therefore fundamental. Note that it is obvious for the standard product measure in the Euclidean space $\mathbb{R}^n$, but not so obvious in a general normed vector space with a general Lebesgue measure.

6 Covering Theorems in Measure Theory

There is a huge variety of covering theorems for measure theory in the literature. Among them, the two most prominent ones are probably the Vitali and Besicovitch covering theorems that we will now describe and that we have formalized for the current project.

Here is a version of the Vitali covering theorem:

Theorem 6.1. *In a metric space X, consider a locally finite measure μ which is doubling: there exists $C > 0$ such that, for any x, r, then $\mu(B(x, 2r)) \leq C\mu(B(x, r))$. Consider a set of balls t with uniformly bounded radii, and a set $s \subseteq X$ at which the family t is fine, i.e., every point of s belongs to balls in t with arbitrarily small radius. Then there exists a disjoint subfamily of t covering almost all s.*

Here is a version of the Besicovitch covering theorem:

Theorem 6.2. *Let E be a finite-dimensional real normed vector space, and s a subset of E. Consider a set t of balls such that, for any $x \in s$, there exist arbitrarily small radii r such that $B(x, r) \in t$. Let μ be a sigma-finite measure. Then there exists a disjoint subfamily of t covering almost all s.*

The assumptions of the Besicovitch theorem on the measure are weaker than those of the Vitali theorem, as they do not assume any doubling property. On the other hand, the former theorem requires stronger geometric assumptions (in this formulation, a finite-dimensional real normed vector space instead of a general metric space). The Besicovitch theorem is harder to prove and often more powerful than the Vitali theorem. The book [3] is a standard reference for these theorems and several powerful extensions, which was used for the formalization.

Both theorems are consequences of purely combinatorial results, from which the measure theoretic versions are then deduced. For instance, the deterministic result leading to the Besicovitch covering theorem is the following statement:

Theorem 6.3. *Let E be a finite-dimensional real normed vector space. There exists a constant $N = N(E)$ with the following property. Consider a set s, and a set of balls t with uniformly bounded radii such that any point in s is the center of some ball in t. Then one may find disjoint subfamilies $t_1, \ldots, t_N$ of t which still cover s.*

To deduce the measure-theoretic version from this deterministic version, one picks one of the subfamilies covering a proportion $1/N$ of s, say t_i, and then works inductively on the subset of s that it still to be covered.

The best value of the constant N has been determined in [6]: it is the maximal number of points one can put inside the unit ball of radius 2 under the condition that their distances are bounded below by 1. This is the version we have formalized in `mathlib`.

Let us mention that this theorem (and the combinatorial theorem implying the Vitali covering theorem) are proved using transfinite induction (i.e., an induction indexed by ordinals). Fortunately, ordinals were already available in `mathlib` before the start of this project.

The book [3] introduces a formalism to deduce consequences of covering theorems in a uniform way, independently of the covering theorem. This formalism is often deemed too abstract by mathematicians, but it turns out to be extremely well suited to formalization. Let us say that a family of sets $(\mathcal{F}_x)_{x \in X}$ in a metric measured space X is a Vitali family if it satisfies the following property: consider a (possibly non-measurable) set s, and for any x in s a subfamily F_x of $\mathcal{F}_x$ containing sets of arbitrarily small diameter. Then one can extract from $\bigcup_{x \in s} F_x$ a disjoint subfamily covering almost all s.

In this language, the Vitali covering theorem says that one gets a Vitali family by taking for $\mathcal{F}_x$ the balls that contain x, in a space where the measure μ is doubling. And the Besicovitch covering theorem states that, in a finite-dimensional real vector space, one gets a Vitali family by taking for $\mathcal{F}_x$ the balls centered at x.

The fundamental theorem on differentiation of measures is the following. On a metric space with a measure μ, consider a Vitali family $\mathcal{F}$ and another measure ρ. Then, for μ-almost every x, the ratio $\rho(a)/\mu(a)$ converges when a shrinks to x along the Vitali family $\mathcal{F}_x$, towards the Radon-Nikodym derivative of ρ with respect to μ. We have formalized this theorem, as follows:

```
theorem vitali_family.ae_tendsto_rn_deriv
[sigma_compact_space α] [borel_space α]
{μ : measure α} [is_locally_finite_measure μ] (v : vitali_family μ)
(ρ : measure α) [is_locally_finite_measure ρ] :
∀ᵐ x ∂μ, tendsto (λ a, ρ a / μ a) (v.filter_at x) (nhds (ρ.rn_deriv μ x))
```

Note that the convergence when a set in a Vitali family shrinks to a point (i.e., its diameter tends to 0) is not one of the standard convergences in mathematics (and Federer discusses it at length and introduces a special notation for it) but it fits perfectly well within the framework of `mathlib` where all notions of convergence are expressed with the single notion of filter, as advocated by [7].

Once this theorem is formalized, versions in the context of the Vitali and the Besicovitch covering theorems readily follow. In these respective contexts, they make it possible to prove that almost every point x of a set s is a Lebesgue density point of s, i.e., $\mu(s \cap B(x,r))/\mu(s) \to 1$ as $r \to 0$. This fact plays a key role in the proof of Theorem 3.2 to prove that $x \mapsto f'(x)$ is almost everywhere measurable (indeed, it is measurable when restricted to the set of Lebesgue density points).

7   Polish Spaces and Descriptive Set Theory

A Polish space is a topological space which is second-countable and on which there exists a complete metric space structure inducing the given topology. A good reference on Polish spaces and descriptive set theory is [8]. This definition may seem strange at first: one could instead require to have a complete second-countable metric space. However, in many applications, there is no natural distance, and the only relevant information is the topology. For instance, the extended reals $[-\infty, \infty]$ are a Polish space, but they don't have a canonical metric to work with.

This slightly awkward definition makes Polish spaces slightly awkward to use in proof assistants, as it refers to the mere existence of a nice metric but without providing it. In the completely classical framework of `mathlib`, this is not a real issue as one can use choice to pick such an arbitrary nice metric. The definition is formalized as follows:

```
class polish_space (α : Type*) [h : topological_space α] : Prop :=
(second_countable [] : second_countable_topology α)
(complete : ∃ m : metric_space α,
  m.to_uniform_space.to_topological_space = h ∧
  @complete_space α m.to_uniform_space)
```

To construct a nice metric space structure on a Polish space, one uses the following incantation in proofs: `letI := upgrade_polish_space α`. It endows the Polish space α with a metric space structure which is registered in the typeclass system as complete and second-countable, and moreover the topology associated to this metric is definitionally equal to the given topology. This makes it possible to work smoothly with Polish spaces as one would do in a non-formalized setting.

An important theme in descriptive set theory is to start with a Polish space and modify its topology to get better behavior while retaining Polishness. For instance, if a map between two Polish spaces is measurable, then one can construct a finer Polish topology on the source space for which the map becomes continuous. This makes it possible to deduce results for measurable maps from results for continuous maps. Unfortunately, this kind of argument does not interact well with the typeclass system, in which each type is supposed to be endowed with at most one typeclass of each kind, and in particular at most one topology. Fortunately, there is a way to override typeclass inference and provide explicitly the topology one would like to use (at the cost of added verbosity and reduced readability), by prefixing a command with @ and then giving explicitly all its arguments, including the implicit and typeclass arguments. For instance, let us give the formalized version of the following basic (but nontrivial) statement: given a Polish space and countably many finer Polish topologies, there exists another Polish topology which is finer than all of them.

```
lemma exists_polish_space_forall_le {ι : Type*} [encodable ι]
[t : topological_space α] [polish_space α]
(m : ι → topological_space α) (hm : ∀ n, m n ⩽ t)
(h'm : ∀ n, @polish_space α (m n)) : ∃ (t' : topological_space α),
  (∀ n, t' ⩽ m n) ∧ (t' ⩽ t) ∧ @polish_space α t'
```

This fact is proved as follows. First, one checks (by constructing a suitable complete metric) that a countable product of Polish spaces is Polish. Consider then the infinite product $Y = α^{\mathbb{N}}$, where the n-th copy of $α$ is endowed with the n-th topology m_n. Then Y is Polish, therefore the pullback of its topology under the diagonal embedding is also Polish and it satisfies all the required properties.

Another trick that proves useful in this kind of argument is to use a type synonym, i.e., a copy of a type with a different name. As typeclass inference does not unfold definitions, the type synonym is not endowed by default with any topology or metric, and one can register new instances that will not conflict with the original ones. As an example, let us sketch the proof that an open subset of a Polish space is Polish:

```
lemma is_open.polish_space [topological_space α] [polish_space α]
{s : set α} (hs : is_open s) : polish_space s
```

One endows $α$ with a distance for which it is complete and second-countable. Then the open subset s of $α$ has the induced topology, which is second-countable, and it also inherits the restricted distance. But in general it is not complete for the restricted distance (think of the interval $(0, 1) \subset \mathbb{R}$). One should therefore use another distance. A suitable formula for it is

$$d'(x, y) = d(x, y) + |1/d(x, α \setminus s) - 1/d(y, α \setminus s)|.$$

As it blows up close to the boundary of s, one can check that this is a complete distance on s, defining the same topology as the original topology and therefore proving that it is Polish. For formalization purposes, we do not put this distance on s (which would then have two competing metric space structures and would force us to use the @ version of statements everywhere), but instead on a type synonym complete_copy s. Then there is no difficulty to check that complete_copy s is Polish. As this new distance defines the same topology as the original one, the identity from s to complete_copy s is a homeomorphism, hence the fact that s itself is Polish follows.

The Lusin-Souslin theorem states that the image of a measurable set under a measurable injective map on a Polish space is still measurable. Here is the formalized statement:

```
theorem measurable_set.image_of_measurable_inj_on
[topological_space γ] [polish_space γ] [measurable_space γ]
[borel_space γ] [topological_space β] [t2_space β] [measurable_space β]
[borel_space β] [second_countable_topology β] {s : set γ} {f : γ → β}
(hs : measurable_set s) (f_meas : measurable f) (f_inj : set.inj_on f s) :
measurable_set (f '' s)
```

Its proof builds on the techniques we have sketched above, but it is considerably more involved. We will not get into more details. Let us just mention that it is first proved assuming that f is continuous, and then generalized to a measurable f by using the trick to modify the topology to turn a measurable map into a continuous map. A key notion in the proof is that of an analytic set, i.e., the image of a Polish space under a continuous map, and a key result is that if two analytic sets are disjoint, then they are contained in disjoint measurable sets (this result is called the Lusin separation theorem). As far as the author knows, there is no known proof of the Lusin-Souslin theorem which does not go through a study of analytic sets, even though these sets do not appear in the conclusion of the theorem.

8 Conclusion

We have described the formalization of the change of variables formula in integrals, in an advanced version. One interest of this theorem from the point of view of formalization is that it involves several areas of mathematics that are often considered quite independent, but that need to interact seamlessly here – as is often the case in advanced mathematics that mix basic results from several areas.

We have explained how the development model of mathlib has made this project reasonable, as well as its general philosophy: things should be done right, in the greatest level of generality, and without taking shortcuts. A lot of refactors are needed to reach this goal. This is possible in mathlib since it is a monorepository project, without a lot of outside users, and would be harder for more mature projects. We may hope that basic definitions stabilize with time, but we are clearly not there yet: the definition of a group was changed less than one year ago to cope with a definitional equality issue in tensor products of abelian groups seen as $\mathbb{Z}$-modules, that showed up when doing advanced mathematics in the liquid tensor experiment. This kind of agile development is clearly a strength of mathlib currently, but once it reaches a critical size other strategies will need to be devised to make sure it can be used in a more stable way by other projects.

The change of variables formula is one of the tools to implement de Rham cohomology. While this project was being done, other necessary tools have been formalized independently, for other projects: homological algebra was developed for the liquid tensor experiment, and vector bundles on manifolds were developed for the sphere eversion project. This means that de Rham cohomology is now a reasonable target!

References

1. van Doorn, F.: Formalized Haar measure. In: 12th International Conference on Interactive Theorem Proving, LIPIcs. Leibniz Int. Proc. Inform. **193**, Art. No. 18, 17. Schloss Dagstuhl. Leibniz-Zent. Inform., Wadern (2021)
2. van Doorn, F., Ebner, G., Lewis, R.Y.: Maintaining a library of formal mathematics. In: Benzmüller, C., Miller, B. (eds.) CICM 2020. LNCS (LNAI), vol. 12236, pp. 251–267. Springer, Cham (2020). https://doi.org/10.1007/978-3-030-53518-6_16
3. Federer, H.: Geometric Measure Theory. Classics in Mathematics. Springer, Heidelberg (1969). https://doi.org/10.1007/978-3-642-62010-2
4. Fremlin, D.H.: Measure Theory: Broad Foundations, vol. 2 (2003). Corrected second printing of the 2001 original
5. Fremlin, D.H.: Measure Theory: Topological Measure Spaces, vol. 4 (2006). Part I, II, Corrected second printing of the 2003 original
6. Füredi, Z., Loeb, P.A.: On the best constant for the Besicovitch covering theorem. Proc. Am. Math. Soc. **121**(4), 1063–1073 (1994). https://doi.org/10.2307/2161215
7. Hölzl, J., Immler, F., Huffman, B.: Type classes and filters for mathematical analysis in Isabelle/HOL. In: Blazy, S., Paulin-Mohring, C., Pichardie, D. (eds.) ITP 2013. LNCS, vol. 7998, pp. 279–294. Springer, Heidelberg (2013). https://doi.org/10.1007/978-3-642-39634-2_21
8. Kechris, A.S.: Classical Descriptive Set Theory, Graduate Texts in Mathematics, vol. 156. Springer, New York (1995). https://doi.org/10.1007/978-1-4612-4190-4
9. de Moura, L., Kong, S., Avigad, J., van Doorn, F., von Raumer, J.: The lean theorem prover (system description). In: Felty, A.P., Middeldorp, A. (eds.) CADE 2015. LNCS (LNAI), vol. 9195, pp. 378–388. Springer, Cham (2015). https://doi.org/10.1007/978-3-319-21401-6_26
10. Raymond, E.S.: The Cathedral and the Bazaar. USENIX Association (1999)
11. The mathlib community: The Lean mathematical library. In: Proceedings of the 9th ACM SIGPLAN International Conference on Certified Programs and Proofs, CPP 2020, pp. 367–381 (2020)

Formalizations

On the Formalization of the Heat Conduction Problem in HOL

Elif Deniz[1]([⊠]), Adnan Rashid[2], Osman Hasan[2], and Sofiène Tahar[1]

[1] Department of Electrical and Computer Engineering, Concordia University,
Montreal, QC, Canada
{e_deniz,tahar}@ece.concordia.ca
[2] School of Electrical Engineering and Computer Science,
National University of Sciences and Technology, Islamabad, Pakistan
{adnan.rashid,osman.hasan}@seecs.nust.edu.pk

Abstract. Partial Differential Equations (PDEs) are widely used for modeling the physical phenomena and analyzing the dynamical behavior of many engineering and physical systems. The heat equation is one of the most well-known PDEs that captures the temperature distribution and diffusion of heat within a body. Due to the wider utility of these equations in various safety-critical applications, such as thermal protection systems, a formal analysis of the heat transfer is of utmost importance. In this paper, we propose to use higher-order-logic (HOL) theorem proving for formally analyzing the heat conduction problem in rectangular coordinates. In particular, we formally model the heat transfer as a one-dimensional heat equation for a rectangular slab using the multivariable calculus theories of the HOL Light theorem prover. This requires the formalization of the heat operator and formal verification of its various properties, such as linearity and scaling. Moreover, we use the separation of variables method for formally verifying the solution of the PDEs, which allows modeling the heat transfer in the slab under various initial and boundary conditions using HOL Light.

Keywords: Heat equation · Partial Differential Equations · Separation of variables · Higher-order logic · Theorem proving · HOL Light

1 Introduction

Partial Differential Equations (PDEs) [1] are commonly used for the mathematical formulation of the physical behavior of many engineering and physical systems. They capture the continuous dynamics of a system by providing a mathematical relationship between various components of the underlying system by incorporating changes in their associated properties. Due to these distinguishing features, they are broadly used in analyzing many physical phenomena such as, heat or sound propagation, electrodynamics, quantum mechanics and fluid dynamics. For example, they play a pivotal role in the thermal analysis of a system by formulating a general heat equation that can be analyzed using various

© The Author(s), under exclusive license to Springer Nature Switzerland AG 2022
K. Buzzard and T. Kutsia (Eds.): CICM 2022, LNAI 13467, pp. 21–37, 2022.
https://doi.org/10.1007/978-3-031-16681-5_2

appropriate boundary and initial conditions [2]. Similarly, this kind of thermal analysis is a foremost step in the design of many safety-critical applications, such as aerospace, nuclear power plants and automobile engines.

The phenomenon of heat transfer/propagation can occur by three different means, namely, heat conduction [2], convection [3], and thermal radiation [4]. Heat conduction or diffusion is the flow of energy in a system/body from the region of high temperature to the region of low temperature by direct collision of molecules. Whereas, convection refers to the transfer of the energy due to the physical movement of a bulk fluid. Thermal radiation is the transfer of energy in the form of electromagnetic wave. Heat conduction is the most important type of heat transfer and it is commonly used to analyze problems arising in the design and operation of industrial appliances, such as heat exchanger and compressors. The first step for analyzing the heat conduction in a given system/body is to construct a mathematical model of the dynamics of the system, such as heat distribution using the heat equation, which is a PDE. These dynamics provide the variation of the temperature as a function of position/space and time within the heat conducting system/body. The heat distribution (temperature field) usually depends on boundary conditions, initial conditions, material properties, and the geometry of the body. The next step in the heat conduction analysis is to find the solution of the heat equation modeled in the first step that can be obtained by determining a temperature distribution that is consistent with the initial and boundary conditions.

Heat equations are generally analyzed using numerical techniques [5] or analytical methods [6]. The two most widely used numerical techniques for analyzing PDE based heat equations are Finite Difference [7] and Finite Element [8] methods. These methods can solve the complex heat conduction problems by providing the closed-from solutions. However, they involve approximation and rounding of the associated mathematical expressions and thus cannot ensure absolute correctness of the results of the associated analysis. Unlike numerical solutions, the analytical methods for analyzing heat conduction do not involve any approximation of the associated mathematical expressions and thus are preferred on numerical methods for ensuring the correctness of the results. Some commonly used analytical techniques for solving the heat conduction problem are separation of variables [9] and transform methods [10].

Conventionally, the heat conduction problem has been analyzed using paper-and-pencil proof and computer based numerical and symbolic methods. However, the former is human-error prone and it is not well-suited for large systems involving extensive human manipulation. Moreover, the required assumptions are not all explicitly mentioned in the analysis, which may lead to inaccurate results. Similarly, the numerical and symbolic methods are based on approximation of the mathematical results due to the finite precision of computer arithmetic. Moreover, the core of the tools involved in the symbolic methods based analysis has a large number of unverified algorithms that puts a question mark on the accuracy of the associated analysis. Given, the safety-critical nature of many systems, these conventional techniques cannot ensure absolute accuracy of the analysis.

As an alternative to related methods and tools, in this paper, we propose to use higher-order-logic theorem proving [11] for formally analyzing the heat conduction problem in rectangular coordinates and thus overcome the above-mentioned inaccuracy limitations. In particular, we formally model the heat equation using the multivariable calculus theories of the HOL Light theorem prover capturing the heat conduction in the system/body. Next, to formally analyze the heat equation, we use the separation of variables method [9] to formally verify the solution of the PDE by incorporating all relevant boundary and initial conditions. One of the primary reasons for choosing HOL Light for the proposed work is the availability of rich theories of multivariable calculus, such as differential, integration, transcendental and real analysis. The HOL Light codes of our formalization is available at [12].

The remainder of the paper is structured as follows: In Sect. 2, we provide an overview of related work on differential equations based formal analysis. Section 3 describes some fundamentals of the multivariate analysis libraries of the HOL Light theorem prover that are necessary for understanding the rest of the paper. We provide the formalization of the heat equation in rectangular coordinates in Sect. 4. Section 5 presents the formal verification of the solution of the heat equation. Finally, Sect. 6 concludes the paper.

2   Related Work

Many higher-order-logic theorem provers, such as HOL Light[1], HOL4[2], Isabelle/HOL[3], Coq[4] and Mizar[5] have been used for the differential equations based formal analysis of the engineering and physical systems. For instance, Immler et al. [13] used Isabelle/HOL for formally verifying the numerical solutions of Ordinary Differential Equation (ODE). The authors formalized the Initial Value Problems (IVPs) and formally verified the existence of a unique solution of the ODE. Moreover, the authors provide an approximation of the solution using the Euler's method. Immler et al. [14] presented a formal reasoning support about the flow of ODEs using Isabelle/HOL. In particular, the authors formally verified a solution of ODEs incorporating various initial conditions. They also formalized the Poincaré map and formally verified its differentiability. However, both these approaches rely on approximating the solutions of differential equations representing the dynamical behavior of the underlying system. Guan et al. [15] used the HOL Light theorem prover to formalize the Euler-Lagrange equation set that is based on Gâutex derivatives. In addition, the authors used their proposed formalization for formally verifying the least resistance problem of gas flow. Similarly, Sanwal et al. [16] formally verified the solutions of the second-order homogeneous linear differential equations using the HOL4 theorem

---

[1] https://www.cl.cam.ac.uk/jrh13/hol-light/.
[2] https://hol-theorem-prover.org/.
[3] https://isabelle.in.tum.de/.
[4] https://coq.inria.fr/.
[5] http://www.mizar.org/.

prover. Moreover, they used their proposed formalization for formally verifying the damped harmonic oscillator and a second-order op-amp circuit. Rashid et al. formalized the Laplace [17] and the Fourier [18] transforms using HOL Light and used these formalization for differential equations based analysis of many systems, such as automobile suspension system [18], unmanned free-swimming submersible vehicle [19] and platoon of automated vehicles [20]. However, the existing formalization of ODEs in HOL4 and HOL Light, respectively, do not provide the formalization of the solution when dealing with separable linear partial differential equations.

Boldo et al. [21] utilized the Coq theorem prover for formally verifying the numerical solution of one-dimensional acoustic wave equation. The authors used the second-order centered finite difference scheme, commonly known as the three-point scheme for convergence of the result. Similarly, Boldo et al. [22] mechanically verified the correctness of a C program implementing numerical scheme for the solution of PDE using both automated and interactive theorem provers. Despite important contributions, both these works approximate the solution of acoustic wave equation and did not provide analytical solution. Otsuki et al. [23] formalized the method of separation of variables and superposition principle and used it for analyzing a one-dimensional wave equation using the Mizar theorem prover. However, they did not extend the solution for the infinite series. In the work we propose in this paper, we provide, for the first time, the formalization in HOL of the heat equation, in the form of a PDE modeling temperature variation for a rectangular solid. We conduct the formal verification in HOL Light of useful properties of the heat equation as well as verify its infinite series solution.

3 Preliminaries

In this section, we provide an overview of some of the fundamental formal definitions and notations of the multivariate calculus theories of HOL Light that are necessary for understanding the rest of the paper. The derivative of a real-valued function is defined in HOL Light as follows:

Definition 1. *Real Derivative*
⊢ ∀f x. real_derivative f x = (@f'. (f has_real_derivative f') (atreal x))

The function `real_derivative` accepts a real valued function f that needs to be differentiated and a real number x, and provides the derivative of f with respect to x. It is formally represented in functional form using the Hilbert choice operator @. The function `has_real_derivative` expresses the same functionality in relational style.

Definition 2. *Higher Real Derivative*
⊢ ∀f x. higher_real_derivative 0 (f:real→real) (x:real) = f x ∧
    (!n. higher_real_derivative (SUC n) (f:real→real) (x:real) =
            (real_derivative (λx. higher_real_derivative n f x) x))

The HOL Light function `higher_real_derivative` accepts an order n of the derivative, a real-valued function f and a real number x, and provides a higher-order derivative of order n for the function f with respect to x.

The infinite summation over a function f: $\mathbb{N} \rightarrow \mathbb{R}$ is formalized in HOL Light as follows:

Definition 3. *Real Sums*
⊢ ∀s f L. real_sums (f real_sums l) s ⇔
        ((λn. sum (s INTER (0..n)) f) → l) sequentially

The HOL Light function `real_sums` accepts a set of natural numbers s: $\mathbb{N} \rightarrow$ bool, a function f: $\mathbb{N} \rightarrow \mathbb{R}$ and a limit value l: $\mathbb{R}$, and returns the traditional mathematical expression $\sum_{k=0}^{\infty} f(k) = L$. Here, INTER captures the intersection of two sets. Similarly, `sequentially` represents a net providing a sequential growth of a function f, i.e., $f(k), f(k+1), f(k+2), ...,$ etc. This is mainly used in modeling the concept of an infinite summation.

We provide the formalization of the summability of a function f: $\mathbb{N} \rightarrow \mathbb{R}$ over s : $\mathbb{N} \rightarrow$ bool, which ensures that there exist some limit value l: $\mathbb{R}$, such that $\sum_{k=0}^{\infty} f(k) = L$ in HOL Light as:

Definition 4. *Real Summability*
⊢ ∀s f. real_summable s f = ∃l. (f real_sums l)

Now, we provide a formalization of an infinite summation, which will be used in the formal analysis of the heat conduction problem in Sect. 5 of the paper.

Definition 5. *Real Infsum*
⊢ ∀s f. real_infsum s f = @l. (f real_sums l) s

where the HOL Light function `real_infsum` accepts s: num $\rightarrow$ bool specifying the starting point and a function f of data-type $\mathbb{N} \rightarrow \mathbb{R}$, and returns a limit value l: $\mathbb{R}$ to which the infinite summation of f converges from the given s.

An infinite summation of a real-valued function Definition 5 can be mathematically expressed in an alternate form as follows:

$$\sum_{w=0}^{\infty} f_w(x) = \lim_{N \to \infty} \sum_{w=0}^{N} f_w(x)$$

We proved this equivalence in HOL Light as follows:

Theorem 1. *Alternate Representation of an Infinite Summation*
⊢ ∀f k s. real_infsum s (λw. f w x) =
        reallim sequentially (λk. sum (s INTER (0..k))(λw. f w x))

4   Formalization of the Heat Conduction Problem

Heat conduction is a phenomenon of energy transfer that occurs due to differences in temperature in adjacent components of a body/system. The heat is transferred from the high-temperature side to the low-temperature side until the body reaches its thermal equilibrium. The heat conduction or temperature variation can be mathematically defined as a function of space and time. Generally, the heat conduction in a body is three dimensional i.e., the conduction is significant in all three dimensions and a temperature variation in a body can be modeled as $T = T(x, y, z, t)$. The heat conduction is said to be two-dimensional when the conduction is significant in two-dimensions and negligible in the third dimension. Similarly, it is one-dimensional when the conduction is significant in one-dimensional only and the temperature variable can be modeled as $T = T(x, t)$. In this paper, we focus on the formalization of the one-dimensional heat conduction problem. In particular, we formally model the temperature variation in a rectangular slab using a PDE as a heat equation and formally verify its analytical solution by the method of separation of variables based on various boundary and initial conditions.

4.1   Heat Conduction Problem Formulation

A heat conduction problem for a rectangular slab having a thickness L is depicted in Fig. 1. We consider it as a one-dimensional heat conduction problem. Here, the function $u(x, t)$ provides the temperature in the slab at a point x and time t [24].

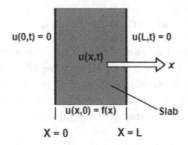

Fig. 1. Heat conduction across thickness of a slab [25]

We can mathematically express the one-dimensional heat conduction (temperature variation) in the rectangular slab as follows [25]:

$$\frac{\partial u(x, t)}{\partial t} = c \frac{\partial^2 u(x, t)}{\partial x^2} \qquad 0 < x < L, \quad t > 0 \tag{1}$$

where c is the thermal diffusivity of the slab that depends on the material used for constructing the slab. Equation (1) can be equivalently written as:

$$\frac{\partial u(x,t)}{\partial t} - c\frac{\partial^2 u(x,t)}{\partial x^2} = 0$$

Moreover, the solution of the heat equation (Eq. (1)) should satisfy the following initial and boundary conditions.

Initial Condition:

$$u(x,t)\,|_{t=0} = u(x,0) = f(x) \tag{2}$$

Boundary Conditions:

$$u(x,t)\,|_{x=0} = u(0,t) = 0 \tag{3}$$

$$u(x,t)\,|_{x=L} = u(L,t) = 0 \tag{4}$$

The heat equation (Eq. (1)) along with Eqs. (2), (3) and (4) is known as the initial boundary-value problem. It becomes an initial-value problem with respect to time that considers the only initial condition represented by Eq. (2). Whereas, in the case of its dependence on space only, it represents a boundary-value problem by incorporating the two boundary conditions expressed as Eqs. (3) and (4). Next, to formally verify the solution of the heat equation, we need to formalize it in higher-order logic.

4.2 Formalization of the Heat Equation

We formalize the heat equation (Eq. (1)) capturing the one-dimensional heat conduction in a rectangular slab in HOL Light as follows:

Definition 6. *The Heat equation*
⊢ heat_equation u(x,t) c ⇔ heat_operator u(x,t) c = &0

where **heat_equation** accepts a function u of type ($\mathbb{R} \times \mathbb{R} \to \mathbb{R}$), a space variable x: $\mathbb{R}$, a time variable t: $\mathbb{R}$ and the thermal diffusivity constant c, and returns the corresponding heat equation. The function **heat_operator** is formalized as follows:

Definition 7. *Heat operator*
⊢ ∀u x t.
heat_operator u(x,t) c = higher_real_derivative 1 (λt. u(x,t)) t -
                         c * higher_real_derivative 2 (λx. u(x,t)) x

Next, we verify a few important properties of the **heat_operator** Definition 7 that are required in formally verifying the solution of the heat equation.

Theorem 2. *Linearity*
⊢ ∀u x t a b.

[A1] (∀t. (λt. u(x,t)) real_differentiable atreal t) ∧
[A2] (∀t. (λt. v(x,t)) real_differentiable atreal t) ∧
[A3] (∀x. (λx. u(x,t)) real_differentiable atreal x) ∧
[A4] (∀x. (λx. v(x,t)) real_differentiable atreal x) ∧
[A5] (∀x. (λx. real_derivative (λx. u(x,t)) x)
            real_differentiable atreal x) ∧
[A6] (∀x. (λx. real_derivative (λx. v(x,t)) x)
            real_differentiable atreal x)

⇒ (heat_operator (λ(x,t). u(x,t) + v(x,t)) (x,t) c =
      heat_operator (λ(x,t). u(x,t)) (x,t) c +
         heat_operator (λ(x,t). v(x,t)) (x,t) c)

Assumptions A1 and A2 ensure that the real-valued functions u and v are differentiable at t, respectively. Assumptions A3 and A4 assert the differentiability of the functions u and v at x, respectively. Similarly, Assumptions A5 and A6 provide the differentiability conditions for the derivatives of the functions u and v at x, respectively. The proof of the above theorem is mainly based on the properties of derivative and differentiability of real-valued functions.

Theorem 3. *Scalar Multiplication*
⊢ ∀u x t a.

[A1] (∀t. (λt. u(x,t)) real_differentiable atreal t) ∧
[A2] (∀x. (λx. u(x,t)) real_differentiable atreal x) ∧
[A3] (∀x. (λx. real_derivative (λx. u(x,t)) x)
            real_differentiable atreal x)

⇒ heat_operator (λ(x,t). a * u(x,t)) (x,t) c =
                    a * heat_operator (λ(x,t). u(x,t))(x,t) c

Assumptions A1 and A2 ensure that the real-valued function u is differentiable at t and x, respectively. Assumption A3 asserts the differentiability condition for the derivative of the function u.

5   Formal Verification of the Solution of the Heat Equation

To find out the solution of the boundary-value problem, i.e., heat equation alongside the boundary conditions Eqs. (1), (3) and (4), we use the method of separation of variables that reduces the problem of solving a partial differential equation to a problem of solving the equivalent ordinary differential equations. By this method, we can mathematically express the solution of the heat equation $u(x, t)$ as a separable equation as follows:

$$u(x, t) = X(x)W(t) \qquad (5)$$

where X and W are functions of x and t, respectively. We formalize Eq. (5) in HOL Light as follows:

Definition 8. *Separable*
⊢ ∀X W t x. separable x t X W = X(x) * W(t)

By using Eq. (5) in the heat equation (Eq. (1)) and after simplification, we obtain the following equation.

$$\frac{1}{c}\frac{\partial[X(x)W(t)]}{\partial t} = \frac{\partial^2[X(x)W(t)]}{\partial x^2} \tag{6}$$

Next, using the property of the partial derivative of a separable function transforms the above equation as follows:

$$\frac{1}{c}\frac{dW(t)}{dt}X(x) = W(t)\frac{d^2X(x)}{dx^2} \tag{7}$$

where the operator $\frac{d}{dt}$ captures the simple derivative with respect to t. We formally verify the equivalence of the left-hand-sides of Eqs. (6) and (7) as the following HOL Light theorem.

Theorem 4. *Equivalence of Partial and Simple Derivatives (Left-hand Side)*
⊢ ∀X x W t.

[A1] (X real_differentiable atreal t) ∧
[A2] (W real_differentiable atreal t)

  ⇒ (real_derivative (λt. separable x t X W) t) =
     real_derivative W t * X x

Assumptions A1 and A2 provide the differentiability of the functions X and W at t, respectively. The proof process of the above theorem is mainly based on the properties of derivatives and differentiability of the real-valued functions along-with some arithmetic reasoning. Similarly, we formally verify the equivalence of the right-hand-sides of Eqs. (6) and (7) as follows:

Theorem 5. *Equivalence of Partial and Simple Derivatives (Right-hand Side)*
⊢ ∀X x W t.

[A1] (∀x. X real_differentiable atreal x) ∧
[A2] (∀x. W real_differentiable atreal x) ∧
[A3] (λx. real_derivative X x) real_differentiable atreal x

⇒ higher_real_derivative 2 (λx. (separable x t X W)) x =
                 W t * higher_real_derivative 2 (λx. X x) x

Assumptions A1 and A2 are very similar to that of Theorem 4. Assumption A3 ensures that the first-order derivative of the real-valued function X is differentiable at x. The verification of Theorem 5 is similar to that of Theorem 4.

Now, after rearranging various terms, Eq. (7) can be expressed as follows:

$$\frac{1}{c}\frac{dW(t)}{dt}\frac{1}{W(t)} = \frac{1}{X(x)}\frac{d^2X(x)}{dx^2} = -\beta^2 \tag{8}$$

where the left- and right-hand sides are functions of only t and x, respectively. The equivalence of these two functions of different variables is only possible when both are equal to some constant, which is represented by $-\beta^2$ in the above equation.

The above equation can be equivalently represented by the following two ordinary differential equations.

$$\frac{d^2 X(x)}{dx^2} + \beta^2 X(x) = 0 \qquad (9)$$

and

$$\frac{dW(t)}{dt} + c.\beta^2 W(t) = 0 \qquad (10)$$

Now, our problem of solving a boundary-value problem Eqs. (1), (3) and (4) has been transformed to solving a set of linear homogenous differential equations with constant coefficients Eqs. (9) and (10). Moreover, the solution of the heat equation Eq. (1) can be obtained by multiplying the solution of these two equations.

The solution of Eq. (9) is mathematically expressed as:

$$X(x) = A cos(\beta x) + B sin(\beta x) \qquad (11)$$

where A and B are the arbitrary constants that can be computed by applying the boundary conditions. Similarly, the solution of the second differential equation Eq. (10) is mathematically described as:

$$W(t) = Ce^{-\beta^2 ct} \qquad (12)$$

where C is the constant of integration and can be computed by applying the boundary conditions.

We formalize the two differential equations Eqs. (9) and (10) in HOL Light as follows:

Definition 9. *Formalization of Eq. (9)*
⊢ ∀X x b. first_equation X x b ⇔
    higher_real_derivative 2 (λx. X(x)) x + b pow (2) * (λx. X(x)) x = 0

Definition 10. *Formalization of Eq. (10)*
⊢ ∀W t b c.
    second_equation W t b c ⇔
        real_derivative (λt. W(t)) t + c * b pow (2) * W(t) = 0

Similarly, we formalized the solutions of these differential equations in HOL Light as:

Definition 11. *Solution of First Differential Equation*
⊢ ∀A B x b. first_equation_sol A B x b = A * cos(b * x) + B * sin(b * x)

Definition 12. *Solution of Second Differential Equation*
⊢ ∀C c b t.
    second_equation_sol C c b t = C * exp (-c * b pow (2) * t)

Next, we formally verify the solution of the first differential equation Eq. (9) as the following HOL Light theorem:

Theorem 6. *Solution of First Differential Equation*
⊢ ∀A B x b.
    (first_equation (λx. first_equation_sol A B x b)) x b

The proof process of the above theorem is based on Definitions 9 and 10 and properties of real derivative alongside some real arithmetic reasoning.

    Similarly, we formally verify the solution of the second differential equation Eq. (10) as follows:

Theorem 7. *Solution of Second Differential Equation*
⊢ ∀C c b t.
    (second_equation (λt. second_equation_sol C c b t))(t) b c

The proof process of the above theorem is based on Definitions 10 and 12 and properties of real derivative alongside some real arithmetic reasoning.

    To find out the values of arbitrary constants A and B of the solution of the ordinary differential equation expressed as Eq. (11), we apply the corresponding boundary conditions. Applying the first boundary condition Eq. (3) results into $A = 0$. Similarly, the application of the second boundary condition Eq. 4 provides $Bsin(\beta L) = 0$. We formally verify values of these arbitrary constants based on the corresponding boundary conditions in HOL Light as follows:

Theorem 8. *Verification of the Arbitrary Constant A*
⊢ ∀A B x b.
    x = &0 ∧ first_equation_sol A B x b = &0
                   ⇒ A = &0

Theorem 9. *Verification of the Arbitrary Constant B*
⊢ ∀A B x b L.
    x = L ∧ A = &0 ∧ first_equation_sol A B x b = &0
    ⇒ first_equation_sol x b A B = B * sin(b * L)

The equation $Bsin(\beta L) = 0$ holds if $B = 0$ or $sin(\beta L) = 0$. In case of $B = 0$ alongside $A = 0$, it results into $X(x) = 0$. This further provides $u(x, t) = 0$ as a solution to the heat equation, which is an uninteresting trivial solution. This means that B is equal to some non-zero value, which implies that $sin(\beta L) = 0$. Since β can have infinitely many values for which $sin(\beta L) = 0$ holds, namely $\beta = \beta_w = \frac{w\pi}{L}$. This results into a non-trivial solution of the boundary-value problem as follows:

$$u(x, t) = u_w(x, t) = \left[B_w sin \left(\frac{w\pi x}{L} \right) \right] e^{-\left(\frac{w\pi}{L} \right)^2 ct} \tag{13}$$

Now, assume that the function $f(x)$ in initial condition Eq. (2) is a linear combination of the function $sin(\frac{w\pi x}{L})$, i.e., Fourier sine series representation as follows:

$$f(x) = \sum_{w=1}^{\infty} B_w sin\left(\frac{w\pi x}{L}\right) \tag{14}$$

We can mathematically express the general solution of the heat equation as the following equation since it is a linear combination of the non-trivial solutions of the boundary-value problem that satisfies the initial condition expressed as Eq. (14).

$$u(x,t) = \sum_{w=1}^{\infty} u_w(x,t) = \sum_{w=1}^{\infty} B_w sin\left(\frac{w\pi x}{L}\right) e^{-\left(\frac{w\pi}{L}\right)^2 ct} \tag{15}$$

The constant B_w of the Fourier sine series representation of $f(x)$ can be determined using the orthogonality property of the sine function and is mathematically expressed as follows:

$$B_w = \frac{2}{L} \int_0^L f(x) sin\left(\frac{w\pi x}{L}\right) dx \qquad w = 1,2,3... \tag{16}$$

We formalize the Fourier sine coefficient in HOL Light as follows:

Definition 13. *Fourier Sine Coefficient*
⊢ ∀f w L.
    fourier_sine_coefficient f w L =
      2 / L * (real_integral (real_interval [0,L])(λx. (f x) *
        sin (&w * pi * x / L)))

where `fourier_sine_coefficient` accepts a function `f` : $\mathbb{R} \to \mathbb{R}$, a number `w` and the width of the slab `L`, and returns a real number representing the Fourier sine coefficient of the function `f`.

Now, the solution of the heat equation capturing the heat conduction in a rectangular slab can be alternatively expressed as:

$$u(x,t) = \sum_{w=1}^{\infty} u_w(x,t) = \sum_{w=1}^{\infty} \left(\frac{2}{L} \int_0^L f(x) sin\left(\frac{w\pi x}{L}\right) dx\right) sin\left(\frac{w\pi x}{L}\right) e^{-\left(\frac{w\pi}{L}\right)^2 ct} \tag{17}$$

We formalize the generalized solution of the heat equation (Eq. (17)) in HOL Light as follows:

Definition 14. *Generalized Solution of the Heat Equation*
⊢ ∀f x t c L.
    heat_solution f x t c L = real_infsum (from 1)
      (λw. (fourier_sine_coefficient f w L) *
        exp (-c * ((&w * pi / L) pow 2) * t) * sin (&w * pi * x / L))

The convergence of the generalized solution of the heat equation depends on the convergence of the infinite series $u_w(x,t)$ and is mathematically expressed as the following bound on $u_w(x,t)$.

$$|u_w(x,t)| \leq M_w \tag{18}$$

where

$$M_w = \left(\frac{2}{L} \int_0^L |f(x)| dx \right) e^{-\left(\frac{w\pi}{L}\right)^2 ct} \tag{19}$$

We compute the upper bound M_w using the upper bound on the Fourier coefficient B_w, and the fact that $\left| sin\left(\frac{w\pi x}{L}\right)\right| \leq 1$, along with the following property of the integral:

$$\left| \int_a^b f(x)dx \right| \leq \int_a^b |f(x)| dx. \tag{20}$$

Next, we formally verify the convergence of the generalized solution of the heat equation as the following HOL Light theorem.

Theorem 10. *Convergence of the Generalized Solution*
⊢ ∀f x c L t.

```
[A1] &0 < L ∧ [A2] &0 < t ∧ [A3] &0 < c ∧
[A4] f absolutely_real_integrable_on real_interval [&0, L]

⇒ ((λw. fourier_sine_coefficient f w L *
          exp (-c * (&w * pi / L) pow 2 * t) sin (&w * pi * x / L))
              real_sums heat_solution f x t c L) (from 1)
```

Assumptions (A1–A3) ensure that the width L, the time t and the constant c are positive real values. Assumption (A4) provides the absolute integrability of the function f over the interval [0, L]. The conclusion presents the convergence of the generalized solution of the heat equation. The verification of above Theorem 10 is mainly based on the following two important lemmas about the summability of the bound M_w and the generalized solution alongside some real arithmetic reasoning.

Lemma 1. *Summability of the Bound* M_w
⊢ ∀f c L t.

```
[A1] &0 < L ∧ [A2] &0 < t ∧ [A3] &0 < c

⇒ real_summable (from 1) (λw. &2 / L * real_integral
                  (real_interval [&0,L]) (λx. abs (f x)) *
                          exp (-c * ((&w * pi / L) pow 2) * t))
```

Assumptions (A1–A3) are the same as those of Theorem 10. The conclusion of the above lemma provides the summability of the upper bound M_w. The verification of Lemma 1 is mainly based on the Ratio test [26] along with some real arithmetic reasoning.

Lemma 2. *Summability of the Generalized Solution*
⊢ ∀f x c L t.

[A1] &0 < L ∧ [A2] &0 < t ∧ [A3] &0 < c ∧
[A4] f absolutely_real_integrable_on real_interval [&0, L]

⇒ real_summable (from 1)(λw. fourier_sine_coefficient f w L *
              exp (-c * (&w * pi / L) pow 2 * t) * sin (&w * pi * x / L))

Assumptions (A1–A4) are the same as those of Theorem 10. The verification of Lemma 2 is mainly based on the Comparison test [26] and Lemma 1 along with some real arithmetic reasoning. More details about the verification of these lemmas and the convergence of the generalized solution of the heat equation can be found in our HOL Light script [12].

Next, we formally verify some interesting properties involving the derivatives of the general solution with respect to position x and time t that capture the heat conduction (variation of temperature) in the rectangular slab with respect to position and time.

Theorem 11. *Derivative of the Generalized Solution with Respect to Time*
⊢ ∀f x t c L u u'.

[A1] (∀t. ((λw. (fourier_sine_coefficient f w L) *
      exp (-c * (&w * pi / L) pow 2 * t) * sin (&w * pi * x / L))
      real_sums u(x,t)) (from 1)) ∧
[A2] (∀t. ((λw. -c * (&w * pi / L) pow 2 * (fourier_sine_coefficient
      f w L) * exp (-c * (&w * pi / L) pow 2 * t) *
      sin (&w * pi * x / L)) real_sums u'(x,t)) (from 1)) ∧
[A3] ((λt. u(x,t)) has_real_derivative u'(x,t)) (atreal t)

⇒ real_derivative (λt. heat_solution f x t c L) t =
    real_infsum (from 1) (λw. -c * (&w * pi / L) pow 2) *
    (fourier_sine_coefficient f w L) *
       exp (-c * (&w * pi / L) pow 2) * t) * sin (&w * pi * x / L))

Assumption A1 provides the condition that the infinite series converges to the function $u(x,t)$. Similarly, Assumption A2 asserts that the derivative of the infinite series with respect to t converges to the derivative of function $u(x,t)$, i.e. $u'(x,t)$. Assumption A3 ensures the function u has derivative $u'(x,t)$ at point t. The verification of the above theorem is mainly based on swapping the operation of differentiation and infinite summation alongwith properties of the infinite summation and derivatives.

Theorem 12. *First Derivative of the Generalized Solution with Respect to Space*
⊢ ∀f x t c L u u'.

[A1] (∀x. ((λw. (fourier_sine_coefficient f w L) * exp (-c * (&w * pi / L)
      pow 2 * t) * sin (w * pi * x / L)) real_sums u(x,t)) (from 1)) ∧
[A2] ((∀x.((λw. (fourier_sine_coefficient f w L) * exp (-c * (&w * pi / L)
      pow 2 * t) * (&w * pi/ L) * cos (&w * pi * x / L)) real_sums
      u'(x,t)))(from 1) ∧

[A3] $((\lambda x.\ u(x,t))$ has_real_derivative u'$(x,t))$ (atreal x)

$\Rightarrow$ real_derivative $(\lambda x.$ heat_solution f x t c L) x =
   real_infsum (from 1)$(\lambda w.$ (fourier_sine_coefficient f w L) *
      exp $(-c * ((\&w * pi\ /\ L)$ pow 2) * t) * $(\&w * pi\ /\ L)$ *
                           cos $(\&w * pi * x\ /\ L))$

The proof process of Theorem 12 is very similar to that of Theorem 11.

Theorem 13. *Second Derivative of the General Solution with Respect to Space*
⊢ ∀f x t c L u u' u''.

[A1] $(\forall x.\ ((\lambda w.$ (fourier_sine_coefficient f w L) * exp $(-c * (\&w * pi\ /\ L)$
   pow 2 * t) * sin $(\&w * pi * x\ /\ L))$ real_sums u$(x,t))$ (from 1)) ∧

[A2] $((\forall x.((\lambda w.$ (fourier_sine_coefficient f w L) * exp $(-c * (\&w * pi\ /\ L)$
   pow 2 * t) * $(\&w * pi/\ L)$ * cos $(\&w * pi * x\ /\ L))$ real_sums
   u'$(x,t)))$(from 1) ∧

[A3] $((\lambda x.\ u(x,t))$ has_real_derivative u'$(x,t))$ (atreal x) ∧

[A4] $((\forall x.\ ((\lambda w.$ (fourier_sine_coefficient f w L) * exp $(-c * (\&w * pi\ /\ L)$
   pow 2 * t) * $(\&w * pi\ /\ L)$ pow 2 * -sin $(\&w * pi * x\ /\ L))$ real_sums
   u''$(x,t)))$ (from 1) ∧

[A5] $((\lambda x.\ u'(x,t))$ has_real_derivative u''$(x,t))$ (atreal x)

$\Rightarrow$ higher_real_derivative 2 $(\lambda x.$ heat_solution f x t c L) x =
   real_infsum (from 1) $(\lambda w.$ (fourier_sine_coefficient f w L) *
   exp $(-c * ((\&w * pi\ /\ L)$ pow 2) * t) * $((\&w * pi\ /\ L)$ pow 2) *
   -sin $(\&w * pi * x\ /\ L))$

The verification of the above theorem is mainly based on Theorem 12 and properties of derivatives along with some arithmetic reasoning.

Discussion

The distinguishing feature of our proposed formal analysis of the heat conduction problem, as compared to traditional analysis techniques, is that all verified theorems are of generic nature, i.e., all functions and variables involved in these theorems are universally quantified and thus can be specialized based on the requirement of the analysis of a rectangular slab with any width and corresponding boundary and initial conditions. Another advantage of our proposed approach is the inherent soundness of the theorem proving technique. It ensures that all the required assumptions are explicitly present along with the theorem, which are often ignored in conventional simulation based analysis and their absence may affect the accuracy of the corresponding analysis. One of the major difficulties in the proposed formalization was the swapping of the infinite summation and the differential operator that is used in the verification of Theorems 11–13. The mathematical proofs available in the literature for this swap operation were very abstract and we developed our own formal reasoning. In addition, to the best of our knowledge, this is the first formal work on the formalization of a one-dimensional heat equation and the verification of its infinite series solution.

6   Conclusion

In this paper, we proposed a HOL theorem proving based approach for formally analyzing the one-dimensional heat conduction in a rectangular slab. We formalized the heat equation and formally verified its linearity and scaling properties. Moreover, we used the separation of variables method for formally verifying the solution of the heat equation incorporating the corresponding boundary and initial conditions. Next, we formally verified convergence of the generalized solution of the heat equation. Finally, we verified some interesting properties regarding the derivatives of the generalized solution of the heat equation that provide useful insights to the variation of the temperature in the body. In future, we plan to formally verify the uniqueness of the generalized solution of the heat equation and its uniform convergence. Another future direction is to formally analyze the heat transfer in composite slabs [27], thermal protection systems [28] and heat transfer through various thermoelectric devices, such as thermoelectric generator and thermocouple [29] that are widely used in many safety-critical systems.

References

1. Strauss, W.A.: Partial Differential Equations: An Introduction. Wiley, Hoboken (2007)
2. Hahn, D.W., Özisik, M.N.: Heat Conduction. Wiley, Hoboken (2012)
3. Jiji, L.M.: Heat Convection. Springer, Heidelberg (2009). https://doi.org/10.1007/978-3-642-02971-4
4. Howell, J.R., Mengüç, M.P., Daun, K., Siegel, R.: Thermal Radiation Heat Transfer. CRC Press, Boca Raton (2020)
5. Minkowycz, W., Sparrow, E.M., Schneider, G.E., Pletcher, R.H.: Handbook of Numerical Heat Transfer. Wiley-Interscience, New York (1988)
6. Han, J.C.: Analytical Heat Transfer. Taylor & Francis, Boca Raton (2012)
7. Smith, G.: Numerical Solution of Partial Differential Equations: Finite Difference Methods. Oxford University Press, Oxford (1985)
8. Hughes, T.J.: The Finite Element Method: Linear Static and Dynamic Finite Element Analysis. Dover Publications, Mineola (2000)
9. Evans, L.: Partial Differential Equations. American Mathematical Society, Berkeley (2010)
10. Andrews, L.C., Shivamoggi, B.K.: Integral Transforms for Engineers. SPIE Press, Bellingham (1999)
11. Harrison, J.: Handbook of Practical Logic and Automated Reasoning. Cambridge University Press, Cambridge (2009)
12. Deniz, E., Rashid, A.: On the Formalization of the Heat Conduction Problem in HOL, HOL Light Script. https://hvg.encs.concordia.ca/code/hol-light/he/heat_conduction.ml
13. Immler, F., Hölzl, J.: Numerical analysis of ordinary differential equations in Isabelle/HOL. In: Beringer, L., Felty, A. (eds.) ITP 2012. LNCS, vol. 7406, pp. 377–392. Springer, Heidelberg (2012). https://doi.org/10.1007/978-3-642-32347-8_26
14. Immler, F., Traut, C.: The flow of ODEs: formalization of variational equation and Poincaré map. J. Autom. Reason. **62**(2), 215–236 (2018). https://doi.org/10.1007/s10817-018-9449-5

15. Guan, Y., Zhang, J., Wang, G., Li, X., Shi, Z., Li, Y.: Formalization of Euler-Lagrange equation set based on variational calculus in HOL light. J. Autom. Reason. **65**, 1–29 (2021)
16. Sanwal, M.U., Hasan, O.: Formal verification of cyber-physical systems: coping with continuous elements. In: Murgante, B., et al. (eds.) ICCSA 2013. LNCS, vol. 7971, pp. 358–371. Springer, Heidelberg (2013). https://doi.org/10.1007/978-3-642-39637-3_29
17. Rashid, A., Hasan, O.: Formalization of transform methods using HOL light. In: Geuvers, H., England, M., Hasan, O., Rabe, F., Teschke, O. (eds.) CICM 2017. LNCS (LNAI), vol. 10383, pp. 319–332. Springer, Cham (2017). https://doi.org/10.1007/978-3-319-62075-6_22
18. Rashid, A., Hasan, O.: On the formalization of Fourier transform in higher-order logic. In: Blanchette, J.C., Merz, S. (eds.) ITP 2016. LNCS, vol. 9807, pp. 483–490. Springer, Cham (2016). https://doi.org/10.1007/978-3-319-43144-4_31
19. Rashid, A., Hasan, O.: Formal analysis of linear control systems using theorem proving. In: Duan, Z., Ong, L. (eds.) ICFEM 2017. LNCS, vol. 10610, pp. 345–361. Springer, Cham (2017). https://doi.org/10.1007/978-3-319-68690-5_21
20. Rashid, A., Siddique, U., Hasan, O.: Formal verification of platoon control strategies. In: Johnsen, E.B., Schaefer, I. (eds.) SEFM 2018. LNCS, vol. 10886, pp. 223–238. Springer, Cham (2018). https://doi.org/10.1007/978-3-319-92970-5_14
21. Boldo, S., Clément, F., Filliâtre, J.-C., Mayero, M., Melquiond, G., Weis, P.: Formal proof of a wave equation resolution scheme: the method error. In: Kaufmann, M., Paulson, L.C. (eds.) ITP 2010. LNCS, vol. 6172, pp. 147–162. Springer, Heidelberg (2010). https://doi.org/10.1007/978-3-642-14052-5_12
22. Boldo, S., Clément, F., Filliâtre, J.C., Mayero, M., Melquiond, G., Weis, P.: Trusting computations: a mechanized proof from partial differential equations to actual program. Comput. Math. Appl. **68**, 325–352 (2014)
23. Otsuki, S., Kawamoto, P.N., Yamazaki, H.: A simple example for linear partial differential equations and its solution using the method of separation of variables. Formalized Math. **27**, 25–34 (2019)
24. Braun, M., Golubitsky, M.: Differential Equations and Their Applications. Springer, New York (1983). https://doi.org/10.1007/978-1-4684-9229-3
25. Hsu, T.R.: Applied Engineering Analysis. Wiley, Hoboken (2018)
26. Kline, M.: Calculus: An Intuitive and Physical Approach. Courier Corporation, North Chelmsford (1998)
27. De Monte, F.: Transient heat conduction in one-dimensional composite slab. A natural analytic approach. Int. J. Heat Mass Transfer **43**, 3607–3619 (2000)
28. Blosser, M.L.: Analytical solution for transient thermal response of an insulated structure. J. Thermophys. Heat Transfer **27**, 422–428 (2013)
29. Montecucco, A., Buckle, J., Knox, A.: Solution to the 1-D unsteady heat conduction equation with internal joule heat generation for thermoelectric devices. Appl. Therm. Eng. **35**, 177–184 (2012)

Isabelle/HOL/GST: A Formal Proof Environment for Generalized Set Theories

Ciarán Dunne[(✉)] and J. B. Wells[(✉)]

Heriot-Watt University, Edinburgh, UK
cmd1@hw.ac.uk
https://www.macs.hw.ac.uk/~jbw/

Abstract. A *generalized set theory* (GST) is like a standard set theory but also can have non-set structured objects that can contain other structured objects including sets. This paper presents Isabelle/HOL support for GSTs, which are treated as type classes that combine *features* that specify kinds of mathematical objects, e.g., sets, ordinal numbers, functions, etc. GSTs can have an exception feature that eases representing partial functions and undefinedness. When assembling a GST, extra axioms are generated following a user-modifiable policy to fill specification gaps. Specialized type-like predicates called *soft types* are used extensively. Although a GST can be used without a model, for confidence in its consistency we build a model for each GST from components that specify each feature's contribution to each tier of a von-Neumann-style cumulative hierarchy defined via ordinal recursion, and we then connect the model to a separate type which the GST occupies.

Keywords: Set theory · Higher-order logic · Soft types · Isabelle

1 Introduction

1.1 Set Theory

Many mathematicians (but not all) have long regarded set theories as suitable foundations of mathematics, in particular Zermelo/Fraenkel set theory (ZF) and other related theories, e.g., ZF plus the Axiom of Choice (ZFC) and Tarski/Grothendieck (TG) set theory which adds a universe axiom. At its core, ZF is a domain type V whose members are called "sets", a predicate $\in$ (membership) of type $V \Rightarrow V \Rightarrow$ bool, and axioms specifying $\in$. Due to cardinality constraints, some operators like $\in$ can not themselves be members of the domain V but must live in higher types. Although just the predicate $\in$ is enough, formalizing ZF is easier with some constants and additional operators at a few further types, e.g., $\mathcal{P}$ (power set) and $\bigcup$ (union) at type $V \Rightarrow V$ and Repl (replacement) at type $V \Rightarrow (V \Rightarrow V \Rightarrow$ bool$) \Rightarrow V$. Nonetheless, ZF needs few types and nearly all interesting mathematical objects live in type V.

Supported by EPSRC [EP/R513040/1 2273715].

ZF is usually specified by "axioms" written in first-order logic (FOL), but ZF can also be given in higher-order logic (HOL). Actively used proof systems implementing ZF or TG in FOL include Isabelle/ZF, Mizar (TG), and Metamath (ZF, when using the set.mm database). Active systems in HOL include the Isabelle/HOL development "ZFC in HOL" [13]. Other systems include: Isabelle/Set [9] (TG, HOL), Isabelle/Mizar [8] (TG), and Egal [3] (TG, HOL).

In all these systems, except for a few key operators like ∈, every mathematical object is a set. Arrangements of sets represent numbers, ordered pairs, functions, and nearly all other kinds of mathematical objects. Because sets are used to represent everything, representation overlaps are unavoidable. With the most commonly used representations, sad coincidences include, e.g., that $\langle 0, 1 \rangle = \{1, 2\}$ where 0, 1, and 2 are natural numbers, and that the successor function on natural numbers is equal to the number 1 as an integer. This troubles philosophers, leads university teachers to choose to deceive students, makes correct definitions more challenging, and adds difficulty to formalization.

1.2 Higher-Order Logic

Although many mathematicians favor set theory as a foundation of mathematics, many computer system implementers prefer formalisms where numerous types are used, rather than set theory's "one big type". Most of these systems extend Church's λ-calculus with simple types. We consider here HOL, which needs at its core only one type constructor $\Rightarrow$ and axioms and inference rules for constants for equality ($=$) and implication ($\rightarrow$). Most connectives ($\wedge$, $\neg$, $\forall$, etc.) are added via simple definitions that provide convenience, compactness, and readability, but no extra power. Sometimes domain-specific axioms are added that do provide extra power (which raises the question of whether these extensions are consistent, which we will address for our systems later in this paper). The systems we present in this paper are developed in Isabelle/HOL, the most active and widely used HOL proof system. Isabelle/HOL adds locale and type class mechanisms that support modularity, some limited type polymorphism, and overloading, and adds type definition features such as semantic subtypes, quotient types, and (co)recursive datatypes. Although Isabelle/HOL can be used just as a logical framework in which to define a set theory (as this paper does), typical uses of Isabelle/HOL generally put different kinds of mathematical objects into numerous fine-grained types.

1.3 Types

Consider ways "types" can be useful. One view of "type" is as an aspect of some object that can be inspected to help determine what you *should or want to* do with it. For example, when storing an incoming parcel, you might store food somewhere cold and jewelry in a safe. This corresponds to using types for overloading. Another view of "type" is as determining what you *meaningfully can* do with some object. For example, if you have two pieces of paper you want to attach, ice cream that you want to eat, a stapler, some staples, a bowl, and a spoon, you will want to use the right tools for each task. This corresponds to using types for avoiding meaningless combinations. There is no firm boundary between these two notions, e.g., you might *want* to store food somewhere cold because otherwise it might rot making the storage *not meaningful*.

Formal proving needs both views of "types" and these views interact. Using a lemma usually requires knowing if an operation is "defined" and the result's "type", and the answers to the same questions for operations on results *ad infinitum*. It is also frequently very useful to make decisions based on "types".

With this in mind, compare the set theory and HOL approaches (excluding HOL used solely as a framework for a set theory). HOL typically has precise and (usually) useful types for each object, obtained automatically with (often) reasonable efficiency, but sometimes it is quite hard to find types that allow fitting everything that is needed together. In set theory, typically the type of nearly everything is set, which conveys little information, but sets are a rich collection of every imaginable predicate and are usable as "types". However, there is (with currently available systems) a lack of automation for finding the right "types" and making them available in the right places at the right times.

1.4   Generalized Set Theories

A *generalized set theory* (GST) [1] specifies a domain type that contains set objects and may also contain other kinds of non-set structured objects, e.g., non-set ordered-pairs, non-set functions, non-set relations, etc. Many mathematicians view their work as based on ZF but their practice is often inconsistent with numbers, tuples, functions, etc., actually being sets. We believe much mathematics is, in effect, actually using a GST.

Subtype distinctions within a GST's domain type avoid representation overlaps. In a GST with non-overlapping sets, numbers, and ordered pairs, the answer to "is the ordered pair $\langle 0, 1 \rangle$ equal to the set $\{1, 2\}$?" is "no", an improvement over "yes" or "maybe" or "how dare you ask that question!". Subtype distinctions also help with showing that operations have meaningful results, and with choosing the most appropriate reasoning for each object.

A GST can have an exception object • (spoken "boom") which can not be confused with any other kind of object, can not lurk hidden inside any other object, and is useful for representing "undefined" results. This is similar to how terms in free logic can be "undefined" or fail to denote, but avoids the costs of actual undefinedness, e.g., inability to use standard HOL proof systems. This also avoids problems of other undefinedness approaches, e.g., the weakness of three-valued logic, the confusion caused by defining 1/0 to be 0, the need for amazingly complicated function domain specifications to handle definedness gaps, etc. Our approach avoids needing separate inner/outer domain quantifiers.

Recent unformalized theoretical work by us with Kamareddine [5] showed how to combine *features* to make a GST. Features include what are traditionally called *structures*, e.g., the natural numbers, and also include "large" (proper-class-sized) concepts like the sets, the ordered pairs, the functions, and the ordinals. This approach, also followed in this paper, is as follows. Each GST has a single domain type d and some features. Each feature specifies the existence of some objects in d, which the feature is usually considered to "own", and can specify relationships among all the objects in d. Additional axioms specify that every object is "owned" by exactly one feature. Features are specified as independently as is feasible so, e.g., the specification of ordered pairs knows nothing about sets and *vice versa*. This is useful because, e.g., for sets and ordered pairs, (1) mathematicians think of these as distinct kinds of things, (2) the independent specification of ordered pairs is easier to comprehend than, e.g., Kuratowski's definition in terms of sets, and (3) this enables custom GSTs with only the desired primitive features. Our recent work also suggested theoretically how to build a model of a GST Q within another GST P and then connect that model to a type containing the GST Q, thereby reducing the question of the consistency of Q to the consistency of P. Our previous work had not, until this paper, been formalized.[1] Aside from this paper and the work mentioned above, the only model building for GSTs seems to be by Aczel and Lunnon [2], and their set theories have the anti-foundation axiom, which is quite different from our work.

1.5   Isabelle/HOL/GST

Building on Isabelle/HOL and earlier work on GSTs, we present a formal proving environment for defining GSTs and reasoning about them, including building models for them. The development consists of much Isabelle/Isar code for high-level formalization, as well as Isabelle/ML for the implementation of machinery that assists GST specification and model building. The source code of our development and instructions for use are available at https://www.macs.hw.ac.uk/~cmd1/isabelle-gst/.

Our development provides: (1) Definitions of a number of *soft type constructors*, operators that construct and manipulate *soft types*, HOL predicates of type $\tau \Rightarrow \star$, where $\star$ is this paper's name for bool, and proofs of many useful properties of these. (2) A formal notion of *feature*, with a corresponding implementation as an Isabelle/ML record type. A feature contains a pointer to an Isabelle (type) class, which manages an abstract specification of the dependencies on other classes, signature, axioms, definitions, theorems, syntax, etc., of a feature, and also contains information used when combining features. (3) Definitions and theorems in Isabelle theories for features for ZF-style sets, functions, ordinals, ordinal recursion, an exception object, ordered pairs, natural numbers and binary relations.[2] (4) Automation for combining features to define GSTs, which are implemented as Isabelle classes with axioms for enforcing: (a) that each object in the domain of individuals is owned by exactly one feature, (b) a policy specifying allowed combinations of objects of different features, and a policy for filling in specification gaps. These two policies primarily support approaches

---

[1] A mostly formalized proof had been given earlier in Isabelle/ZF of the existence of a model of a much simpler system with just two fixed features [4].

[2] Natural numbers and binary relations can be found in the Isabelle source code, as can ordered pairs, which also have a LaTeX presentation in this paper's long version.

to handling "undefined" operations, including the use of an exception object. (5) Generic definitions and reasoning for models of GSTs, which are cumulative hierarchies defined using ordinal recursion. (6) A notion of *model component*, with a corresponding implementation as an ML record type. Each model component specifies a schematic description of a part of a model, primarily constraints placed on the zero, successor, and limit cases of the ordinal recursion used in defining models. (7) Automatic generation of terms, theorems and proof states that assist in implementing GST models. (8) Automatic lifting and generation of transfer rules for constants on types obtained by Isabelle/HOL type definitions on GST models, gained by interfacing Lifting and Transfer. (9) A bootstrap of our development from Paulson's "ZFC in HOL", carried out by instantiating our GZF (set), Ordinal, OrdinalRec (ordinal recursion), OPair (ordered pair), and Function classes at the type V of "ZFC in HOL", which makes a GST in which every object is a set. (10) An example GST called ZF^+ (defined as a class) with sets and all of the following as non-set objects: functions, ordinals, and the exception object. (11) An Isabelle type d_0 which instantiates the ZF^+ class, obtained by type definition on a model built using our model-building kit in V, justifying confidence in using ZF^+, provided you have faith in Isabelle/HOL and the axioms of ZFC that were added at V.

Items (5), (6), (7), and (8) with model building details are not presented in this paper, but are described in the appendix of the long version of this paper. Full details can be found in the `ModelKit/` directory of the Isabelle source.

1.6   Summary of Contributions

This paper presents a formalization of generalized set theories in Isabelle/HOL. Section 2 formulates a logical framework with classes like those of Isabelle. Section 3.1 defines GST features as classes with associated data. Section 3.2 presents example features for ZF-style sets, ordinals, functions, and an exception element. Section 3.3 defines how to combine features to create a GST and shows how to combine our example features to make the GST ZF^+ which has all these features within one domain type. Section 4 presents examples of working in ZF^+. Section 5 presents a methodology for building models of GSTs in Isabelle and how we use the Lifting and Transfer packages to create types that instantiate GSTs. Section 5 also discusses how we justify confidence in ZF^+ by building a model in the type V of "ZFC in HOL", defining a type d_0 isomorphic to this model, and instantiating ZF^+ at d_0.

2   Mathematical Definitions and Logical Framework

We assume a set-theoretic meta-level with at least: equality $\equiv$; definitions $:=$; an empty set $\varnothing$; set membership ϵ, binary union $\uplus$, set literals $\{\!\{\,\square_1, \ldots, \square_n\,\}\!\}$, and comprehensions $\{\!\{\,\square \mid \square\,\}\!\}$; an empty list $\diamond$, tuple/list literals $[\square_1, \ldots, \square_n]$, membership ϵ:, cons #, append @, and comprehensions $[\square \mid \square]$; and a set of natural numbers $\mathbb{N}$. Uses of $\dot{\epsilon}$ declare that symbols on the left side of $\dot{\epsilon}$ are metavariables ranging over the set on the right side.

2.1 Syntax and Types

Our logical framework is close to and inspired by the Isabelle/HOL formulation of Kunčar and Popescu [12]. Figure 1 defines the meta-level sets TVar of *type variables*, RVar of *raw term variables*, and Type, Var, and Const. Type consists of type variables, the type $\star$ of truth claims/assumptions, *domain types* d_i, and *operator types* $\sigma \Rightarrow \tau$.[3] The constructor $\Rightarrow$ is right associative so $(\tau_1 \Rightarrow \tau_2 \Rightarrow \tau_3) \equiv (\tau_1 \Rightarrow (\tau_2 \Rightarrow \tau_3))$. $\mathsf{TV_{typ}}(\sigma)$ yields the set of type variables occurring in σ. Var is the set of *term variables*, which are pairs of raw term variables and types. A term variable $[\bar{x}, \sigma]$ may be written as $\bar{x}_\sigma$. Const is a set of *constants*. The metavariable $\mathcal{K}$ ranges over lists of constants. The fixed meta-level function $\mathsf{ctyp} : \mathsf{Const} \to \mathsf{Type}$ assigns each constant a type. We write $\bar{\kappa} \ \overline{::} \ \tau$ for $\mathsf{ctyp}(\bar{\kappa}) \equiv \tau$. Figure 1 assigns some constants some types.

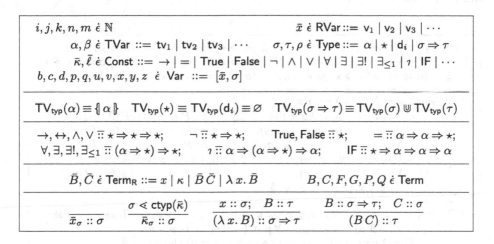

Fig. 1. Base syntax, type variables, constants, terms, and the typing relation.

The notation $\sigma[\alpha := \tau]$ denotes the *type substitution* replacing occurrences of the type variable α in σ by τ. A type σ is an *instance* of τ (written $\sigma \ll \tau$) iff $\sigma \equiv \tau[\alpha_1 := \rho] \ldots [\alpha_n := \rho_n]$ for some $\rho_1, \ldots, \rho_n$. For example, $(d_1 \Rightarrow d_1) \ll (\alpha \Rightarrow \alpha)$. A *constant instance* is a pair $[\bar{\kappa}, \sigma]$ such that $\sigma \ll \mathsf{ctyp}(\bar{\kappa})$ (e.g., $[\mathcal{P}, d_0 \Rightarrow d_0]$ and $[\in, d_1 \Rightarrow d_1]$ are constant instances for the constants $\mathcal{P} \ \overline{::} \ \alpha \Rightarrow \alpha$ and $\in \ \overline{::} \ \alpha \Rightarrow \alpha \Rightarrow \star$ respectively). Constant instances $[\bar{\kappa}, \sigma]$ may be written as $\bar{\kappa}_\sigma$, and if $\mathsf{ctyp}(\bar{\kappa})$ has no type variables, then we may write $\bar{\kappa}$ for the constant instance $\bar{\kappa}_{\mathsf{ctyp}(\bar{\kappa})}$. Let θ, κ range over constant instances.

Figure 1 defines the set $\mathsf{Term_R}$ of *raw terms*. A raw term is a variable x, a constant instance κ, an *application* $\bar{B}\,\bar{C}$, or an *abstraction* $\lambda\,x.\bar{B}$. Define the *free variable* meta-level function $\mathsf{FV_{trm}}$ and α-equivalence on $\mathsf{Term_R}$ as usual. Let $\mathsf{Term_\alpha}$ be $\mathsf{Term_R}$ modulo α-equivalence. Define *term substitution* and *type*

---

[3] At object level, "function" means an object satisfying the Fun predicate of Fig. 4.

substitution on Term_α as usual with the notation $\bar{B}[x := \bar{C}]$ and $\bar{B}[\alpha := \sigma]$. Define β-equivalence on Term_α as usual. Let Term be Term_α modulo β-equivalence. Lift $\mathsf{FV}_{\mathrm{trm}}$ and term and type substitution to Term.

Let the typing relation :: between Term and Type be the least relation satisfying the rules in Fig. 1. Let a term B be a *formula* iff $B :: \star$. A *simple definition* is a formula of the form $\kappa =_\sigma B$.[4] Let φ, ψ, and γ range over formulas, let Δ, Γ, and Θ range over sets of formulas, and let Φ range over formula lists.

We adopt the following notation. Given $x \equiv \bar{x}_\sigma$, the expression $\lambda\, x :: \sigma.\, B$ stands for $\lambda\, \bar{x}_\sigma.\, B$. Inside term expressions $B =_\sigma C$ and $\lambda\, x :: \sigma.\, B$, we allow omitting the type σ (and the ::), provided that σ can be uniquely determined by the typing rules and other type information in or about B and C. The notation $\lambda x_1 \cdots x_n.\, B$ stands for the nested abstractions $\lambda x_1.\, \cdots (\lambda x_n.\, B)$. The notation $[\, \kappa_1 :: \tau_1, \ldots, \kappa_n :: \tau_n \,]$ denotes the list $[\, \kappa_1, \ldots, \kappa_n \,]$ and asserts that $\kappa_i \,\epsilon\, \mathsf{Const}$ and $\kappa_i \mathbin{\overset{..}{\cdot}} \tau_i$ for $1 \le i \le n$.

$$(\colon) \mathbin{\overset{..}{\cdot}} \alpha \Rightarrow (\alpha \Rightarrow \star) \Rightarrow \star \qquad\qquad \to \mathbin{\overset{..}{\cdot}} (\alpha \Rightarrow \star) \Rightarrow (\beta \Rightarrow \star) \Rightarrow ((\alpha \Rightarrow \beta) \Rightarrow \star)$$

$$\top, \bot \mathbin{\overset{..}{\cdot}} \alpha \Rightarrow \star \qquad\qquad \Pi \mathbin{\overset{..}{\cdot}} (\alpha \Rightarrow \star) \Rightarrow (\alpha \Rightarrow \beta \Rightarrow \star) \Rightarrow ((\alpha \Rightarrow \beta) \Rightarrow \star)$$

$$\sqsubseteq \mathbin{\overset{..}{\cdot}} (\alpha \Rightarrow \star) \Rightarrow (\alpha \Rightarrow \star) \Rightarrow \star \qquad \sqcap, \sqcup \mathbin{\overset{..}{\cdot}} (\alpha \Rightarrow \star) \Rightarrow (\alpha \Rightarrow \star) \Rightarrow (\alpha \Rightarrow \star)$$

$\mathsf{SoftTypeOps} := \{\!| \ (\colon) = \lambda\, x\, p.\, p\, x,$

$\qquad\qquad \top = \lambda\, x.\, \mathsf{True}, \quad \bot = \lambda\, x.\, \mathsf{False}, \qquad\qquad\qquad \Pi\, x : P.\, Q := \Pi\, P\, (\lambda\, x.\, Q)$

$\qquad\qquad \to = \lambda\, p\, q\, f.\, \forall\, x.\, (x : p \to f\, x : q), \qquad\qquad\quad \forall\, x : P.\, \varphi := \forall\, x.\, (x : P \to \varphi)$

$\qquad\qquad \Pi = \lambda\, p\, q\, f.\, \forall\, x.\, (x : p \to f\, x : q\, x), \qquad\qquad \exists\, x : P.\, \varphi := \exists\, x.\, (x : P \wedge \varphi)$

$\qquad\qquad \sqcap = \lambda\, p\, q\, x.\, (x : p \wedge x : q), \qquad\qquad\qquad\quad \exists_{\le 1}\, x : P.\, \varphi := \exists_{\le 1}\, x.\, (x : P \wedge \varphi)$

$\qquad\qquad \sqcup = \lambda\, p\, q\, x.\, (x : p \vee x : q), \qquad\qquad\qquad\quad \exists!\, x : P.\, \varphi := \exists!\, x.\, (x : P \wedge \varphi)$

$\qquad\qquad \sqsubseteq = \lambda\, p\, q.\, \forall\, x.\, (x : p \to x : q) \ |\!\}$

$\mathsf{HOL} := \{\!| \ \forall\, x.\, x =_\alpha x, \quad \forall\, p, x, y.\, x =_\alpha y \to p\, x \to p\, y, \quad \forall\, p.\, p = \mathsf{True} \vee p = \mathsf{False},$

$\qquad\qquad \forall\, p, d.\, (\exists!\, p) \to p\, (\imath\, d\, p), \quad \forall\, p, d.\, (\neg\, \exists!\, p) \to \imath\, d\, p = d,$

$\qquad\qquad \mathsf{IF} = \lambda\, b\, x\, y.\, \imath\, c.\, (b \to c = x) \wedge (\neg\, b \to c = y) \ \text{else}\ x, \ \ldots\ |\!\}$

(ASSM)	If $\varphi\,\epsilon\,\mathsf{HOL} \uplus \mathsf{ZFCinHOL} \uplus \Delta \uplus \Gamma$, then $\Delta; \Gamma \vdash \varphi$	
(IMPI)	If $\Delta; \Gamma \uplus \{\!| \varphi |\!\} \vdash \psi$, then $\Delta; \Gamma \vdash \varphi \to \psi$	
(IMPE)	If $\Delta; \Gamma \vdash \varphi \to \psi$ and $\Delta; \Gamma \vdash \varphi$, then $\Delta; \Gamma \vdash \psi$	
(TYP-INST)	If $\Delta; \Gamma \vdash \varphi$ and $\alpha \notin \mathsf{TV}_{\mathsf{set}}(\Gamma)$, then $\Delta; \Gamma \vdash \varphi[\alpha := \sigma]$	
(TRM-INST)	If $\Delta; \Gamma \vdash \varphi$, $B :: \sigma$, and $\bar{x}_\sigma \notin \mathsf{FV}_{\mathsf{set}}(\Delta \uplus \Gamma)$, then $\Delta; \Gamma \vdash \varphi[\bar{x}_\sigma := B]$	
(EXT)	If $\Delta; \Gamma \vdash F\, \bar{x}_\sigma =_\tau G\, \bar{x}_\sigma$, then $\Delta; \Gamma \vdash F =_{\sigma \Rightarrow \tau} G$	

Fig. 2. Types, definitions and notation for soft types, axioms, and inference rules.

If $\bar{\kappa}$ is *infix*, an application $(\bar{\kappa}_\sigma\, B)\, C$ is written as $B\, \bar{\kappa}_\sigma\, C$. The constants $=$, $\wedge$, $\vee$, and $\to$ are all infix, and listed here in descending order of precedence. Negation and operator application take precedence over infix operators, e.g., $\neg\, P \wedge \neg\, Q$ is $(\neg\, P) \wedge (\neg\, Q)$ and $F\, x =_\sigma G\, x$ is $(F\, x) =_\sigma (G\, x)$. If $\bar{\kappa}$ is a *binder*, an application

---

[4] B must not refer to κ, including via a chain of other definitions.

$\bar{\kappa}_\sigma \,(\lambda x :: \sigma.\,B)$ is written as $(\bar{\kappa}x :: \sigma.\,B)$, and $\bar{\kappa}x_1,\ldots,x_n.\,B$ stands for the nested applications of quantifiers and abstractions $\bar{\kappa}_\sigma \,(\lambda\,x_1 :: \sigma.\,\cdots(\bar{\kappa}_\sigma \,(\lambda\,x_n :: \sigma.\,B)))$. The constants $\forall$, $\exists$, $\exists!$, and $\exists_{\leq 1}$ are all binders.

2.2 Soft Types

A *soft type* is an operator of type $\tau \Rightarrow \star$ for some τ. A key difference from "hard" types is that each object will satisfy many (often infinitely many) soft types. Figure 2 gives types and simple definitions for operators for building and using soft types. Our soft type constructors are mostly the same as those used by Kappelmann, Chen, and Krauss in Isabelle/Set [9], which in turn are along the lines suggested much earlier by Krauss [10]. To help the reader think "types", and also to allow proof tactics to follow soft-type-specific strategies, we write "$F : P$" as a *soft typing* which means the same thing as "$P\,F$", i.e., F satisfies the predicate P. We also have a non-dependent constructor $\rightarrowtail$ and a dependent constructor Π for soft types on operators, soft intersection and union type constructors $\sqcap$ and $\sqcup$, and soft subtyping $\sqsubseteq$. Our development derives the standard introduction and elimination rules for each of these concepts. The constants $(:)$, $\rightarrowtail$, $\sqcap$, $\sqcup$, and $\sqsubseteq$ are all infix, and $\rightarrowtail$ is right-associative. The figure gives notation for dependent operator soft types, restricted quantification, and restricted binding of λ-expressions. Let SoftTypeOps be the set of simple definitions as defined in Fig. 2.

For example, using the soft operator type constructor $\rightarrowtail$, given a predicate $P :: \sigma \Rightarrow \star$ and a predicate $Q :: \tau \Rightarrow \star$, the term $P \rightarrowtail Q$ is a predicate of type $(\sigma \Rightarrow \tau) \Rightarrow \star$ such that if $P \rightarrowtail Q$ is true of $x :: \sigma \Rightarrow \tau$ and P is true of $y :: \sigma$ then Q is true of $x\,y :: \tau$. Precisely, $F : P \rightarrowtail Q$ means $\forall\,b.\;b : P \to F\,b : Q$.

2.3 Inference Rules and Axioms

Let $\mathsf{TV}_{\mathsf{trm}}$ be the extension of $\mathsf{TV}_{\mathsf{typ}}$ to terms. Let $\mathsf{FV}_{\mathsf{set}}(\varGamma)$ be the union of all $\mathsf{FV}_{\mathsf{trm}}(\varphi)$ for all $\varphi \,\epsilon\, \varGamma$. Let $\mathsf{TV}_{\mathsf{set}}(\varGamma)$ be the union of all $\mathsf{TV}_{\mathsf{trm}}(\varphi)$ for all $\varphi \,\epsilon\, \varGamma$. Let $\varGamma[x := B]$ be the set of all $\varphi[x := B]$ for all $\varphi \,\epsilon\, \varGamma$.

HOL is the set of axioms (formulas) given in Fig. 2 that implement reflexivity of equality, indiscernability of equal objects, the law of the excluded middle, a definite description operator with a default ($\imath$), a conditional operator (IF), and simple definitions (not shown) for True, $\forall$, $\exists$, False, $\neg$, $\wedge$, $\vee$, $\leftrightarrow$, $\exists_{\leq 1}$, and $\exists!$. Nice notation for $\imath$ and IF are given thus: $(\imath\,x.\,\varphi\,\text{else}\,D) := \imath\,D\,(\lambda\,x.\,\varphi)$, (if P then B else C) := IF $P\,B\,C$. Let ZFCinHOL be the set of axioms used in Paulson's "ZFC in HOL" [13]. Let the *deduction relation* $\vdash$ be the least relation satisfying the inference rules in Fig. 2. Our normal use will be to derive judgements of the form HOL $\uplus$ SoftTypeOps $\uplus\,\Delta;\varGamma \vdash \varphi$ where Δ contains additional definitions specific to the topic of the proof and $\varGamma$ contains local assumptions.

2.4 Classes

A *(type) class* is a tuple $\mathcal{C} \equiv [\mathcal{D},\mathcal{K},\Phi,\Theta]$. $\mathcal{D}$ is a list of classes called the *dependencies* of $\mathcal{C}$. $\mathcal{K}$ is a list $[\kappa_1,\ldots,\kappa_n]$ of pairwise distinct constants called the *parameters* of $\mathcal{C}$, such that $\mathsf{TV}_{\mathsf{trm}}(\kappa_1) \uplus \ldots \uplus \mathsf{TV}_{\mathsf{trm}}(\kappa_n) \equiv \{\!| \alpha |\!\}$ for some α, i.e., exactly one type variable occurs in the parameters of $\mathcal{C}$, which we refer to as $\mathsf{tv}(\mathcal{C})$. Φ is a list of formulas and Θ is a list of simple definitions, called the *axioms* and *definitions* of $\mathcal{C}$ respectively.

We write $\sigma[\![d_i]\!]$ and $B[\![d_i]\!]$ for the results of the type substitutions $\sigma[\mathsf{tv}(\mathcal{C}) := d_i]$ and $B[\mathsf{tv}(\mathcal{C}) := d_i]$ respectively. A *parameter instantiation* of $\mathcal{C}$ at d_i is a set of formulas $\{\!| \kappa_1[\![d_i]\!] = B_1, \ldots, \kappa_n[\![d_i]\!] = B_n |\!\}$, where $\mathsf{TV}(B_j) \equiv \varnothing$ for $j \in \{\!| 1, \ldots, n |\!\}$. Relative to a set of hypothesis Γ and a set of definitions Δ, we say that $\mathcal{C}$ is *instantiated* at d_i if we have $\Delta; \Gamma \vdash \varphi$ for any $\varphi \in \Phi[\![d_i]\!]$. Typically Δ will contain parameter instantiations for all of the dependencies of $\mathcal{C}$.

3   GSTs as Type Classes

3.1   GST Features

A *feature* is a tuple $\mathcal{F} \equiv [\mathcal{C}, P_{\mathsf{logo}}, P_{\mathsf{cargo}}, \kappa_{\mathsf{default}}]$, where $\mathcal{C}$ is a class, $P_{\mathsf{logo}}, P_{\mathsf{cargo}} :: \mathsf{tv}(\mathcal{C}) \Rightarrow \star$ are the *cargo* and *logo* soft types of $\mathcal{F}$, and κ is a constant called the *default parameter* of $\mathcal{F}$. The terms $P_{\mathsf{logo}}, P_{\mathsf{cargo}}, \kappa_{\mathsf{default}}$ keep track of information used when combining features to create GSTs. P_{logo} (a.k.a. $\mathsf{logo}(\mathcal{F})$) should be chosen as the soft type of all objects contributed to the domain by a feature's axioms. For features that do not contribute any objects to the domain, $\perp$ should be used. P_{cargo} (a.k.a. $\mathsf{cargo}(\mathcal{F})$) should be chosen as the soft type satisfied of all objects contained in the internal structure of some object $X : P_{\mathsf{logo}}$. Keeping track of cargo types allows preventing the exception object $\bullet$ from being contained in sets and functions, allowing the benefits of a free logic, i.e., terms can be "undefined", without the need for anything to actually really be undefined and without the need for separate quantifiers for an "inner" and "outer" domain. The $\bullet$ object is in the spirit of concepts like the number 0 and the empty set $\emptyset$, i.e., it is a object representing what would otherwise be the lack of an object. Each feature's class has its default parameter $\kappa_{\mathsf{default}}$ (a.k.a. $\mathsf{default}(\mathcal{F})$) in its list of parameters, given as the symbol $\circ$ decorated with a subscript. The default parameter is intended to be a placeholder that can be used in a feature's axioms and definitions for exceptional results and normally it will be axiomatized to be equal to some specific object (typically $\bullet$) when features are combined into a GST.

3.2   Example Features

We now define features for sets, ordinals, ordinal recursion, ordered pairs, functions, and an exception ($\bullet$). Figures 3 and 4 define constants, axioms, and definitions for each feature's class. Figure 5 defines the classes and their features (in boldface). The Isabelle/HOL development can be found in `GST_Features.thy`.

We have formulated some of our features' axioms as soft typings. This is more useful in a GST than in set-only ZF, because even basic set operations like $\bigcup$ and $\mathcal{P}$ can be "undefined" because, e.g., the argument might be a non-set.

$\text{GZF}_{\text{consts}} := [\ \circ_{\text{GZF}} :: \alpha,\ \text{Set} :: \alpha \Rightarrow \star,\ \in\ ::\ \alpha \Rightarrow \alpha \Rightarrow \star,\ \bigcup :: \alpha \Rightarrow \alpha,\ \mathcal{P} :: \alpha \Rightarrow \alpha,$
$\qquad\qquad \emptyset :: \alpha,\ \text{Succ} :: \alpha \Rightarrow \alpha,\ \text{Inf} :: \alpha,\ \text{Repl} :: \alpha \Rightarrow (\alpha \Rightarrow \alpha \Rightarrow \star) \Rightarrow \alpha\ ]$

$\text{GZF}_{\text{axioms}} := [\ \bigcup : \text{SetOf Set} \rightarrow \text{Set},\quad \mathcal{P} : \text{Set} \rightarrow \text{SetOf Set},$
$\qquad\qquad \emptyset : \text{Set},\quad \text{Succ} : \text{Set} \rightarrow \text{Set},\quad \text{Inf} : \text{Set},$
$\qquad\qquad \text{Repl} : (\Pi\, x : \text{Set}.\, \text{ReplPred}\, x \rightarrow \text{Set}),$
$\qquad\qquad \forall\, x, y : \text{Set}.\, \forall\, b.\, (b \in x \leftrightarrow b \in y) \rightarrow x = y,$
$\qquad\qquad \forall\, x : \text{SetOf Set}.\, \forall\, b.\, b \in \bigcup x \leftrightarrow (\exists\, y.\, y \in x \land b \in y),$
$\qquad\qquad \forall\, x, y : \text{Set}.\, y \in \mathcal{P}\, x \leftrightarrow y \subseteq x,$
$\qquad\qquad \forall\, b.\, \neg b \in \emptyset,\quad \forall\, x : \text{Set}.\, b \in \text{Succ}\, x \leftrightarrow (b \in x \lor b = x),$
$\qquad\qquad \emptyset \in \text{Inf} \land (\forall\, b.\, b \in \text{Inf} \rightarrow \text{Succ}\, b \in \text{Inf}),$
$\qquad\qquad \forall\, x : \text{Set}.\, \forall\, p : \text{ReplPred}\, x.$
$\qquad\qquad\qquad \forall\, c.\, c \in \text{Repl}\, x\, p \leftrightarrow (\exists\, b.\, b \in x \land p\, b\, c \land c : \text{SetMem})\ ]$

$\text{GZF}_{\text{defs}} := [\ \subseteq\ =\ (\lambda\, x\, y.\, \forall\, b.\, b \in x \rightarrow b \in y),\quad \text{SetMem} = (\lambda\, b.\, \exists\, y : \text{Set}.\, b \in y),$
$\qquad\qquad \text{SetOf} = (\lambda\, p\, x.\, x : \text{Set} \land \forall\, b \in x.\, b : p),$
$\qquad\qquad \text{ReplPred} = (\lambda\, x\, p.\, \forall\, b \in x.\, \exists_{\leq 1}\, c : \text{SetMem}.\, p\, b\, c)\ ]$

$\text{Ord}_{\text{consts}} := [\ \circ_{\text{Ord}} :: \alpha,\ \text{Ord} :: \alpha \Rightarrow \star,\ <\ ::\ \alpha \Rightarrow \alpha \Rightarrow \star,\ 0 :: \alpha,\ \text{succ} :: \alpha \Rightarrow \alpha,\ \omega :: \alpha\ ]$

$\text{Ord}_{\text{axioms}} := [\ 0 : \text{Ord},\quad \text{succ} : \text{Ord} \rightarrow \text{Ord},\quad \omega : \text{Limit},$
$\qquad\qquad \forall\, u : \text{Ord}.\, \neg u < 0,$
$\qquad\qquad \forall\, u, v : \text{Ord}.\, u < \text{succ}\, v \leftrightarrow (u < v \lor u = v),$
$\qquad\qquad \forall\, u : \text{Limit}.\, u = \omega \lor \omega < u,$
$\qquad\qquad \forall\, u, v, w : \text{Ord}.\, u < v \rightarrow v < w \rightarrow u < w \quad \text{(transitivity)},$
$\qquad\qquad \forall\, u, v : \text{Ord}.\, u < v \rightarrow \neg v < u \quad \text{(antisymmetry)},$
$\qquad\qquad \forall\, u, v : \text{Ord}.\, u < v \lor u = v \lor v < u \quad \text{(trichotomy)},$
$\qquad\qquad \forall\, p.\, (\forall\, u : \text{Ord}.\, (\forall\, v : \text{Ord}.\, v < u \rightarrow p\, v) \rightarrow p\, u) \rightarrow (\forall\, w : \text{Ord}.\, p\, w)\ ]$

$\text{Ord}_{\text{defs}} := [\ \text{Limit} = (\lambda\, u.\, u : \text{Ord} \land 0 < u \land (\forall\, v : \text{Ord}.\, v < u \rightarrow \text{succ}\, v < u))\ ]$

$\text{OrdRec}_{\text{consts}} := [\ \circ_{\text{OrdRec}} :: \alpha,\ \text{predSet} :: \alpha \Rightarrow \alpha,\ \text{supOrd} :: \alpha \Rightarrow \alpha,$
$\qquad\qquad \text{OrdRec} :: (\alpha \Rightarrow (\alpha \Rightarrow \alpha) \Rightarrow \alpha) \Rightarrow (\alpha \Rightarrow \alpha \Rightarrow \alpha) \Rightarrow \alpha \Rightarrow \alpha \Rightarrow \alpha\ ]$

$\text{OrdRec}_{\text{axioms}} := [\ \text{predSet} : \text{Ord} \rightarrow \text{SetOf Ord},\quad \text{supOrd} : \text{SetOf Ord} \rightarrow \text{Ord},$
$\qquad\qquad \forall\, u, v : \text{Ord}.\, u \in \text{predSet}\, v \leftrightarrow u < v,$
$\qquad\qquad \forall\, x : \text{SetOf Ord}.\, \forall\, u.\, u \in x \rightarrow u < \text{succ}\, (\text{supOrd}\, x),$
$\qquad\qquad \forall\, g, f, x.\, \text{OrdRec}\, g\, f\, x\, 0 = x,$
$\qquad\qquad \forall\, g, f, x.\, \forall\, u : \text{Ord}.\, \text{OrdRec}\, g\, f\, x\, (\text{succ}\, u) = f\, (\text{succ}\, u)\, (\text{OrdRec}\, g\, f\, x\, u),$
$\qquad\qquad \forall\, g, f, x.\, \forall\, u : \text{Limit}.\, \text{OrdRec}\, g\, f\, x\, u =$
$\qquad\qquad\qquad g\, u\, (\lambda\, v.\, \text{if}\ v : \text{Ord} \land v < u\ \text{then}\ \text{OrdRec}\, g\, f\, x\, v\ \text{else}\ \circ_{\text{OrdRec}})\ ]$

Fig. 3. Constants, axioms, and definitions for the GZF, Ordinal, and OrdinalRec classes

We use a number of soft types that classify objects in a GST's domain, e.g., Set, Function, and Pair. For example, $\emptyset : \text{Set}$. An example of using the soft operator type constructor $\rightarrow$ is that using the axioms dom : Fun $\rightarrow$ Set and $\mathcal{P} : \text{Set} \rightarrow \text{SetOf Set}$ and $\bigcup : \text{SetOf Set} \rightarrow \text{Set}$, we can deduce that if f is a function (i.e., $f : \text{Fun}$), then $T = \bigcup (\mathcal{P}\, (\text{dom}\, f))$ is a set (i.e., $T : \text{Set}$) and is not "undefined" (i.e., $T \neq \bullet$). Our soft types are similar in spirit to the soft types of Mizar [15], but differ in many details.

$$\begin{aligned}
\mathsf{Fun}_{\mathsf{consts}} := \big[\, &\circ_{\mathsf{Fun}} :: \alpha,\ \mathsf{Fun} :: \alpha \Rightarrow \star,\ \mathsf{app} :: \alpha \Rightarrow \alpha \Rightarrow \alpha \Rightarrow \star,\ \twoheadrightarrow :: \alpha \Rightarrow \alpha \Rightarrow \alpha, \\
&\mathsf{mkFun} :: \alpha \Rightarrow (\alpha \Rightarrow \alpha \Rightarrow \star) \Rightarrow \alpha,\ \mathsf{dom} :: \alpha \Rightarrow \alpha,\ \mathsf{ran} :: \alpha \Rightarrow \alpha \,\big]
\end{aligned}$$

$$\begin{aligned}
\mathsf{Fun}_{\mathsf{axioms}} := \big[\, &\mathsf{mkFun} : (\Pi\, x : \mathsf{Set}.\, \mathsf{FunPred}\, x \twoheadrightarrow \mathsf{Fun}), \\
&\mathsf{dom} : \mathsf{Fun} \twoheadrightarrow \mathsf{Set},\ \mathsf{ran} : \mathsf{Fun} \twoheadrightarrow \mathsf{Set}, \\
&\twoheadrightarrow\, :\, \mathsf{Set} \twoheadrightarrow \mathsf{Set} \twoheadrightarrow \mathsf{SetOf\, Fun}, \\
&\forall\, f : \mathsf{Fun}.\, \forall\, b,c,d.\, \mathsf{app}\, f\, b\, c \wedge \mathsf{app}\, f\, b\, d \rightarrow c = d \\
&\forall\, f,g : \mathsf{Fun}.\, (\forall\, b,c.\, \mathsf{app}\, f\, b\, c \leftrightarrow \mathsf{app}\, g\, b\, c) \rightarrow f = g \\
&\forall\, f : \mathsf{Fun}.\, \forall\, b.\, b \in \mathsf{dom}\, f \leftrightarrow (\exists\, c.\, \mathsf{app}\, f\, b\, c) \\
&\forall\, f : \mathsf{Fun}.\, \forall\, c.\, c \in \mathsf{ran}\, f \leftrightarrow (\exists\, b.\, \mathsf{app}\, f\, b\, c) \\
&\forall\, x,y : \mathsf{Set}.\, \forall\, f : \mathsf{Fun}.\, (f \in x \twoheadrightarrow y) \leftrightarrow (\mathsf{dom}\, f \subseteq x \wedge \mathsf{ran}\, f \subseteq y) \\
&\forall\, x : \mathsf{Set}.\, \forall\, p : \mathsf{FunPred}\, x.\, \forall\, b,c. \\
&\quad \mathsf{app}\, (\mathsf{mkFun}\, x\, p)\, b\, c \leftrightarrow (b \in x \wedge p\, b\, c \wedge b : \mathsf{FunMem} \wedge c : \mathsf{FunMem}) \,\big]
\end{aligned}$$

$$\begin{aligned}
\mathsf{Fun}_{\mathsf{defs}} := \big[\, &\mathsf{FunMem} = (\lambda\, b.\, \exists\, f : \mathsf{Fun}.\, b \in \mathsf{dom}\, f \vee b \in \mathsf{ran}\, f), \\
&\mathsf{FunPred} = (\lambda\, x\, p.\, \forall\, b : \mathsf{FunMem}.\, b \in x \rightarrow (\exists_{\leq 1}\, c : \mathsf{FunMem}.\, p\, b\, c)) \,\big]
\end{aligned}$$

Fig. 4. Constants, axioms, and definitions for the Function class

$$\begin{aligned}
\mathsf{GZF} &:= [\,\diamond, \mathsf{GZF}_{\mathsf{consts}}, \mathsf{GZF}_{\mathsf{axioms}}, \mathsf{GZF}_{\mathsf{defs}}\,] \\
\mathsf{Ordinal} &:= [\,\diamond, \mathsf{Ord}_{\mathsf{consts}}, \mathsf{Ord}_{\mathsf{axioms}}, \mathsf{Ord}_{\mathsf{defs}}\,] \\
\mathsf{OrdinalRec} &:= [\,[\,\mathsf{GZF}, \mathsf{Ordinal}\,], \mathsf{OrdRec}_{\mathsf{consts}}, \mathsf{OrdRec}_{\mathsf{axioms}}, \mathsf{OrdRec}_{\mathsf{defs}}\,] \\
\mathsf{Function} &:= [\,[\,\mathsf{GZF}\,], \mathsf{Fun}_{\mathsf{consts}}, \mathsf{Fun}_{\mathsf{axioms}}, \mathsf{Fun}_{\mathsf{defs}}\,] \\
\mathsf{Exception} &:= [\,\diamond, [\,\circ_{\mathsf{Exc}} :: \alpha, \mathsf{Exc} :: \alpha \Rightarrow \star, \bullet :: \alpha\,], \diamond, \diamond\,]
\end{aligned}$$

$$\begin{aligned}
\mathbf{GZF} &:= [\,\mathsf{GZF}, \mathsf{Set}, \mathsf{SetMem}, \circ_{\mathsf{GZF}}\,] \\
\mathbf{Ordinal} &:= [\,\mathsf{Ordinal}, \mathsf{Ord}, \bot, \circ_{\mathsf{Ord}}\,] \\
\mathbf{OrdRec} &:= [\,\mathsf{OrdinalRec}, \bot, \bot, \circ_{\mathsf{OrdRec}}\,] \\
\mathbf{Function} &:= [\,\mathsf{Function}, \mathsf{Fun}, \mathsf{FunMem}, \circ_{\mathsf{Fun}}\,] \\
\mathbf{Exc} &:= [\,\mathsf{Exception}, \mathsf{Exc}, \bot, \circ_{\mathsf{Exc}}\,]
\end{aligned}$$

Fig. 5. Definition of our example features and their classes.

We define specialized soft type constructors for our features, e.g., using the SetOf soft type constructor we can build the soft type SetOf Set of those sets that contain only sets. For example, $\{0\}$: SetOf Ord, because $\{0\}$ is a set containing only ordinals, and $\{\{0\}\}$: SetOf Set, and also $\{\{0\}\}$: SetOf (SetOf Ord).

The feature **GZF** provides *generalized Zermelo/Fraenkel* sets. $\mathsf{GZF}_{\mathsf{defs}}$ defines the cargo soft type SetMem, a soft type for objects that belong to some set in the domain. **Ordinal** provides ordinal numbers. **OrdRec** provides an operator for recursion on ordinals, which is crucial for building models of GSTs. Adding the **Ordinal** and **OrdRec** features saved us at least a month of development time because it allows us to pass the development of ordinal recursion from a type implementing ZFC_in_HOL to a GST whose model is built within that type. **Function** provides functions with a function application relation app, operators to find the domain and range of a function, a partial function space operator $\twoheadrightarrow$, and a function-building operator mkFun. **Exc** provides the *exception object* • (spoken "boom"). The important behavior of • is given by axioms generated when combining features that use cargo soft types (e.g., SetMem, FunMem) to ensure that • can not occur within container objects (e.g., sets, functions).

Our current design requires each object in a GST's domain to be "owned" by exactly one feature. However, classes associated with a feature may be used independently of that feature and can be instantiated by any GST's domain provided some collections of objects in the domain can be found that satisfy the requirements the class places on parts of the type. For example, Function can be instantiated by sets of ordered pairs (cf. `GZF/SetRel.thy`), and we do this to build a GST at type **V** (our founder domain).

3.3 Feature Combination

A *feature configuration* is a triple $[\mathcal{F}, D, \mathcal{B}]$, where D is a term to be identified with default($\mathcal{F}$), and $\mathcal{B}$ is a *blacklist* of features whose objects will be excluded from the internal structure of objects of feature $\mathcal{F}$. GSTs are classes defined by combining feature configurations. Figure 6 defines operations that generate extra axioms.

Most of our features use soft typing axioms constraining operator behavior, e.g., dom : Fun $\rightharpoonup$ Set specifies that dom yields a set when applied to a function. These soft typing judgements have a special status: they are also used to generate axioms specifying what operators do in other cases, e.g., specifying that dom B will yield • when B is not a function. The formula otherwise($\kappa, [b_1 : P_1, \ldots, b_n : P_n], D$) produces an axiom that identifies D with the result of applying κ to arguments that do not satisfy at least one of the $P_1, \ldots, P_n$. The formula list allOtherwise($\mathcal{F}, D$) calls otherwise on each parameter κ with an axiom in $\mathcal{F}$ of the form $\kappa : P \rightharpoonup Q$ or $\kappa : (\Pi\, x : P.\, Q)$. Hence, allOtherwise(**Function**, •) generates the axiom $\forall b.\, \neg\, b : \mathsf{Fun} \rightarrow \mathsf{dom}\, b = \bullet$, among others. We expect some users might prefer other policies and we aim to make this more flexible.

$$
\begin{aligned}
\mathsf{typList}_i(P \rightharpoonup Q) &:= b_i : P \mathbin{\#} \mathsf{typList}_{i+1}(Q)\\
\mathsf{typList}_i(\Pi\, P\, Q) &:= b_i : P \mathbin{\#} \mathsf{typList}_{i+1}(Q\, b_i)\\
\mathsf{typList}_i(R) &:= \diamond
\end{aligned}
$$

$$
\begin{aligned}
&\mathsf{otherwise}(\kappa, [b_1 : P_1, \ldots, b_n : P_n], D) :=\\
&\qquad \forall\, b_1, \ldots, b_n.\, (\neg\, b_1 : P_1 \vee \ldots \vee \neg\, b_n : P_n) \rightarrow (\kappa\, b_1 \ldots b_n = D)
\end{aligned}
$$

$$
\begin{aligned}
\mathsf{allOtherwise}(\mathcal{F}, D) := [\,&\mathsf{otherwise}(\kappa, \mathsf{typList}(R), D) \mid (\kappa : R) \in: \mathsf{axioms}(\mathcal{F}),\\
&R \equiv (P \rightharpoonup Q) \text{ or } R \equiv (\Pi\, x : P.\, Q)\,]
\end{aligned}
$$

$$
\begin{aligned}
\mathsf{cover}([P_1, \ldots, P_n]) &:= (P_1 \sqcup \ldots \sqcup P_n = \top)\\
\mathsf{disjoint}([P_1, \ldots, P_n]) &:= [\,(P_1 \sqcap P_2 = \bot), \ldots, (P_1 \sqcap P_n = \bot), \ldots, (P_{n-1} \sqcap P_n = \bot)\,]
\end{aligned}
$$

$$
\begin{aligned}
\mathsf{admitCargo}(P, [\,Q_1, \ldots, Q_n\,]) &:= (Q_1 \sqcup \ldots \sqcup Q_n \sqsubseteq P)\\
\mathsf{restrictCargo}(P, [\,Q_1, \ldots, Q_n\,]) &:= ((Q_1 \sqcup \ldots \sqcup Q_n) \sqcap P = \bot)\\
\mathsf{cargoAx}(\mathcal{F}, \mathcal{W}, \mathcal{B}) &:= [\,\mathsf{admitCargo}(\mathsf{cargo}(\mathcal{F}), [\,\mathsf{logo}(\mathcal{G}) \mid \mathcal{G} \in: \mathcal{W}, \mathcal{G} \not\in: \mathcal{B}]),\\
&\qquad \mathsf{restrictCargo}(\mathsf{cargo}(\mathcal{F}), [\,\mathsf{logo}(\mathcal{G}) \mid \mathcal{G} \in: \mathcal{B}])\,]
\end{aligned}
$$

Fig. 6. 'Otherwise' axioms for operators, logo axioms, and cargo axioms.

The formula $\mathsf{cover}([P_1, \ldots, P_n])$ ensures that $B : P_i$ holds for some P_i, and $\mathsf{disjoint}([P_1, \ldots, P_n])$ ensures that both $B : P_i$ and $B : P_j$ cannot hold for $i \neq j$. The formula $\mathsf{admitCargo}(P, [Q_1, \ldots, Q_n])$ states that $Q_1, \ldots, Q_n$ are all subtypes of P, and $\mathsf{restrictCargo}(P, [Q_1, \ldots, Q_m])$ states that if $B : Q_i$ for any $i \leq m$, then $\neg B : P$. The formulas of $\mathsf{cargoAx}$ constrain the internal structure of the objects of a feature, e.g., $\mathsf{cargoAx}(\mathbf{GZF}, [\mathbf{GZF}, \mathbf{Exc}], [\mathbf{Exc}])$ stands for: $[(\mathsf{Set} \sqsubseteq \mathsf{SetMem}), (\mathsf{Exc} \sqcap \mathsf{SetMem} = \bot)]$.

Figure 7 defines mkGST and gives an example GST named ZF^+. The operation $\mathsf{mkGST}(\mathsf{spec})$ defines a class $\mathcal{C}$ from a list spec of feature configurations. The dependencies of $\mathcal{C}$ are the classes of the features in spec, the axioms of $\mathcal{C}$ are given by $\mathsf{GST}_{\mathsf{axioms}}(\mathsf{spec})$, and the definitions of $\mathcal{C}$ are a list of simple definitions of the default parameter of each feature in spec given by $\mathsf{GST}_{\mathsf{defs}}(\mathsf{spec})$. Our example GST ZF^+ has sets, non-set ordinals, non-set functions, and a distinguished non-set $\bullet$. Each feature is configured to use $\bullet$ as its default value, and $\bullet$ is blacklisted from the internal structure of sets, and functions.

4   Examples of Working in a GST

These examples assume parameter instantiations of the class ZF^+ and all the classes in its dependencies (i.e., GZF, Ordinal, Function, and Exception) at the type d_0. This means that from the class ZF^+ and each class in its dependencies deductions can use definitions of the constants in the class parameters as well as the class definitions and the class axioms. When we say "define X", we mean "add the indicated simple definition for X to the definitions used in deductions".

Ordinal Left-Subtraction. Suppose we already have an infix ordinal addition operator such that if $i, j : \mathsf{Ord}$ then $i + j : \mathsf{Ord}$, and otherwise $i + j = \bullet$. Define

---

Given $\mathsf{spec} \equiv [[\mathcal{F}_1, \mathcal{D}_1, \mathcal{B}_1], \ldots, [\mathcal{F}_n, \mathcal{D}_n, \mathcal{B}_n]]$, then:

$$\mathsf{GST}_{\mathsf{axioms}}(\mathsf{spec}) := \mathsf{allOtherwise}(\mathcal{F}_1, \mathcal{D}_1) \, @ \ldots @ \, \mathsf{allOtherwise}(\mathcal{F}_n, \mathcal{D}_n)$$
$$@ \; \mathsf{disjoint}\,[\,\mathsf{logo}(\mathcal{F}_1), \ldots, \mathsf{logo}(\mathcal{F}_n)\,]$$
$$@ \; [\,\mathsf{cover}\,[\,\mathsf{logo}(\mathcal{F}_1), \ldots, \mathsf{logo}(\mathcal{F}_n)\,]\,]$$
$$@ \; \mathsf{cargoAx}(\mathcal{F}_1, [\mathcal{F}_1, \ldots, \mathcal{F}_n], \mathcal{B}_1)$$
$$@ \; \ldots @ \; \mathsf{cargoAx}(\mathcal{F}_n, [\mathcal{F}_1, \ldots, \mathcal{F}_n], \mathcal{B}_n)$$
$$\mathsf{GST}_{\mathsf{defs}}(\mathsf{spec}) := [\,\mathsf{default}(\mathcal{F}_1) = \mathcal{D}_1, \ldots, \mathsf{default}(\mathcal{F}_n) = \mathcal{D}_n\,]$$

$$\mathsf{mkGST}(\mathsf{spec}) := [\,[\,\mathsf{class}(\mathcal{F}_1), \ldots, \mathsf{class}(\mathcal{F}_n)\,], \diamond, \mathsf{GST}_{\mathsf{axioms}}(\mathsf{spec}), \mathsf{GST}_{\mathsf{defs}}(\mathsf{spec})\,]$$

$$\mathsf{ZF}^+_{\mathsf{spec}} := [\,[\,\mathbf{GZF}, \bullet, [\,\mathbf{Exc}\,]\,], [\,\mathbf{Ordinal}, \bullet, \diamond\,], [\,\mathbf{Function}, \bullet, [\,\mathbf{Exc}\,]\,], [\,\mathbf{Exc}, \bullet, \diamond\,]\,]$$
$$\mathsf{ZF}^+ := \mathsf{mkGST}(\mathsf{ZF}^+_{\mathsf{spec}})$$

Fig. 7. Generating GSTs from specifications and ZF^+, an example GST.

the *left-subtraction* operator $-_{\mathsf{left}} :: \mathsf{d}_0 \Rightarrow \mathsf{d}_0 \Rightarrow \mathsf{d}_0$ on ordinals:

$$-_{\mathsf{left}} = (\lambda\, i\, j.\, \imath\, k.\, i = j + k\, \mathsf{else} \bullet)$$

Given ordinals i, j : Ord where $j < i$ or $j = i$, the term $i -_{\mathsf{left}} j$ is the unique ordinal k such that $i = j + k$. If $i < j$ or if either i or j are not ordinals, then $i -_{\mathsf{left}} j = \bullet$. For example:

$$5 -_{\mathsf{left}} 3 = (\imath\, k.\, 5 = 3 + k\, \mathsf{else} \bullet) = 2$$
$$3 -_{\mathsf{left}} 5 = (\imath\, k.\, 3 = 5 + k\, \mathsf{else} \bullet) = \bullet$$
$$3 -_{\mathsf{left}} \emptyset = (\imath\, k.\, 3 = \emptyset + k\, \mathsf{else} \bullet) = \bullet$$

Function Application and Some Other Function Operators. Define the *function application* infix operator (') $:: \mathsf{d}_0 \Rightarrow \mathsf{d}_0 \Rightarrow \mathsf{d}_0$ such that if f : Fun and $x \in \mathsf{dom}\, f$, then $f\, `\, x$ is the value of $f$ at $x$, and otherwise $f\, `\, x = \bullet$:

$$(`) = (\lambda\, f\, x.\, \imath\, y.\, \mathsf{app}\, f\, x\, y\, \mathsf{else} \bullet)$$

Define an operator $(\dot\lambda)$ that when given a domain x : Set extracts from an operator $t :: \mathsf{d}_0 \Rightarrow \mathsf{d}_0$ a function of type d_0:

$$\dot\lambda = (\lambda\, x\, t.\, \mathsf{mkFun}\, x\, (\lambda\, b\, c.\, b \in x \wedge c = t\, b))$$

We furthermore adopt the following nice notation: $(\dot\lambda\, b \in x.\, B) := (\dot\lambda\, x\, (\lambda\, b.\, B))$. Define an operator to lift a binary operator on objects of type d_0 to a binary operator on functions:

$$\mathsf{lift} = \lambda\, o\, f\, g.\, (\dot\lambda\, b \in \mathsf{dom}\, f \cap \mathsf{dom}\, g.\, o\, (f\, `\, b)\, (g\, `\, b))$$

Define an operator FunRet $:: (\mathsf{d}_0 \Rightarrow \star) \Rightarrow \mathsf{d}_0 \Rightarrow \star$ that takes a soft type P and yields the soft type of functions whose return values always satisfy P:

$$\mathsf{FunRet} = (\lambda\, p\, f.\, f : \mathsf{Fun} \wedge \mathsf{ran}\, f : \mathsf{SetOf}\, p)$$

Pointwise Left-Subtraction on Ordinal-Valued Functions. A function f : Fun is *ordinal-valued* iff f : FunRet Ord. Thus, $\mathsf{lift}\, (-_{\mathsf{left}})\, f\, g$ computes the pointwise left-subtraction of ordinal-valued functions f, g : FunRet Ord. For example, consider functions f, g with $\mathsf{dom}\, f$ and $\mathsf{dom}\, g$ being $\mathsf{predSet}\, \omega$ (the set of all ordinals less than ω), where $f\, `\, j = j + 2$ and $g\, `\, j = j + 1$ for all $j \in \mathsf{predSet}\, \omega$. Clearly f, g are ordinal-valued. Hence for any ordinal $j < \omega$,

$$\begin{aligned}
(\mathsf{lift}\, (-_{\mathsf{left}})\, f\, g)\, `\, j &= (\dot\lambda\, b \in \mathsf{dom}\, f \cap \mathsf{dom}\, g.\, (f\, `\, b) -_{\mathsf{left}} (g\, `\, b))\, `\, j \\
&= (f\, `\, j -_{\mathsf{left}} g\, `\, j) = (j + 2) -_{\mathsf{left}} (j + 1) \\
&= 2 -_{\mathsf{left}} 1 = 1
\end{aligned}$$

Combining Operators. Define an operator override that takes two binary operators $o_1, o_2 :: d_0 \Rightarrow d_0 \Rightarrow d_0$ and two predicates $t_1, t_2 :: d_0 \Rightarrow d_0 \Rightarrow \star$ saying when to use each operator and builds a new operator by combining them:

$$\text{override } t_1 \, o_1 \, t_2 \, o_2 = (\lambda \, x \, y. \, \text{if } (t_1 \, x \, y) \text{ then } (o_1 \, x \, y)$$
$$\text{else if } (t_2 \, x \, y) \text{ then } (o_2 \, x \, y)$$
$$\text{else } \bullet)$$

Define an infix operator $(-)$ that combines ordinal left-subtraction with ordinal left-subtraction lifted to ordinal-valued functions:

$$(-) = \text{override} \, (\lambda \, x \, y. \ x, y : \text{Ord}) \, (-_{\text{left}}) \, (\lambda \, x \, y. \ x, y : \text{FunRet Ord}) \, (\text{lift} \, (-_{\text{left}}))$$

If i, j are ordinals, then $i - j = i -_{\text{left}} j$. If f, g are ordinal-valued functions, then $f - g$ is the function such that if $j \in \text{dom } f$ and $j \in \text{dom } g$, then $(f - g) \, ' \, j = (f \, ' \, j) -_{\text{left}} (g \, ' \, j)$, and otherwise $(f - g) \, ' \, j = \bullet$. Because $\text{Ord} \sqcap \text{Fun} = \bot$ is an axiom of ZF^+, we know that at most one of $x, y : \text{Ord}$ and $x, y : \text{FunRet Ord}$ can be true, so there is no possibility of $(-)$ using $(-_{\text{left}})$ where the user might have intended instead for $(-)$ to use $(\text{lift} \, (-_{\text{left}}))$.

5   Building a Model for ZF$^+$ in ZFC

Normally, using our example GST ZF^+ is done in a type that it is instantiated at. Such a type could be gained by including the axioms of ZF^+ in the hypotheses of deductions (i.e., $\text{ZF}^+_{\text{axioms}}[\![d_0]\!] \subseteq \Gamma$). For confidence in ZF^+'s consistency, we opt to *define* d_0 in terms of a model built in **V**, a type axiomatized by Paulson's ZFCinHOL.[5] Models of GSTs are cumulative hierarchies of sets defined via ordinal recursion. We build models of GSTs within other GSTs using classes called *model components* that correspond to features. We use Isabelle/ML code to generate terms, theorems, and proof states that assist in instantiating these classes and using them as interpretations of GST features. Details of our approach may be seen in the appendix of the longer version of this paper.

The only axioms (i.e., excluding simple definitions) used for this are those of HOL and ZFCinHOL. From this point, only *definitional mechanisms* (as defined by Kunčar and Popescu [11]) are used to define ZF^+ in d_0 (i.e., class instantiation, `typedef`, `lift_definition`). Hence, if HOL and ZFCinHOL are consistent, then so are the axioms of ZF^+.

5.1   Building a Model in V

To build a model in **V**, our method needs the GZF, Ordinal, OrdRec, and Function classes to be instantiated at **V**.

---

[5] Interested readers can see the definition of the ZF^+ class in `GST_Features.thy`, and the type definition of d_0 and instantiation of the ZF^+ class in `Founder/Test.thy`.

ZFCinHOL provides axioms for set theory, and supports von Neumann ordinals and transfinite recursion. Hence instantiations for GZF, Ordinal, OrdRec are achieved by choosing correct definitions for each parameter, and then using automatic proof methods to discharge proof obligations. For Function, we have a generic result that this class may be instantiated in any GST with the GZF feature, using sets of Kuratowski ordered pairs.

We define a model in V and a soft type $M :: V \Rightarrow \star$ of all objects in the model. We then do an Isabelle/HOL type definition to define d_0 as a type isomorphic to the model of ZF^+ built in V (i.e., $\texttt{typedef}\, d_0 = \mathsf{Collect}\, M$).[6]

5.2 Instantiating ZF^+ in d_0

To instantiate the ZF^+ class at d_0, we must provide instantiations for each parameter of ZF^+, and prove each axiom. This is achieved with help from the Lifting and Transfer [7] packages. The parameters of each feature of ZF^+—namely GZF, Ordinal, Function and Exc—are instantiated using Isabelle/ML code calling the Lifting package to lift constants acting on the model in V to d_0. To prove the axioms of each feature, the Transfer package is used to transform each subgoal into an equivalent statement about the model. Our approach only needs the transferred versions of a feature's axioms to be proved once—the proofs may be reused when building a model of any GST using that feature. The axioms of ZF^+ generated in the feature combination process (e.g., cover, disjoint) must also be proved on d_0. This is also achieved with Transfer, but these statements are proved manually.

6 Related and Future Work

Other Related Work. Two bodies of work are notable in being similar in spirit to the way we assemble GSTs, specifically work by Farmer et al. on "little theories" [6], which emphasizes small theories that are then connected together by morphisms, and work led by Rabe on the MMT framework [14], which emphasizes combining even smaller theory bits with morphisms.

Future Work. We aim to further develop generalized set theories and our Isabelle/HOL formalization in the following directions. *Importing features from Isabelle/HOL.* The "ZFC in HOL" development that we bootstrap from provides support (packaged up in two type classes) for obtaining structures within the set theory with the same behavior as other Isabelle/HOL types. We aim to be able to pass such structures on to GSTs we build in the form of extra features. *Foundation.* We have not yet implemented the axiom of foundation, which states that there are no infinite chains going by steps from an object to any of its "children". Many use cases do not need this, so it is not crucial, but some of our future work will depend on it. We plan to handle this as described in previous work on formalizing GSTs [5]. *Overloading.* Isabelle supports overloading (via

---

[6] The operator Collect converts a predicate to a HOL-set as required by $\texttt{typedef}$.

classes or *ad hoc* commands) so symbols can have different meanings at different types. However, we want to give symbols different meanings for different features within one type, a GST's domain. For example, we want machinery that will support overloading ∈ to work both with Set objects and groups within a GST. We also want to allow symbols for set operations like ∈ to be used both with soft types (which are close to Isabelle/HOL sets) and Set objects in a GST. *Universes.* To support some applications in category theory and also for building models of complicated type theories to embed them within a GST, we aim to add a feature that adds some kind of universe axiom like in TG, so those who need it could have a much larger and more powerful GST. *Automation.* Others are working on automation of finding and using soft types and we aim to build on any progress they make. Our model-building process currently needs lots of manual proving, but much can be automated and/or given a nicer interface.

References

1. Aczel, P.: Generalised set theory. In: Logic, Language and Computation, Volume 1. CSLI Publications (1996)
2. Aczel, P., Lunnon, R.: Universes and parameters. In: Situation Theory and Its Applications, Volume 2. CSLI Publications (1991)
3. Brown, C.E., Kaliszyk, C., Pąk, K.: Higher-order Tarski Grothendieck as a foundation for formal proof. In: 10th International Conference Interactive Theorem Proving (ITP 2019). Dagstuhl Publishing (2019)
4. Dunne, C., Wells, J.B., Kamareddine, F.: Adding an abstraction barrier to ZF set theory. In: Benzmüller, C., Miller, B. (eds.) CICM 2020. LNCS (LNAI), vol. 12236, pp. 89–104. Springer, Cham (2020). https://doi.org/10.1007/978-3-030-53518-6_6
5. Dunne, C., Wells, J.B., Kamareddine, F.: Generating custom set theories with non-set structured objects. In: Kamareddine, F., Sacerdoti Coen, C. (eds.) CICM 2021. LNCS (LNAI), vol. 12833, pp. 228–244. Springer, Cham (2021). https://doi.org/10.1007/978-3-030-81097-9_19
6. Farmer, W.M., Guttman, J.D., Javier Thayer, F.: Little theories. In: Kapur, D. (ed.) CADE 1992. LNCS, vol. 607, pp. 567–581. Springer, Heidelberg (1992). https://doi.org/10.1007/3-540-55602-8_192
7. Huffman, B., Kunčar, O.: Lifting and transfer: a modular design for quotients in Isabelle/HOL. In: Gonthier, G., Norrish, M. (eds.) CPP 2013. LNCS, vol. 8307, pp. 131–146. Springer, Cham (2013). https://doi.org/10.1007/978-3-319-03545-1_9
8. Kaliszyk, C., Pąk, K.: Semantics of Mizar as an Isabelle object logic. J. Autom. Reasoning **63** (2019). https://doi.org/10.1007/s10817-018-9479-z
9. Kappelmann, K., Chen, J., Krauss, A.: Isabelle/Set (2022). Git repository on github.com. Committers include also Cezary Kaliszyk and Karol Pąk. https://github.com/kappelmann/Isabelle-Set/
10. Krauss, A.: Adding soft types to Isabelle (2010). Unpublished paper on author's web page. https://www21.in.tum.de/~krauss/publication/2010-soft-types-note/
11. Kunčar, O., Popescu, A.: Safety and conservativity of definitions in HOL and Isabelle/HOL. Proc. ACM Program. Lang. **2**, 1–26 (2017)
12. Kunčar, O., Popescu, A.: A consistent foundation for Isabelle/HOL. J. Autom. Reasoning **62**(4), 531–555 (2019)

13. Paulson, L.C.: Zermelo Fraenkel set theory in higher-order logic. Archive of Formal Proofs (2019). Formal proof development available at https://isa-afp.org/entries/ZFC_in_HOL.html
14. Rabe, F.: The future of logic: foundation-independence. Log. Univers. **10**, 1–20 (2015)
15. Wiedijk, F.: Mizar's soft type system. In: Schneider, K., Brandt, J. (eds.) TPHOLs 2007. LNCS, vol. 4732, pp. 383–399. Springer, Heidelberg (2007). https://doi.org/10.1007/978-3-540-74591-4_28

Formalising Basic Topology for Computational Logic in Simple Type Theory

David Fuenmayor[1,2]([✉]) [iD] and Fabián Fernando Serrano Suárez[3] [iD]

[1] University of Bamberg, Bamberg, Germany
`david.fuenmayor@uni-bamberg.de`
[2] University of Luxembourg, Esch-sur-Alzette, Luxembourg
[3] Universidad Nacional de Colombia, Manizales, Colombia
`ffserranos@unal.edu.co`

Abstract. We present a formalisation of basic topology in simple type theory encoded using the Isabelle/HOL proof assistant. In contrast to related formalisation work, which follows more 'traditional' approaches, our work builds upon closure algebras, encoded as Boolean algebras of (characteristic functions of) sets featuring an axiomatised closure operator (cf. seminal work by Kuratowski and McKinsey & Tarski). With this approach we primarily address students of computational logic, for whom we bring a different focus, closer to lattice theory and logic than to set theory or analysis. This approach has allowed us to better leverage the automated tools integrated into Isabelle/HOL (model finder Nitpick and Sledgehammer) to do most of the proof and refutation heavy-lifting, thus allowing for assumption-minimality and less-verbose interactive proofs.

Keywords: Closure algebras · Topology · Simple type theory · Isabelle/HOL

1 Introduction

Inspired by Kevin Buzzard's project on formalising undergraduate mathematics in dependent type theory (using the Lean proof assistant),[1] we have started an effort on formalising *basic* topology in simple type theory (using Isabelle/HOL). With this work we aim at addressing students of (logic in) computer science, for whom we bring a different focus, closer to logic and lattice theory.

It is well known that formal logic has a fundamental role in computer science, e.g., in areas like knowledge representation, automated reasoning, programming languages, computability & complexity theory. Less known is the close relationship between logic and topology [15,19], and its many applications, e.g., in the semantics of logical systems (e.g. [1,8]) and programs (e.g. [11]), duality theory (e.g. [12]), as well as type theory (e.g. [18]), to mention only a few. In this

---

[1] Cf. Xena Project (https://xenaproject.wordpress.com/).

K. Buzzard and T. Kutsia (Eds.): CICM 2022, LNAI 13467, pp. 56–74, 2022.
https://doi.org/10.1007/978-3-031-16681-5_4

respect, we argue, students of logic in computer science, in particular *computational logic* [17], should consider learning some basic topological notions, and preferably using a conceptual framework that feels more 'native' to logic, instead of subordinating the study of topology to areas like analysis or geometry, which may be more suited to the 'standard' mathematician's agenda, but arguably not the computer scientist's.[2]

We investigate an (admittedly not mainstream) approach for presenting topology, which is based upon closure algebras, as famously introduced by Kuratowski [13], and further developed in the seminal work by McKinsey & Tarski's 'The Algebra of Topology' [16]. This has several motivations: Firstly, it is a relatively unexplored approach,[3] and its founding conceptual framework is closer to lattice theory and logic than to analysis or set theory, and so, arguably, better suited to computational logicians.

Secondly, this approach lends itself to a formalisation in (extensional) simple type theory [5], referred to as HOL,[4] by encoding (extensions of) Boolean algebras by means of their (Stone) representation as algebras of sets. This is, arguably, a most natural approach towards their *shallow embedding* in a proof assistant like Isabelle/HOL, which nicely dovetails with previous work on the shallow embedding of modal logics (e.g. [4]). This way we can directly reuse the Boolean connectives of the meta-language (HOL) without relying on any background theory of orderings, sets, lattices, etc. In fact, as argued in [2], such a shallow embedding allows us to harness existing automated theorem provers for reasoning with many different sorts of non-classical logical connectives in a very efficient way (avoiding e.g. inductive definitions and proofs). In fact, a related Isabelle/HOL formalisation has previously employed this approach for automating to a considerable extent reasoning in paraconsistent and paracomplete logics [9]. In the present work, we conveniently make use of Isabelle/HOL's 'prefix' polymorphism to introduce type variables in our formalisation. This aims at fostering future reusability.

Thirdly, we have striven to make our formalisation completely library independent.[5] As a side-product, this makes it quite close to being 'proof assistant agnostic'. Even more, our formalisation can feasibly be reused *mutatis mutandis* for reasoning using stand-alone automated higher-order provers.[6] This has being

---

[2] Unless, of course, she happens to be working on a particular *application* area related to those.

[3] It could be further argued that this fact alone already legitimises their study.

[4] HOL is in fact an acronym for (classical) higher-order logic. It is traditionally employed to refer to the sort of simple type theory instantiated in proof assistants like Isabelle/HOL.

[5] 'Main' is imported for technical reasons to enable the execution of Sledgehammer and Nitpick.

[6] At the time of writing these include LEO-II, Leo-III, Satallax, Zipperposition, and more recently CVC5 and E (starting with release 3.0); as well as model finders Nitpick and Nunchaku. Most of them are available via *System-on-TPTP* (https://tptp.org/cgi-bin/SystemOnTPTP).

motivated by ongoing efforts towards embedding symbolic reasoning capabilities in future AI systems [3,10].

Last but not least, Isabelle's powerful facilities for introducing custom notation, as well as its structured proof language, *Isar* [20], are harnessed to provide a nicely readable, and students' friendly formalisation.[7] We shall remark that the formulas shown in this paper appear in essentially the same way as in our Isabelle/HOL sources,[8] which have been made available under: https://github.com/davfuenmayor/basic-topology/.

Paper Structure: In Sect. 2 we discuss the encoding of topological Boolean algebras, which serves as the starting point of our formalisation of basic topology, to be succinctly summarised in Sect. 3. In Sect. 4 we provide some methodological reflections upon the presented formalisation, followed by a discussion of related work in Sect. 5. We conclude in Sect. 6.

2   Topological Boolean Algebras

In a nutshell, topological Boolean algebras are Boolean algebras extended with additional unary operations, where the latter are given some topological interpretation (in a broad sense). These additional unary operations are often referred to as (topological) *operators*. Paradigmatic cases of such operators are the topological (Kuratowski-) *closure* and its dual the *interior* operator [13]. Moreover, other operations can also be interpreted as topological, in particular the *border*, *frontier*, and *derived-set* operators [14,21,22].

It is important to remark that, in the present work, we will abuse terminology and employ the denominations 'closure', 'interior', etc. quite liberally, so that they apply to all operations that are *intended* to act as such, independently of the currently imposed constraints. For example, while it is known that closure operators are, among others, additive (distribute over unions/joins) and idempotent, we will still conveniently refer to a unary operation as a 'closure', even if it is not necessarily additive or idempotent, in case it is treated as such in the current context and this does not lead to confusions.

2.1   Boolean Algebras

A distinctive feature of our approach consists in encoding Boolean algebras via their (Stone) representation as algebras of sets [12]. This means that each element of (the carrier of) the algebra will be a set (of points). Inspired by the 'propositions as sets of worlds' paradigm from modal logic, we may think of points as 'worlds', and thus of the elements of our Boolean algebras as 'propositions'.

---

[7] To be fair, this feature is in fact shared to a great extent with other Isabelle/HOL formalisations.

[8] We didn't embed any automatically extracted LATEX sources because of space constraints.

We interpret a type variable $'w$ as corresponding to a domain of points (worlds), so that the $'w$-parametric type $'w \Rightarrow bool$ (aliased $'w\,\sigma)$[9] can be interpreted as the type of sets of points (propositions), encoded using their respective characteristic functions.

We start by encoding the lattice-ordering relation and derived equality, observing that the latter corresponds to its counterpart in HOL because of extensionality:

$$A \preceq B \stackrel{\text{def}}{=} \forall w.\ (A\ w) \longrightarrow (B\ w) \qquad A \approx B \stackrel{\text{def}}{=} A \preceq B \wedge B \preceq A$$
$$i.e.\ A \approx B = \forall w.\ (A\ w) \longleftrightarrow (B\ w) = (A = B)$$

We employ **boldface** for the algebra connectives $(\wedge, \vee, \rightarrow, -, \perp, \top, \text{etc.})$. We define meet $(\wedge)$ and join $(\vee)$ as binary operations, i.e., as *curried* terms of type $'w\,\sigma \Rightarrow {}'w\,\sigma \Rightarrow {}'w\,\sigma$, by reusing HOL's conjunction and disjunction, what makes them tantamount to set intersection and union, respectively. Similarly the top and bottom elements are encoded as zero-ary connectives (of type $'w\,\sigma)$ by reusing the meta-logical *True* and *False* terms:

$$A \wedge B \stackrel{\text{def}}{=} \lambda w.\ (A\ w) \wedge (B\ w) \qquad\qquad \top \stackrel{\text{def}}{=} \lambda w.\ True$$
$$A \vee B \stackrel{\text{def}}{=} \lambda w.\ (A\ w) \vee (B\ w) \qquad\qquad \perp \stackrel{\text{def}}{=} \lambda w.\ False$$

Drawing upon the classicality of our meta-logic (HOL) we can define an implication operation analogously by reusing HOL's classical counterpart. Moreover the Boolean complement $(A \rightarrow \perp)$ can be equivalently defined by reusing HOL's negation.

$$A \rightarrow B \stackrel{\text{def}}{=} \lambda w.\ (A\ w) \longrightarrow (B\ w) \qquad -A \stackrel{\text{def}}{=} \lambda w.\ \neg(A\ w)$$

Other related operations, such as difference $(\leftharpoondown)$, symmetric difference $(\triangle)$, and double implication $(\leftrightarrow)$ have been introduced for convenience in the expected way. It is intuitively evident that the domain set for type $'w$, together with the operations defined above, satisfies the axioms of a lattice, and, more specifically, of a Boolean algebra. This was verified automatically using Sledgehammer [6].

Observe that this approach towards encoding (extensions of) Boolean algebras by drawing upon their (Stone) representation as algebras of sets allows us to directly reuse the Boolean connectives of the meta-language (HOL). As argued before, this technique allows us to harness automated tools in a very efficient way [2].

We also introduce the infinitary operations $\bigwedge$ and $\bigvee$ acting on sets of elements (i.e. sets of sets of points), thus having type $('w\,\sigma \Rightarrow bool) \Rightarrow {}'w\,\sigma$,

$$\bigwedge S \stackrel{\text{def}}{=} \lambda w.\ \forall X.\ (S\ X) \longrightarrow (X\ w) \qquad\qquad \bigvee S \stackrel{\text{def}}{=} \lambda w.\ \exists X.\ (S\ X) \wedge (X\ w)$$

---

[9] Here we follow Isabelle/HOL's convention of writing type variables with a leading apostrophe as well as omitting parentheses for parameterised types, so that $('w)\,\sigma$ becomes $'w\,\sigma$.

and automatically prove that $\bigvee S$ and $\bigwedge S$ correspond indeed to the supremum and infimum of S respectively; i.e. our encoded Boolean algebras are lattice-complete.

We also encode some useful notions corresponding to a set of propositions being 'closed under meets/joins'. We do analogously for the corresponding infinitary variant, whereby a set of propositions can be said to be 'closed under infima/suprema'.

$$\texttt{meet\_closed}\ S \overset{\text{def}}{=} \forall X\, Y.\ (S\ X) \wedge (S\ Y) \longrightarrow S\ (X \wedge Y)$$

$$\texttt{join\_closed}\ S \overset{\text{def}}{=} \forall X\, Y.\ (S\ X) \wedge (S\ Y) \longrightarrow S\ (X \vee Y)$$

$$\texttt{infimum\_closed}\ S \overset{\text{def}}{=} \forall D.\ D \sqsubseteq S \longrightarrow S\ (\textstyle\bigwedge D)$$

$$\texttt{supremum\_closed}\ S \overset{\text{def}}{=} \forall D.\ D \sqsubseteq S \longrightarrow S\ (\textstyle\bigvee D)$$

where $D \sqsubseteq S \overset{\text{def}}{=} \forall X.\ (D\ X) \longrightarrow (S\ X)$. We have also encoded two variants `infimum_closed'` and `supremum_closed'` where the subset D is assumed non-empty.

Moreover, for a given operation φ we define the set of its fixed points $\texttt{fp}\ \varphi$ as:

$$\texttt{fp}\ \varphi \overset{\text{def}}{=} \lambda X.\ (\varphi\ X) \approx X$$

Finally, we introduce some convenient second-order operations (acting on unary operations, thus having type $(\text{'}w\,\sigma \Rightarrow \text{'}w\,\sigma) \Rightarrow (\text{'}w\,\sigma \Rightarrow \text{'}w\,\sigma))$. One of them can be seen as a sort of 'operationalised' fixed-point construction $(\cdot)^{\texttt{fp}}$:

$$\varphi^{\texttt{fp}} \overset{\text{def}}{=} \lambda X.\ (\varphi\, X) \leftrightarrow X$$

Others return the *complement* $(\cdot)^{\text{c}}$ and the *dual* $(\cdot)^{\text{d}}$ of a unary operation:

$$\varphi^{\text{c}} \overset{\text{def}}{=} \lambda X.\ -(\varphi\ X) \qquad \varphi^{\text{d}} \overset{\text{def}}{=} \lambda X.\ -(\varphi -X)$$

We note that several interesting properties and interrelations for these second-order constructions have been easily proven (fully automatically), such as e.g.

Proposition 1

1. the involutivity of $(\cdot)^{\text{c}}$, $(\cdot)^{\text{d}}$ and $(\cdot)^{\texttt{fp}}$;
2. the main relation between fixed point constructions: $(\texttt{fp}\ \varphi)\ X \longleftrightarrow (\varphi^{\texttt{fp}}\ X) \approx \top$.

2.2   Topological Conditions

We can extend the encoding of Boolean algebras by adding further unary operations (with type $\text{'}w\,\sigma \Rightarrow \text{'}w\,\sigma$) in order to obtain different sorts of topological Boolean algebras (cf. [9]). Here we encode and interrelate some useful axiomatic conditions on them. For example, we can say of a unary operation that it is:

additive (ADDI), multiplicative (MULT), expansive (EXPN), contractive (CNTR), normal (NORM), dual normal (DNRM), or idempotent (IDEM). We present below the corresponding definitions, noting that, in the formalisation sources, definitions involving equalities have often been stated separately for the $\preceq$ direction (with suffix 'a') and for the $\succeq$ direction (suffix 'b'). Also note that to avoid clutter all free variables (A, B, S, etc.) shall be taken as universally quantified.

$$\text{ADDI } \varphi \stackrel{\text{def}}{=} \varphi(A \vee B) \approx (\varphi\, A) \vee (\varphi\, B) \quad \text{dually } \text{MULT } \varphi \stackrel{\text{def}}{=} \varphi(A \wedge B) \approx (\varphi\, A) \wedge (\varphi\, B)$$

$$\text{EXPN } \varphi \stackrel{\text{def}}{=} A \preceq (\varphi\, A) \quad \text{dually } \text{CNTR } \varphi \stackrel{\text{def}}{=} (\varphi\, A) \preceq A$$

$$\text{NORM } \varphi \stackrel{\text{def}}{=} (\varphi \perp) \approx \perp \quad \text{dually } \text{DNRM } \varphi \stackrel{\text{def}}{=} (\varphi \top) \approx \top$$

$$\text{IDEM } \varphi \stackrel{\text{def}}{=} (\varphi\, A) \approx \varphi(\varphi\, A)$$

The four conditions on the left (ADDI, EXPN, NORM, IDEM) correspond to the well-known axiomatic conditions on topological *closure* operators originally introduced by Kuratowski [13]. Likewise, the dual conditions (MULT, CNTR, DNRM, IDEM) serve to axiomatise an *interior* operator. Boolean algebras extended with a unary operation satisfying the first (resp. second) set of conditions are thus called closure (resp. interior) algebras. It is well known that closure/interior algebras and topologies are two sides of the same coin. This fact has been exploited, at least since the seminal work by McKinsey & Tarski in the 1940's [16], to provide topological semantics for intuitionistic and modal logics (e.g. [8]). Previous Isabelle/HOL formalisation work has build upon these insights to provide semantics for paraconsistent and paracomplete logics [9].

Additional related conditions are: monotonic (MONO), infinitely additive (iADDI) and infinitely multiplicative (iMULT).[10] Note that $[\![\varphi\, S]\!]$ stands for the image of S under φ.

$$\text{MONO } \varphi \stackrel{\text{def}}{=} A \preceq B \longrightarrow (\varphi\, A) \preceq (\varphi\, B)$$

$$\text{iADDI } \varphi \stackrel{\text{def}}{=} (\varphi \bigvee S) \approx \bigvee [\![\varphi\, S]\!] \quad \text{dually } \text{iMULT } \varphi \stackrel{\text{def}}{=} (\varphi \bigwedge S) \approx \bigwedge [\![\varphi\, S]\!]$$

We have verified several useful interrelations between the conditions above. For them, as in the rest of the presented results in this paper, we explicitly say under which (minimal) assumptions they hold (using Nitpick [7] to find countermodels otherwise).

Proposition 2

1. MONO φ iff MONO φ^{d}; EXPN φ iff CNTR φ^{d}; NORM φ iff DNRM φ^{d}; IDEM^a φ iff IDEM^b φ^{d};
2. EXPN φ implies both DNRM φ and IDEM^a φ; CNTR φ implies both NORM φ and IDEM^b φ;

---

[10] Note that by employing iADDI (iMULT) in place of ADDI (MULT) to axiomatise a closure (interior) algebra, we obtain a so-called Alexandrov topology (cf. Sect. 3.8 for a discussion).

3. the following are equivalent: $\texttt{MONO}\,\varphi, \texttt{MULT}^a\,\varphi, \texttt{ADDI}^b\,\varphi, \texttt{iMULT}^a\,\varphi, \texttt{iADDI}^b\,\varphi$;
4. $\texttt{(i)ADDI}^a\,\varphi$ iff $\texttt{(i)MULT}^b\,\varphi^d$; $\texttt{(i)MULT}^a\,\varphi$ iff $\texttt{(i)ADDI}^b\,\varphi^d$;
5. $\texttt{MULT}\,\varphi$ implies $\texttt{meet\_closed}(\texttt{fp}\,\varphi)$ and $\texttt{iMULT}\,\varphi$ implies $\texttt{infimum\_closed}$ $(\texttt{fp}\,\varphi)$; the converse holds under the additional assumptions $\texttt{MONO}\,\varphi$, $\texttt{CNTR}\,\varphi$ and $\texttt{IDEM}^a\,\varphi$;
6. $\texttt{ADDI}\,\varphi$ implies $\texttt{join\_closed}(\texttt{fp}\,\varphi)$ and $\texttt{iADDI}\,\varphi$ implies $\texttt{supremum\_closed}$ $(\texttt{fp}\,\varphi)$; the converse holds under the additional assumptions $\texttt{MONO}\,\varphi$, $\texttt{EXPN}\,\varphi$ and $\texttt{IDEM}^b\,\varphi$;
7. assuming $\texttt{MONO}\,\varphi$, we have that $\texttt{EXPN}\,\varphi$ implies $\texttt{infimum\_closed}(\texttt{fp}\,\varphi)$ and that $\texttt{CNTR}\,\varphi$ implies $\texttt{supremum\_closed}(\texttt{fp}\,\varphi)$.

2.3   Closure Algebras

In the previous section we remarked that the four conditions: $\texttt{ADDI}$, $\texttt{EXPN}$, $\texttt{NORM}$ and $\texttt{IDEM}$, correspond to the well-known Kuratowski conditions on *closure* operators. Hence, from now on they will be aliased (in that order) as $\texttt{Cl\_1}$ to $\texttt{Cl\_4}$, and we will speak of a Boolean algebra with an operator satisfying some 'meaningful' subset of the conditions $\texttt{Cl\_1}$ to $\texttt{Cl\_4}$ as a *closure algebra*.[11] Dually, the conditions: $\texttt{MULT}$, $\texttt{CNTR}$, $\texttt{DNRM}$, and $\texttt{IDEM}$, have been aliased (in that order) as $\texttt{Int\_1}$ to $\texttt{Int\_4}$ and the corresponding algebras called *interior algebras*. Moreover, we will introduce the aliases $\texttt{Cl\_1'}$ for $\texttt{ADDI}^a$ and $\texttt{Cl\_4'}$ for $\texttt{IDEM}^b$. Recalling the results in the previous section, we have that, given $\texttt{MONO}$, $\texttt{Cl\_1'}$ encodes the gist of $\texttt{Cl\_1}$; and given $\texttt{EXPN}$, $\texttt{Cl\_4'}$ encodes the gist of $\texttt{Cl\_4}$.

In fact, several additional topological operators can be defined in terms of a closure operator together with the Boolean operations. In doing this we can conveniently extend closure algebras by further defined operations. The following terms, having type $(`w\,\sigma \Rightarrow `w\,\sigma) \Rightarrow (`w\,\sigma \Rightarrow `w\,\sigma)$, transform a given operator $\mathcal{C}$ (intended as a closure) into

- an *interior* operator $\mathcal{I}[\mathcal{C}]$ defined as $\mathcal{C}$'s dual, i.e. $\mathcal{I}[\mathcal{C}] \overset{\text{def}}{=} \mathcal{C}^d$;
- a *frontier* (aka. boundary) operator $\mathcal{F}[\mathcal{C}] \overset{\text{def}}{=} \lambda A.\ (\mathcal{C}\ A) \wedge \mathcal{C}\ (-A)$;
- a *border* operator $\mathcal{B}[\mathcal{C}] \overset{\text{def}}{=} \lambda A.\ A \wedge \mathcal{C}\ (-A)$;
- a *residue* operator $\mathcal{R}[\mathcal{C}] \overset{\text{def}}{=} \lambda A.\ \mathcal{B}[\mathcal{C}]\ (\mathcal{B}[\mathcal{C}]^d\ A)$.

We have automatically verified several relations among these operators, for example:

Proposition 3. Assuming $\texttt{Cl\_2}\ \mathcal{C}$ only, we have:

1. $\mathcal{I}[\mathcal{C}]^{\texttt{fp}} = \mathcal{B}[\mathcal{C}]^c$ and $\mathcal{B}[\mathcal{C}]^{\texttt{fp}} = \mathcal{I}[\mathcal{C}]^c$;
2. $\mathcal{C}^{\texttt{fp}} = \mathcal{B}[\mathcal{C}]^d$ and $(\mathcal{B}[\mathcal{C}]^d)^{\texttt{fp}} = \mathcal{C}$;

---

[11] As noted before, we are abusing terminology, as we do not fix a minimal set of conditions that an operator needs to satisfy to deserve being called a 'closure'; e.g. Moore/hull closures only satisfy $\texttt{MONO}$, $\texttt{EXPN}$, and $\texttt{IDEM}$, while Čech closure operators may not satisfy $\texttt{IDEM}$ (they are sometimes called 'preclosures'). Anyhow, we let the context dictate how operators are called.

3. $\mathcal{F}[\mathcal{C}]^{\mathrm{fp}} = \lambda X.\ (\mathcal{B}[\mathcal{C}]\ X) \vee (\mathcal{C}^{\mathrm{c}}\ X)$.

Moreover, we have conveniently aliased the following fixed-point expressions:

- $\mathrm{Cl}[\mathcal{C}] \stackrel{\mathrm{def}}{=} \mathrm{fp}\ \mathcal{C}$, such that $\mathrm{Cl}[\mathcal{C}]\ A$ reads 'A is closed';
- $\mathrm{Op}[\mathcal{C}] \stackrel{\mathrm{def}}{=} \mathrm{fp}\ \mathcal{I}[\mathcal{C}]$, such that $\mathrm{Op}[\mathcal{C}]\ A$ reads 'A is open';
- and similarly: $\mathrm{Br}[\mathcal{C}] \stackrel{\mathrm{def}}{=} \mathrm{fp}\ \mathcal{B}[\mathcal{C}]$; $\mathrm{Fr}[\mathcal{C}] \stackrel{\mathrm{def}}{=} \mathrm{fp}\ \mathcal{F}[\mathcal{C}]$; $\mathrm{Rs}[\mathcal{C}] \stackrel{\mathrm{def}}{=} \mathrm{fp}\ \mathcal{R}[\mathcal{C}]$.

We have verified (mostly automatically) several properties, among which we count:

Proposition 4

1. $\mathrm{Cl}[\mathcal{C}]\ A$ iff $\mathrm{Op}[\mathcal{C}]\ -A$, i.e. the complement of a closed set is open (and viceversa);
2. given $\mathrm{Cl\_4}\ \mathcal{C}$ (i.e. $\mathcal{C}$ is idempotent), we have that $\mathrm{Op}[\mathcal{C}]\ (\mathcal{I}[\mathcal{C}]\ A)$ and $\mathrm{Cl}[\mathcal{C}]\ (\mathcal{C}\ A)$, i.e. interiors (of sets/propositions) are open and closures are closed;
3. given $\mathrm{MONO}\ \mathcal{C}$, $\mathrm{Cl\_2}\ \mathcal{C}$ and $\mathrm{Cl\_4'}\ \mathcal{C}$, we have $\mathrm{Cl}[\mathcal{C}]\ (\mathcal{F}\ A)$, i.e. frontiers are closed;
4. given $\mathrm{MONO}\ \mathcal{C}$, we have that $\mathrm{Br}[\mathcal{C}]\ (\mathcal{B}\ A)$, i.e. borders are fixed-points of the border;
5. given $\mathrm{Cl\_2}\ \mathcal{C}$, we have both: (i) $\mathrm{Op}[\mathcal{C}]\ A$ iff $\mathcal{B}[\mathcal{C}]\ A \approx \bot$, i.e. A is open iff its border is empty; and (ii) $\mathrm{Br}[\mathcal{C}]\ A$ iff $\mathcal{I}[\mathcal{C}]\ A \approx \bot$, i.e. A is a fixed-point of the border iff its interior is empty (some literature would then call A a 'boundary set');
6. given $\mathrm{MONO}\ \mathcal{C}$, $\mathrm{Cl\_2}\ \mathcal{C}$ and $\mathrm{Cl\_4'}\ \mathcal{C}$, we have that closure and interior can be characterised in terms of their fixed points, i.e. $\mathcal{C}\ A \approx \bigwedge(\lambda S.\ \mathrm{Cl}[\mathcal{C}]\ S \wedge A \preceq S)$ and $\mathcal{I}[\mathcal{C}]\ A \approx \bigvee(\lambda S.\ \mathrm{Op}[\mathcal{C}]\ S \wedge S \preceq A)$.

3 Formalising Basic Topology

We encode some basic topological notions and verify several properties and relationships. Most of them have been proven fully automatically (via Sledgehammer [6]), yet some others required a (manual) decomposition into subgoals by Isar [20] proof. This methodology will be further described in Sect. 4. In the subsections below we will only showcase a (somewhat representative) subset of the definitions and results in the corresponding Isabelle/HOL sources. As an interesting remark, we were able to harness Nitpick [7] to find countermodels to all finitely countersatisfiable conjectures (not shown).

3.1 Properties of Sets

We have started by encoding some well-known topological properties of sets. As an illustration, sets with an empty interior are called 'boundary' sets in the literature, and we just saw in Proposition 4 (5) that they can be equivalently characterised (given $\mathrm{Cl\_2}$) as the fixed-points of the border operator. In fact there are more connections between fixed-points of topological operators and well-known topological properties of sets which are worth exploring. We list some of those automatically verified results below. We introduce first some relevant definitions:

- *Boundary* sets are those having an empty interior: $\texttt{boundary}[\mathcal{C}]\ A \overset{\text{def}}{=} \mathcal{I}[\mathcal{C}]\ A \approx \bot$.
- A set is *dense* iff its closure is the whole domain: $\texttt{dense}[\mathcal{C}]\ A \overset{\text{def}}{=} \mathcal{C}\ A \approx \top$, and
- *nowhere-dense* iff its closure is boundary: $\texttt{nowhereDense}[\mathcal{C}]\ A \overset{\text{def}}{=} \texttt{boundary}[\mathcal{C}]\ (\mathcal{C}\ A)$.

Proposition 5

1. A set is dense iff its complement is boundary: $\texttt{dense}[\mathcal{C}]\ A = \texttt{boundary}[\mathcal{C}]\ {-}A$.
2. Assuming Cl_2, boundary (resp. dense) sets can be equivalently characterised as the fixed-points of the border (resp. dual-border) operator: $\texttt{boundary}[\mathcal{C}] = \texttt{Br}[\mathcal{C}] = \texttt{fp}\ \mathcal{B}[\mathcal{C}]$; resp. $\texttt{dense}[\mathcal{C}] = \texttt{fp}\ \mathcal{B}[\mathcal{C}]^{\text{d}}$.
3. Assuming Cl_2, nowhere-dense closed sets can be equivalently characterised as the fixed-points of the frontier operator: $\texttt{Fr}[\mathcal{C}]\ A = (\texttt{nowhereDense}[\mathcal{C}]\ A \land \texttt{Cl}[\mathcal{C}]\ A)$.

3.2   Relativisation

A notion of *relativisation* can be defined for topological operators. By seeing topology through the lens of closure algebras this notion plays a role analogous to that of subspace.

Observe that the domain of points has been left implicit as the semantic domain of the type variable 'w, but in fact we can define *relativised* closure operators wrt. a given subset/proposition S (of type '$w\ \sigma$). For this we will define a function $\downarrow_S$ that takes an operator $\mathcal{C}$ (intended as closure) and returns an S-relativised variant of $\mathcal{C}$ as follows:

$$\mathcal{C}{\downarrow}_S \overset{\text{def}}{=} \lambda A.\ S \land \mathcal{C}\ A.$$

The Kuratowski closure conditions can also be relativised wrt. a set S, e.g. we have:

$$\texttt{Cl\_1}{\downarrow}_S\ \varphi \overset{\text{def}}{=} \texttt{ADDI}{\downarrow}_S\ \varphi \overset{\text{def}}{=} S \land \varphi(A \lor B) \approx (S \land \varphi\ A) \lor (S \land \varphi\ B)$$

doing analogously for the other conditions. Importantly, we need to take extra care when working with the interior operator, which needs to be redefined:

$$\mathcal{I}{\downarrow}_S[\mathcal{C}] \overset{\text{def}}{=} \lambda X.\ S \leftharpoonup \mathcal{C}\,(S \leftharpoonup X).$$

Likewise, the notions of closed and (specially) open sets need to be revisited:

$$\texttt{Cl}{\downarrow}_S[\mathcal{C}] \overset{\text{def}}{=} \texttt{fp}\ \mathcal{C}{\downarrow}_S\ (= \texttt{Cl}[\mathcal{C}{\downarrow}_S])\quad \text{dually}\quad \texttt{Op}{\downarrow}_S[\mathcal{C}] \overset{\text{def}}{=} \texttt{fp}\ \mathcal{I}{\downarrow}_S[\mathcal{C}].$$

After these precautions, we proceed to verify automatically several properties, e.g.

Proposition 6

1. relativised closure operators satisfy the relativised Kuratowski conditions, i.e., for N = 1..4 we have that if $\text{Cl\_N}(\mathcal{C})$ then $\text{Cl\_N}{\downarrow}_S(\mathcal{C}{\downarrow}_S)$, noting that only for the case N = 4 we need to further assume MONO and Cl_2 (for the operation $\mathcal{C}$);
2. given MONO, Cl_2 and Cl_4, we have that a set is closed relative to S iff it is the intersection of S with a closed set: $\text{Cl}{\downarrow}_S[\mathcal{C}]\ A$ iff $\exists E.\ \text{Cl}[\mathcal{C}]\ E \wedge A \approx S \wedge E$;
3. given MONO, if S is closed, then the property of being closed relative to S implies being closed in the absolute sense: $\text{Cl}[\mathcal{C}]\ S \longrightarrow \text{Cl}{\downarrow}_S[\mathcal{C}] \sqsubseteq \text{Cl}[\mathcal{C}]$;
4. given MONO, 'relatively closed' is transitive: $\text{Cl}{\downarrow}_B[\mathcal{C}]\ A \wedge \text{Cl}{\downarrow}_C[\mathcal{C}]\ B \longrightarrow \text{Cl}{\downarrow}_C[\mathcal{C}]\ A$.

The duals of some of the statements above can also be proven (e.g. exchanging closure by interior and closed by open). Moreover, several topological properties of sets (like those studied in the previous subsection) are preserved when relativising. The proof reconstruction of these problems can be an interesting exercise for students.

3.3   Neighborhoods and Limits

We call a *neighbourhood function* a term of type $'w \Rightarrow 'w\,\sigma \Rightarrow bool$ that given a point ($'w$) returns a set of 'neighbourhoods' ($'w\,\sigma \Rightarrow bool$) such that the latter satisfies four special conditions, Nbhd_1 to Nbhd_4, characterising sets of neighbourhoods as proper filters that are *centered* (i.e. contain the point in question). The conditions are

$$\text{Nbhd\_1 } F \overset{\text{def}}{=} \forall x.\ \texttt{meet\_closed}\ (F\ x)$$

$$\text{Nbhd\_2 } F \overset{\text{def}}{=} \forall x.\ \texttt{upwards\_closed}\ (F\ x)$$

$$\text{Nbhd\_3 } F \overset{\text{def}}{=} \forall x.\ \texttt{nonEmpty}\ (F\ x)$$

$$\text{Nbhd\_4 } F \overset{\text{def}}{=} \forall x.\ \forall N.\ (F\ x)\ N \longrightarrow N\ x$$

with $\texttt{upwards\_closed}\ S \overset{\text{def}}{=} \forall X Y.\ S\ X \wedge X \preceq Y \longrightarrow S\ Y$ and $\texttt{nonEmpty}\ S \overset{\text{def}}{=} \exists x.\ S\ x$. Moreover, neighbourhood functions are useful to introduce the notions of adherent (aka. closure) and accumulation (aka. limit) points of a set A wrt. a function F.

$$\text{ADH}[F]\ A \overset{\text{def}}{=} \lambda p.\ \forall N.\ (F\ p)\ N \longrightarrow \neg(\texttt{Disj}\ N\ A) \qquad \text{(adherent points)}$$

$$\text{ACC}[F]\ A \overset{\text{def}}{=} \lambda p.\ \forall N.\ (F\ p)\ N \longrightarrow \neg(\texttt{Disj}\ N\ (A \leftharpoonup \{p\})) \qquad \text{(accumulation points)}$$

with $\texttt{Disj}\ A\ B \overset{\text{def}}{=} \forall x.\ \neg A\ x \vee \neg B\ x\ (= A \wedge B \approx \bot)$. In fact, given a closure operator $\mathcal{C}$ we can define a *neighbourhood* function $\mathcal{N}[\mathcal{C}]$ as an interior operator with 'swapped' parameters, and we can also characterise a notion of *limit point*:

$$\mathcal{N}[\mathcal{C}]\ x\ N \overset{\text{def}}{=} \mathcal{I}[\mathcal{C}]\ N\ x \qquad \text{(read as: N is a neighbourhood of x wrt. } \mathcal{C})$$

$$\texttt{limit}[\mathcal{C}]\ A \overset{\text{def}}{=} \lambda w.\ \mathcal{C}(A \leftharpoonup \{w\})\ w \qquad \text{(limit points of A wrt. } \mathcal{C})$$

Again, we have proven (fully automatically) several interesting properties, e.g.

Proposition 7

1. Cl_1' $\mathcal{C}$ implies Nbhd_1 $\mathcal{N}[\mathcal{C}]$ (and analogously for other Nbhd_X conditions);
2. given MONO $\mathcal{C}$, Cl_2 $\mathcal{C}$ and Cl_4' $\mathcal{C}$, the characterisation of neighbourhoods in terms of open sets obtains, i.e., $\mathcal{N}[\mathcal{C}]\ p = (\lambda N.\ (\exists E.\ \mathrm{Op}[\mathcal{C}]\ E \wedge E \preceq N \wedge E\ p))$;
3. given MONO $\mathcal{C}$ and Cl_2 $\mathcal{C}$, the characterisation of open sets in terms of neighbourhoods obtains, i.e., $\mathrm{Op}[\mathcal{C}]\ N = (\forall x.\ N\ x \longrightarrow \mathcal{N}[\mathcal{C}]\ x\ N)$;
4. given MONO $\mathcal{C}$, limit and accumulation points coincide, i.e., $\mathrm{limit}[\mathcal{C}] = \mathrm{ACC}[\mathcal{N}[\mathcal{C}]]$.

3.4 Separation

We have formalised several notions of separation for sets and verified some relationships.

$$\mathtt{Sep}[\mathcal{C}]\ A\ B \overset{\mathrm{def}}{=} \mathtt{Disj}(\mathcal{C}\ A)\ B \wedge \mathtt{Disj}(\mathcal{C}\ B)\ A$$

$$\mathtt{ClosedDisj}[\mathcal{C}]\ A\ B \overset{\mathrm{def}}{=} \mathrm{Cl}[\mathcal{C}]\ A \wedge \mathrm{Cl}[\mathcal{C}]\ B \wedge \mathtt{Disj}\ A\ B$$

$$\mathtt{OpenDisj}[\mathcal{C}]\ A\ B \overset{\mathrm{def}}{=} \mathrm{Op}[\mathcal{C}]\ A \wedge \mathrm{Op}[\mathcal{C}]\ B \wedge \mathtt{Disj}\ A\ B$$

$$\mathtt{SepByOpens}[\mathcal{C}]\ A\ B \overset{\mathrm{def}}{=} \exists G\ H.\ \mathtt{OpenDisj}[\mathcal{C}]\ G\ H \wedge A \preceq G \wedge B \preceq H$$

Proposition 8. We have verified, among others,

1. several properties of Sep (involving set union, difference, etc.);
2. that $\mathtt{ClosedDisj}[\mathcal{C}]$ implies $\mathtt{Sep}[\mathcal{C}]$; and
3. that, given MONO $\mathcal{C}$, $\mathtt{SepByOpens}[\mathcal{C}]$ (and therefore $\mathtt{OpenDisj}[\mathcal{C}]$) implies $\mathtt{Sep}[\mathcal{C}]$.

Employing the notions above we can formalise some of the so-called *separation axioms*:

$$\mathtt{T0}\ \mathcal{C} \overset{\mathrm{def}}{=} \forall p\ q.\ p \neq q \longrightarrow \exists G.\ \mathrm{Op}[\mathcal{C}]\ G \wedge \neg(G\ p \longleftrightarrow G\ q))$$

$$\mathtt{T1}\ \mathcal{C} \overset{\mathrm{def}}{=} \forall p\ q.\ p \neq q \longrightarrow \exists G\ H.\ \mathrm{Op}[\mathcal{C}]\ G \wedge \mathrm{Op}[\mathcal{C}]\ H \wedge G\ p \wedge \neg G\ q \wedge H\ q \wedge \neg H\ p$$

$$\mathtt{T2}\ \mathcal{C} \overset{\mathrm{def}}{=} \forall p\ q.\ p \neq q \longrightarrow \mathtt{SepByOpens}[\mathcal{C}]\ \{p\}\ \{q\}$$

$$\mathtt{Regular}\ \mathcal{C} \overset{\mathrm{def}}{=} \forall p\ Q.\ \mathrm{Cl}[\mathcal{C}]\ Q \wedge \neg Q\ p \longrightarrow \mathtt{SepByOpens}[\mathcal{C}]\ \{p\}\ Q$$

$$\mathtt{Normal}\ \mathcal{C} \overset{\mathrm{def}}{=} \forall P\ Q.\ \mathtt{ClosedDisj}\ P\ Q \longrightarrow \mathtt{SepByOpens}[\mathcal{C}]\ P\ Q$$

$$\mathtt{T3}\ \mathcal{C} \overset{\mathrm{def}}{=} \mathtt{T0}\ \mathcal{C} \wedge \mathtt{Regular}\ \mathcal{C}$$

$$\mathtt{T4}\ \mathcal{C} \overset{\mathrm{def}}{=} \mathtt{T1}\ \mathcal{C} \wedge \mathtt{Normal}\ \mathcal{C}$$

We have verified several well-known relationships between these separation axioms, e.g.

Proposition 9

1. several alternative definitions for T0 and T1 (under different minimal conditions);
2. that T3 implies T2, which implies T1, which implies T0;
3. that, given MONO and Cl_2, T4 implies T3;
4. that Regular collapses T0, T1 and T2 together;
5. and that, given MONO, Cl_2 and T1, we have that Normal implies Regular.

3.5   Connectedness

We formalise the notion of connectedness by first introducing a notion of *separation*, whereby two non-empty separated sets A and B constitute a separation of a set S (wrt. closure $\mathcal{C}$) when S is the union of A and B. A set S is *connected* if it has no separation.

$$\texttt{Separation}[\mathcal{C}]\ S\ A\ B \stackrel{\text{def}}{=} \texttt{nonEmpty}\ A \wedge \texttt{nonEmpty}\ B \wedge \texttt{Sep}[\mathcal{C}]\ A\ B \wedge S \approx A \vee B$$

$$\texttt{Separated}[\mathcal{C}]\ S \stackrel{\text{def}}{=} \exists A\ B.\ \texttt{Separation}[\mathcal{C}]\ S\ A\ B$$

$$\texttt{Connected}[\mathcal{C}]\ S \stackrel{\text{def}}{=} \neg\texttt{Separated}[\mathcal{C}]\ S$$

Employing these definitions we have proved, among others, that

Proposition 10

1. empty set and singletons are connected: $\texttt{Connected}\ [\mathcal{C}]\ \bot$, $\forall x.\ \texttt{Connected}[\mathcal{C}]\ \{x\}$;
2. given MONO, $\texttt{Connected}[\mathcal{C}]\ S \wedge S \preceq X \longrightarrow (\forall A\ B.\ \texttt{Separation}[\mathcal{C}]\ X\ A\ B \longrightarrow S \preceq A \vee S \preceq B)$;
3. given MONO, if S is connected and $S \preceq X \preceq \mathcal{C}(S)$ then X is connected too;
4. given MONO and Cl_2, we have that if every two points in a set X are contained in some connected subset of X then X itself is connected; i.e. $\forall p\ q.\ (X\ p \wedge X\ q) \longrightarrow (\exists S.\ S \preceq X \wedge \texttt{Connected}[\mathcal{C}]\ S \wedge S\ p \wedge S\ q)$ implies $\texttt{Connected}[\mathcal{C}]\ X$.

In particular, item 4 above provides an exemplary instance of a proposition that could not (yet) be proven fully automatically using Sledgehammer. Therefore, we have had recourse to interactive Isar proof reconstruction as illustrated in Fig. 1.

3.6   Compactness

We encode two equivalent definitions of compactness, based upon closed sets and open sets respectively. The first one builds upon the so-called *finite intersection property* (FIP). The second is a conveniently dualised variant of the first, which corresponds to the traditional 'open covers' definition in the literature. We start

```
lemma assumes mono: ‹MONO C› and cl2: ‹Cl_2 C› shows
 ‹(∀p q. X p ∧ X q ∧ (∃S. S ⪯ X ∧ Connected[C] S ∧ S p ∧ S q)) ⟶ Connected[C] X›
proof -
{ assume premise: ‹∀p q. X p ∧ X q ∧ (∃S. S ⪯ X ∧ Connected[C] S ∧ S p ∧ S q)›
  have ‹Connected[C] X› proof
    assume ‹Separated[C] X›
    then obtain A and B where sepXAB: ‹Separation[C] X A B›
      using Separated_def by blast
    hence nonempty: ‹nonEmpty A ∧ nonEmpty B›
      by (simp add: Separation_def)
    let ?p = ‹SOME a. A a› and ?q = ‹SOME b. B b› (*since A and B are non-empty*)
    from nonempty have ‹X ?p ∧ X ?q›
      by (simp add: premise)
    hence ‹∃S. S ⪯ X ∧ Connected[C] S ∧ S ?p ∧ S ?q›
      by (simp add: premise)
    then obtain S where aux: ‹S ⪯ X ∧ Connected[C] S ∧ S ?p ∧ S ?q› ..
    from aux nonempty have ‹¬Disj S A ∧ ¬Disj S B›
      by (metis Disj_def someI)
    moreover from aux mono sepXAB have ‹S ⪯ A ∨ S ⪯ B›
      using conn_prop3 by blast
    moreover from cl2 sepXAB have ‹Disj A B›
      by (simp add: Sep_disj Separation_def)
    ultimately show False
      by (metis (mono_tags, lifting) Disj_def subset_def)
  qed
} thus ?thesis ..
qed
```

Fig. 1. Isar proof reconstruction for Proposition 10 (4) in Isabelle/HOL.

by defining the FIP which we then dualise towards what we dubbed 'finite union property' (FUP).

$$\text{FIP } S \stackrel{\text{def}}{=} \forall D. \text{ nonEmpty } D \wedge D \sqsubseteq S \wedge \text{finite } D \longrightarrow \neg \bigwedge D \approx \bot$$

$$\text{FUP } S \stackrel{\text{def}}{=} \exists D. \text{ nonEmpty } D \wedge D \sqsubseteq S \wedge \text{finite } D \quad \wedge \quad \bigvee D \approx \top$$

where the predicate finite draws on Dedekind's definition of finite/infinite sets:[12]

$$\text{finite } A \stackrel{\text{def}}{=} \forall f. \text{ mapping } f \, A \, A \wedge \text{injective } f \, A \longrightarrow \text{surjective } f \, A \, A$$

After verifying some properties of FIP and FUP we prove the statement below.

Proposition 11. The following are equivalent:

$$\text{compact}^{cl} \, C \stackrel{\text{def}}{=} \forall S. \, S \sqsubseteq \text{Cl}[C] \wedge \text{FIP } S \longrightarrow \neg \bigwedge S \approx \bot$$

$$\text{compact}^{op} \, C \stackrel{\text{def}}{=} \forall S. \, S \sqsubseteq \text{Op}[C] \wedge \bigvee S \approx \top \longrightarrow \text{FUP } S$$

An interesting exercise for students is to extend these results to others known in the literature, and to use compactness to further relate different separation axioms (cf. Sect. 3.4).

---

[12] The predicates mapping, injective and surjective are formalised in the usual way; we refer to the Isabelle/HOL sources for these and other miscellaneous definitions and lemmata.

3.7   Continuity and Homeomorphism

We have formalised different notions of continuity using closure/interior operators as well as their fixed-points (closed/open sets). We present two of them as an illustration. ($[\![\varphi \ S]\!]$ and $[\![\varphi \ S]\!]^{-1}$ correspond to the direct and inverse image of S under φ, resp.)

$$\mathrm{Cont}^{C}[\mathcal{C}_1, \mathcal{C}_2] \ \varphi \ \stackrel{\mathrm{def}}{=} \ \forall U. \ [\![\varphi \ (\mathcal{C}_1 \ U)]\!] \preceq \mathcal{C}_2 \ [\![\varphi \ U]\!]$$

$$\mathrm{Cont}^{cl}[\mathcal{C}_1, \mathcal{C}_2] \ \varphi \ \stackrel{\mathrm{def}}{=} \ \forall V. \ \mathrm{Cl}[\mathcal{C}_2] \ V \longrightarrow \mathrm{Cl}[\mathcal{C}_1] \ [\![\varphi \ V]\!]^{-1}$$

They basically say that a mapping φ (type $'u \Rightarrow 'v$) is continuous iff (i) its direct image distributes (increasingly) over the closure, (ii) the inverse image of closed sets are closed.

We also give definitions based on point-continuity, using both interior operators and neighbourhood functions, and automatically prove that all of them are equivalent, e.g.

$$\mathrm{pCont}_1^{I}[\mathcal{C}_1, \mathcal{C}_2] \ \varphi \ x \ \stackrel{\mathrm{def}}{=} \ \forall V. \ \mathcal{I}[\mathcal{C}_2] \ V \ (\varphi \ x) \longrightarrow \exists U. \ \mathcal{I}[\mathcal{C}_1] \ U \ x \wedge [\![\varphi \ U]\!] \preceq V$$

$$\mathrm{pCont}_2^{N}[\mathcal{C}_1, \mathcal{C}_2] \ \varphi \ x \ \stackrel{\mathrm{def}}{=} \ \forall V. \ \mathcal{N}[\mathcal{C}_2] \ (\varphi \ x) \ V \longrightarrow \exists U. \ \mathcal{N}[\mathcal{C}_1] \ x \ U \wedge U \preceq [\![\varphi \ V]\!]^{-1}$$

Proposition 12

1. assuming MONO $\mathcal{C}_1$, MONO $\mathcal{C}_2$, Cl_2 $\mathcal{C}_1$, Cl_2 $\mathcal{C}_2$ and Cl_4' $\mathcal{C}_2$, we have in fact that $\mathrm{Cont}^{cl}[\mathcal{C}_1, \mathcal{C}_2] \ \varphi$ iff $\mathrm{Cont}^{C}[\mathcal{C}_1, \mathcal{C}_2] \ \varphi$ (other definitions being also equivalent);
2. we also have that $\mathrm{pCont}_1^{I}[\mathcal{C}_1, \mathcal{C}_2] \ \varphi \ x$ iff $\mathrm{pCont}_2^{N}[\mathcal{C}_1, \mathcal{C}_2] \ \varphi \ x$;
3. and, moreover, $\mathrm{Cont}^{C}[\mathcal{C}_1, \mathcal{C}_2] \ \varphi$ iff $\forall x. \ \mathrm{pCont}_1^{I}[\mathcal{C}_1, \mathcal{C}_2] \ \varphi \ x$.

Moreover, we have formalised notions of closed-maps, open-maps, and homeomorphism, interrelating them with continuity (not shown). The preservation of properties such us denseness, separability/connectedness and compactness are interesting exercises for students to verify (as structured Isar proofs with the help of Sledgehammer, cf. Sect. 4).

3.8   Specialisation Orderings and Relations

A topology is called 'Alexandrov' (or 'finitely generated') when the intersection (union) of any arbitrary family of open (closed) sets is open (closed); in more algebraic terms, this means that the set of fixed points of the closure (interior) operation is closed under infinite suprema (infima). These topologies have interesting properties relating them to the semantics of modal logic, since, assuming the Kuratowski conditions, their closure operators are in one-to-one correspondence with relations, called 'specialisation preorders' (i.e. modal **S4** accessibility relations). As usual we weaken our assumptions and harness automated tools to explore minimal conditions under which several properties and interrelations obtain. We first interdefine (Alexandrov) closures and their corresponding (specialisation) relations:

$$\mathcal{C}[R] \overset{\text{def}}{=} \lambda A.\ \lambda w.\ \exists v.\ R\ w\ v \wedge A\ v \quad \text{conversely} \quad \mathbf{sp}[\mathcal{C}] \overset{\text{def}}{=} \lambda w.\ \lambda v.\ \mathcal{C}\ \{v\}\ w$$

We explore the minimal conditions under which: (i) closure conditions for the operator $\mathcal{C}[R]$ obtain, and (ii) properties for the relation $\mathbf{sp}[\mathcal{C}]$ obtain.

Proposition 13

1. $\mathcal{C}[R]$ satisfies (directly) iCl_1 and Cl_3; moreover iCl_1 $\mathcal{C}$ implies $\mathcal{C} = \mathcal{C}[\mathbf{sp}[\mathcal{C}]]$;
2. assuming R reflexive, then Cl_2 $\mathcal{C}[R]$; if, moreover, R is transitive, then Cl_4 $\mathcal{C}[R]$;
3. $\mathbf{sp}[\mathcal{C}]$ is reflexive assuming Cl_2 $\mathcal{C}$, and transitive assuming MONO $\mathcal{C}$ and Cl_4 $\mathcal{C}$;
4. assuming MONO $\mathcal{C}$, Cl_2 $\mathcal{C}$ and Cl_4' $\mathcal{C}$, then $\mathbf{sp}[\mathcal{C}]$ is antisymmetric iff TO $\mathcal{C}$;
5. assuming MONO $\mathcal{C}$, Cl_2 $\mathcal{C}$, and TO $\mathcal{C}$, then $\mathbf{sp}[\mathcal{C}]$ is symmetric iff T1 $\mathcal{C}$;

4   Methodological Remarks

The formalisation discussed in the previous section is a snapshot of an ongoing bottom-up theory development effort. Just enough material has been formalised to justify the claim that our approach is a promising one towards formalising basic topology for (students of) computational logic. We aim at further developing this work, and thus have set up a repository with the Isabelle/HOL sources (https://github.com/davfuenmayor/basic-topology). We want to encourage interested readers to contribute to this effort.

As the reader may have noticed, we made use of little to none Isabelle/HOL-specific constructs like type classes, locales, typedefs, etc. In particular, datatypes and inductive definitions/proofs have not been needed (so far...). It is argued (cf. [2]) that such an approach readily leverages the resources of automated theorem provers. As readers can verify for themselves, Sledgehammer [6] and Nitpick [7] perform very well on the provided sources, with only a small caveat worth mentioning: Since we have made generous use of (non-inductive) definitions, they still need to be carefully crafted and selectively unfolded to best leverage Sledgehammer's performance. There were actually several cases where a couple of definitions needed to be manually unfolded before successfully invoking Sledgehammer. In fact, this didn't happen in the first stages of our formalisation work when we worked mostly with abbreviations. Still, we think that, as a formalisation grows and definitions become more complex, building a layered, hierarchical structure of definitions becomes necessary. We have made some first steps in this direction, but we are yet to figure out the optimal way to build such a definitions' hierarchy, in order to get the most out of Sledgehammer and Isabelle's rewriting machinery.

The Nitpick model finder has also been very useful in quickly computing counterexamples to candidate theorems in case of missing assumptions, thus allowing for assumption-minimal proofs (wrt. a known set of relevant conditions, as happens in topology, e.g., with Kuratowski's closure axioms). Our approach also has the advantage that models generated by Nitpick are relatively good legible (after some initial training). It is also worth noting that Nitpick has found countermodels of small cardinalities (spaces with less than 4 points) for every single finitely countersatisfiable conjecture that we have tried in our formalisation (most of those have been deleted from the sources, but readers can easily reproduce this for themselves). Moreover, Sledgehammer's suggestions are also helpful in the quest for proof-minimality (wrt. required lemmata or definitions), though we have not insisted much on optimising proofs in this regard. A quite surprising benefit of our approach being based on lattice theory (via Stone-type representations) is that lattice dualities can be exploited to create dual definitions (e.g. FIP & FUP in Sect. 3.6). Beyond the evident aesthetic appeal of dualities, they also have an important pragmatic value, as they provide us with two ways to articulate and solve any given problem. This has proven very useful in the present formalisation work, as well as in previous ones based on a similar approach [9].

In its current form, this material can be employed as part of a course on (semantics for) computational logic. In our opinion, the present approach, harnessing hammers and model finders to do most of the proof and refutation heavy-lifting, is particularly suited for teaching when combined with the expressive capabilities of a structured proof language such as Isabelle's *Isar* [20]. From a pedagogical point of view, we conceive the proof workflow for a conjecture as follows:[13]

1. Students first employ model finders to find out whether the conjecture is countersatisfiable (avoiding the Sisyphean task of trying to prove non-theorems). This helps uncover formalisation errors, as well as missing (implicit) assumptions early on.
2. Students can then try hammers as 'oracles' directly on the conjecture, thus receiving suggestions for sets of definitions and lemmata required for a proof. If automatic kernel reconstruction succeeds we might finish at this point.
3. In case an explicit proof is desired, students should start a structured proof, where hammers (and model finders) can again be employed as 'oracles' at any point during the interactive proof reconstruction, eventually starting nested proofs in a structured fashion. Thus, the level of proof granularity can be suitably chosen, allowing for less-verbose and, arguably, more enlightening proofs.

We think that similar approaches to ours (whenever they are possible) can drastically improve the learning curve for employing proof assistants in the classroom,

---

[13] We have followed this very same workflow during the formalisation work, noting that thanks to Sledgehammer's good performance, at the time of writing (May 2022) we have had recourse to interactive proofs (as shown in Fig. 1) only as a fallback in a few cases (hence finishing already at step 2). In a teaching context, students are surely expected to go the extra mile.

since only a basic knowledge of simple type theory and (very rudimentary) lattice theory is needed to work with these formalisations. We believe that this is a rather safe requirement for nowadays students in computer science, who are probably already familiarised with some sort of functional programming or even programming language semantics.

5 Related Work

An important source for definitions and lemmata, as well as for methodological inspiration, has been the seminal paper by Kazimierz Kuratowski [13], and particularly his textbook on topology [14]. Special mention deserve here the relatively unknown papers by Miron Zarycki [21, 22].[14] We find it quite unfortunate, especially for logicians, that this way of introducing topology has not met the recognition it deserves.

Seminal work in this area is McKinsey & Tarski's 'The Algebra of Topology' [16], from which we also took inspiration. The same holds for later work by Leo Esakia and colleagues (e.g. [8]), as well as for work on topological semantics for logics (e.g. [1]). Worth mentioning here are [19] and [15], whose philosophy and approach has strongly influenced us, even if our methods differ to some extent.

Several topology libraries have been developed for proof assistants. In Lean's *mathlib*[15] extensive use is made of filters in the formalisation of results on general topology, topological algebra, etc. In Coq's library[16] some basic notions and results of general topology are introduced in a more conventional fashion. In Isabelle's *AFP* we find the entry *Topology*[17] also drawing upon traditional textbook approaches. A similar observation can be made of the library `HOL/Topological_Spaces.thy`, serving as basis for formalisations in areas of analysis and geometry. Finally, our work can be considered as a 'spin-off' of selected parts of [9], by taking as starting point the (simplified) formalisation of closure algebras upon which we further develop notions specific to topology.

6 Further Work and Conclusions

The approach introduced in the present work intends to encourage the use of automated reasoning tools, as integrated in proof assistants, in order to facilitate the comprehension of fundamental mathematical concepts, starting with topology at present.

Beyond pedagogical ambitions (cf. the discussion in Sect. 4), with this work we want to give impetus to the study, using formal methods, of the scope and

---

[14] English translations of Zarycki's works are now available on the web thanks to Mark Bowron (see https://www.researchgate.net/scientific-contributions/Miron-Zarycki-2016157096).

[15] https://leanprover-community.github.io/theories/topology.html.

[16] https://github.com/coq-community/topology.

[17] https://www.isa-afp.org/browser_info/current/AFP/Topology/index.html.

applications of topological methods for (logic in) computer science, especially by drawing upon lattice theory. In particular, we aim at investigating applications in computational logic for the sort of topological Boolean algebras presented here, e.g., in the study of formal semantics and elementary model theory (as explored in [9]). The present work makes a first step in this direction by explicating in lattice-theoretical terms some basic notions in topology that are well-suited to this undertaking.

As a distinctive feature, this formalisation employs only 'vanilla' simple type theory, and is thus suitable for encoding using different proof assistants (and even higher order automated theorem provers) much in the spirit of *shallow semantical embeddings* [2], where it has been argued that this readily leverages the performance of automated tools.

Using Isabelle's integrated model finder (Nitpick) and hammer (Sledgehammer) in tandem, we have been able to easily uncover minimal conditions under which several results hold. This is not only useful pedagogically, but can also support a future research program on 'reverse mathematics' for topology and related areas. In particular, Nitpick was of enormous help in catching typos and other formalisation mistakes early on, thus sparing us the Sisyphean task of searching for non-existent proofs. Moreover, we have relied on Sledgehammer for most of the proof heavy-lifting, having recourse to interactive Isar proofs only in a few cases. Based on this experience, we have sketched a proposal for introducing proof assistants in the classroom by employing our approach, which we expect to be implementing very soon.

Acknowledgements. The first author acknowledges financial support from the Luxembourg National Research Fund (FNR), under grant CORE C20/IS/14616644.

References

1. Aiello, M., Pratt-Hartmann, I., Van Benthem, J., et al.: Handbook of Spatial Logics. Springer, Dordrecht (2007). https://doi.org/10.1007/978-1-4020-5587-4
2. Benzmüller, C.: Universal (meta-)logical reasoning: recent successes. Sci. Comput. Program. **172**, 48–62 (2019)
3. Benzmüller, C., Parent, X., van der Torre, L.: Designing normative theories for ethical and legal reasoning: LogiKEy framework, methodology, and tool support. Artif. Intell. **287**, 103348 (2020)
4. Benzmüller, C., Paulson, L.C.: Quantified multimodal logics in simple type theory. Logica Universalis (Spec. Issue Multimodal Log.) **7**(1), 7–20 (2013)
5. Benzmüller, C., Andrews, P.: Church's type theory. In: Zalta, E.N. (ed.) The Stanford Encyclopedia of Philosophy. Metaphysics Research Lab, Stanford University, Summer 2019 edn. (2019). https://plato.stanford.edu/archives/sum2019/entries/type-theory-church/
6. Blanchette, J.C., Kaliszyk, C., Paulson, L.C., Urban, J.: Hammering towards QED. J. Formalized Reasoning **9**(1), 101–148 (2016)
7. Blanchette, J.C., Nipkow, T.: Nitpick: a counterexample generator for higher-order logic based on a relational model finder. In: Kaufmann, M., Paulson, L.C. (eds.) ITP 2010. LNCS, vol. 6172, pp. 131–146. Springer, Heidelberg (2010). https://doi.org/10.1007/978-3-642-14052-5_11

8. Esakia, L.: Intuitionistic logic and modality via topology. Ann. Pure Appl. Logic **127**(1–3), 155–170 (2004)
9. Fuenmayor, D.: Topological semantics for paraconsistent and paracomplete logics. Archive of Formal Proofs (2020). https://isa-afp.org/entries/Topological_Semantics.html
10. Fuenmayor, D., Benzmüller, C.: Normative reasoning with expressive logic combinations. In: De Giacomo, G., et al. (eds.) ECAI 2020–24th European Conference on Artificial Intelligence, 8–12 June, Santiago de Compostela, Spain. Frontiers in Artificial Intelligence and Applications, vol. 325, pp. 2903–2904. IOS Press (2020)
11. Gierz, G., Hofmann, K.H., Keimel, K., Lawson, J.D., Mislove, M., Scott, D.S.: Continuous Lattices and Domains, vol. 93. Cambridge University Press, Cambridge (2003)
12. Givant, S.: Duality Theories for Boolean Algebras with Operators. Springer, Cham (2014). https://doi.org/10.1007/978-3-319-06743-8
13. Kuratowski, K.: Sur l'opération A de l'analysis situs. Fundam. Math. **3**(1), 182–199 (1922)
14. Kuratowski, K.: Topology: Volume I. Academic Press (1966)
15. Martin, N.M., Pollard, S.: Closure Spaces and Logic, Mathematics and Its Applications, vol. 369. Springer, New York (1996). https://doi.org/10.1007/978-1-4757-2506-3
16. McKinsey, J.C., Tarski, A.: The algebra of topology. Ann. Math. **45**, 141–191 (1944)
17. Paulson, L.C.: Computational logic: its origins and applications. Proc. Roy. Soc. A Math. Phys. Eng. Sci. **474**(2210), 20170872 (2018)
18. The Univalent Foundations Program: Homotopy Type Theory: Univalent Foundations of Mathematics (2013). https://homotopytypetheory.org/book, Institute for Advanced Study
19. Vickers, S.: Topology via Logic. Cambridge University Press, Cambridge (1996)
20. Wenzel, M.: Isabelle/Isar-a generic framework for human-readable proof documents. From Insight to Proof-Festschrift in Honour of Andrzej Trybulec **10**(23), 277–298 (2007)
21. Zarycki, M.: Quelques notions fondamentales de l'analysis situs au point de vue de l'algèbre de la logique. Fundam. Math. **9**(1), 3–15 (1927)
22. Zarycki, M.: Some properties of the derived set operation in abstract spaces. Nauk. Zap. Ser. Fiz.-Mat. **5**, 22–33 (1947)

Formalising the Kruskal-Katona Theorem in Lean

Bhavik Mehta[✉][iD]

Department of Pure Mathematics and Mathematical Statistics,
Centre for Mathematical Sciences, Wilberforce Road, Cambridge CB3 0WB, UK
bm489@cam.ac.uk

Abstract. The Kruskal-Katona theorem is a celebrated result of extremal combinatorics providing precise cardinality bounds on the 'shadow' of a family of finite sets: the family given by removing an element from each set of the original. We describe a formalisation of the Kruskal-Katona theorem in the Lean theorem prover, including a computable implementation of the shadow as well as standard inequalities about it, and a definition of the colexicographic ordering on finite sets. In addition, we apply these results to other classical combinatorial theorems: Sperner's theorem on antichains and the Erdős-Ko-Rado theorem on intersecting families.

Keywords: Formalisation · Lean Theorem Prover · Extremal combinatorics

1 Introduction

Extremal Combinatorics is a modern and rapidly developing area of discrete mathematics [1], with problems that are often motivated by questions in other areas, such as Geometry, Number Theory, Theoretical Computer Science and Game Theory. It studies the maximal (or minimal) size of a collection of objects (such as natural numbers, edges in a graph or subsets of a finite set), subject to particular restrictions.

One of the earliest results in this area is Sperner's theorem [19], which shows that if you wish to find a large collection of subsets of an n-element set such that none is a subset of another, your collection can be no larger than $\binom{n}{\lfloor \frac{n}{2} \rfloor}$. This result has been reproved and generalised many times over the years [2]. One such direction, given by Lubell [15], Yamamoto [21], Meshalkin [17] and Bollobás [3], sometimes known as the LYM (Lubell-Yamamoto-Meshalkin) inequality, immediately implies Sperner's theorem.

One proof of this inequality uses a 'local' version which concerns an operator on families of finite sets, known as the *shadow*. In brief, the shadow consists of those sets formed by removing any one element from any set in the original family. The local LYM inequality gives an elegant bound on the size of the shadow in terms of the size of the original family, but can be vastly improved upon by

© The Author(s), under exclusive license to Springer Nature Switzerland AG 2022
K. Buzzard and T. Kutsia (Eds.): CICM 2022, LNAI 13467, pp. 75-91, 2022.
https://doi.org/10.1007/978-3-031-16681-5_5

the Kruskal-Katona theorem [13]: while the local LYM inequality is tight for only very specific set families, the Kruskal-Katona theorem can in principle be used to give an explicit expression for the minimum possible shadow size for a given family size.

In this paper we describe our formal proof in the Lean theorem prover [18] of each of these named results, concluding with a slight generalisation of the Kruskal-Katona theorem and an application of it to the Erdős-Ko-Rado theorem [9]. Our source for this formalisation is a lecture series given by Imre Leader on Combinatorics at the University of Cambridge in 2018 [14]. We use results from the Lean mathematical library mathlib [20], which is characterised by a distributed and decentralised community of contributors, and ubiquitous classical reasoning. Some of the more general results formalised have already been accepted into mathlib, and the remainder are in the process of being merged. We provide a snapshot of our code online[1]. The majority of this formalisation process took place when the author was new to the Lean theorem prover, in late 2019. As a consequence, the results presented here use a relatively old version of Lean and mathlib from early 2020, so the experienced Lean reader may notice some slightly surprising notation, though the proofs of course still compile.

In Sect. 2 we discuss related work, focusing on other formalisations of combinatorics. Section 3 presents the mathematical background of the proofs, to demonstrate the ideas relevant in this area of combinatorics. Sections 4 and 5 describe the formalisation of these results, first discussing the shadow and LYM inequalities, then moving onto the statement and proof of the Kruskal-Katona theorem. Finally, Sect. 6 concludes.

2   Related Work

Frankl's conjecture on union-closed families is an open problem in extremal combinatorics with close ties to many of the statements relevant here - work in its direction were formalised in Isabelle/HOL in 2012 [16], which verified some existing partial results and confirmed existing union-closed families. Also in Isabelle/HOL, Edmonds and Paulson [8] have formalised a number of results on set systems and hypergraphs, with similar themes to the Erdős-Ko-Rado theorem, however the focus of their results was on the use of linear algebraic techniques in combinatorics.

Within the field of combinatorics, the cap-set problem [6], the regularity lemma [7], and Hall's marriage theorem [12] have been formalised in Lean. The cap-set problem also utilises linear algebraic techniques, and the proof for the latter theorem is particularly relevant to this work, as it allows for an alternate proof of Sperner's theorem to be derived.

3   Mathematical Background

We now give the results which are formalised in this paper. While we endeavour to communicate the flavour of all the relevant concepts, we will occasionally omit

---

[1] https://github.com/b-mehta/combinatorics.

some motivation as may be given in a textbook, and sketch proofs rather than giving them in full detail. The interested reader may consult [4] or [2] for a more thorough exposition.

We write $n \in \mathbb{N}$ and define $X = [n] := \{1, 2, \ldots, n\}$ throughout, referred to as the *ground set*. Many of the results in this paper will concern set families, also known as set systems, which are simply (finite) families of subsets of $[n]$, i.e. subsets $\mathcal{A}$ with $\mathcal{A} \subseteq \mathcal{P}(X)$.

Definition 1 (r-set). *For a set $B \subseteq X$ and natural r, we write*

$$B^{(r)} := \{A \subseteq B : |A| = r\}.$$

In particular, $X^{(r)}$ as r varies in $0 \leq r \leq n$ form a partition of $\mathcal{P}(X)$, and $|X^{(r)}| = \binom{n}{r}$. A visualisation of these two facts is given in Fig. 1, noting that both $X^{(n)}$ and $X^{(0)}$ have exactly one member.

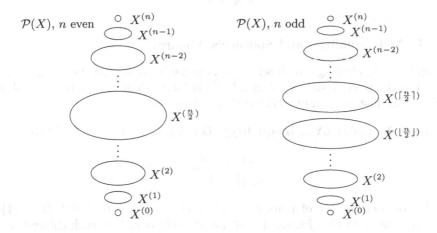

Fig. 1. The partition structure of $P([n])$

Definition 2 (Shadow). *Let $\mathcal{A} \subseteq X^{(r)}$, for some $1 \leq r \leq n$. The shadow of $\mathcal{A}$ (see Fig. 2a) is*

$$\partial \mathcal{A} := \{A \setminus \{i\} : A \in \mathcal{A}, \ i \in A\} \subseteq X^{(r-1)}.$$

Example 1. If $\mathcal{A} = \{\{1, 2, 3\}, \{1, 2, 4\}, \{1, 3, 4\}, \{1, 3, 5\}\} \subseteq [5]^{(3)}$, then

$$\partial \mathcal{A} = \{\{1, 2\}, \{1, 3\}, \{1, 4\}, \{1, 5\}, \{2, 3\}, \{2, 4\}, \{3, 4\}, \{3, 5\}\} \subseteq [5]^{(2)}.$$

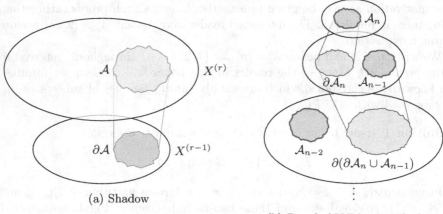

(a) Shadow

(b) Proof of LYM using shadows

Fig. 2. .

3.1   LYM Inequalities and Sperner's Theorem

In Fig. 2a, a simple diagram showing the shadow is given, and the picture suggests that the relative size of $\partial\mathcal{A}$ in $X^{(r-1)}$ is larger than the relative size of $\mathcal{A}$ in $X^{(r)}$. Our first inequality confirms this in general.

Theorem 1 (Local LYM inequality). *Let* $\mathcal{A} \subseteq X^{(r)}$, $1 \leq r \leq n$. *Then*

$$\frac{|\partial\mathcal{A}|}{\binom{n}{r-1}} \geq \frac{|\mathcal{A}|}{\binom{n}{r}}.$$

Proof. Consider the set of pairs $P = \{(A, B) : A \in \mathcal{A}, B \in \partial\mathcal{A}, B \subseteq A\}$, and count this set twice. Fixing $A \in \mathcal{A}$, we have $|P| = r|\mathcal{A}|$ as each choice has r corresponding choices of $B \in \partial\mathcal{A}$ with $B \subseteq A$. On the other hand, fixing $B \in \partial\mathcal{A}$, there are at most $n - r + 1$ available choices of A, so $|P| \leq (n - r + 1)|\partial\mathcal{A}|$. The result immediately follows. □

The local LYM inequality can be used to prove the LYM inequality about antichains.

Definition 3 (Antichain). *A family* $\mathcal{A} \subseteq \mathcal{P}(X)$ *is an* antichain *if* $\forall A, B \in \mathcal{A}$, $A \neq B \Rightarrow A \nsubseteq B$.

Example 2. The set family $\{\{1\}, \{4, 6, 7\}, \{2, 4, 5, 6\}\} \subseteq \mathcal{P}([7])$ is an antichain.

Theorem 2 (LYM). *Let* $\mathcal{A} \subseteq \mathcal{P}(X)$ *be an antichain. Then*

$$\sum_{r=0}^{n} \frac{|\mathcal{A} \cap X^{(r)}|}{\binom{n}{r}} \leq 1.$$

Proof (Sketch). Write $\mathcal{A}_r = \mathcal{A} \cap X^{(r)}$. Clearly $\frac{|\mathcal{A}_n|}{\binom{n}{n}} \leq 1$. As $\mathcal{A}$ is an antichain, $\partial \mathcal{A}_n$ is disjoint from $\mathcal{A}_{n-1}$, so we can write

$$\frac{|\mathcal{A}_n|}{\binom{n}{n}} + \frac{|\mathcal{A}_{n-1}|}{\binom{n}{n-1}} \leq \frac{|\partial \mathcal{A}_n|}{\binom{n}{n-1}} + \frac{|\mathcal{A}_{n-1}|}{\binom{n}{n-1}} = \frac{|\partial \mathcal{A}_n \cup \mathcal{A}_{n-1}|}{\binom{n}{n-1}} \leq 1$$

using the local LYM inequality for the first step. Similarly, $\partial(\partial \mathcal{A}_n \cup \mathcal{A}_{n-1})$ is disjoint from $\mathcal{A}_{n-2}$ since $\mathcal{A}$ is an antichain, hence

$$\frac{|\mathcal{A}_n|}{\binom{n}{n}} + \frac{|\mathcal{A}_{n-1}|}{\binom{n}{n-1}} + \frac{|\mathcal{A}_{n-2}|}{\binom{n}{n-2}} \leq \frac{|\partial \mathcal{A}_n \cup \mathcal{A}_{n-1}|}{\binom{n}{n-1}} + \frac{|\mathcal{A}_{n-2}|}{\binom{n}{n-2}}$$

$$\leq \frac{|\partial(\partial \mathcal{A}_n \cup \mathcal{A}_{n-1})| + |\mathcal{A}_{n-2}|}{\binom{n}{n-2}} \leq 1$$

where we have used our previous inequality, then the local LYM inequality and disjointness in that order. A visual interpretation of this proof is given in Fig. 2b. Continuing this process inductively, we are left with the required result.   □

The LYM inequality allows us to quickly prove Sperner's theorem.

Theorem 3 (Sperner, [19]). *Let $\mathcal{A} \subseteq \mathcal{P}(X)$ be an antichain. Then $|\mathcal{A}| \leq \binom{n}{\lfloor \frac{n}{2} \rfloor}$.*

Proof. For each r, $\binom{n}{r} \leq \binom{n}{\lfloor \frac{n}{2} \rfloor}$, and the $X^{(r)}$ partition $\mathcal{P}(X)$, so we are done from LYM.   □

3.2   Kruskal-Katona

While the local LYM inequality gives a sharp lower bound on the size of the shadow (in particular, it is attained for $\mathcal{A} = \varnothing$ and $\mathcal{A} = X^{(r)}$), we can do even better and give a precise lower bound on $|\partial \mathcal{A}|$ for fixed $\mathcal{A}$, which is the content of the Kruskal-Katona theorem. To state the theorem, we must first define the colexicographic ordering, an important linear (total) order on $X^{(r)}$.

Definition 4 (Colexicographic order). *For $A, B \in X^{(r)}$, say $A < B$ in the colexicographic order (or colex order) if $A \neq B$ and $\max(A \triangle B) \in B$. (Recall that $\triangle$ refers to the symmetric difference, and note that $A \neq B$ implies $A \triangle B \neq \varnothing$ so the maximum is well-defined).*

Example 3. The colex ordering on $[4]^{(2)}$ is

$$\{1,2\} < \{1,3\} < \{2,3\} < \{1,4\} < \{2,4\} < \{3,4\}.$$

Theorem 4 (Kruskal-Katona theorem). *Let $\mathcal{A} \subseteq X^{(r)}$ for $1 \leq r \leq n$ and $\mathcal{C}$ the initial segment of colex on $X^{(r)}$ with $|\mathcal{C}| = |\mathcal{A}|$. Then $|\partial \mathcal{A}| \geq |\partial \mathcal{C}|$.*

In particular, among all subsets of $X^{(r)}$ of a fixed size, the minimum possible shadow size is the one given by the initial segment of colex. This strengthens the local LYM inequality: instead of giving a lower bound on $|\partial \mathcal{A}|$ uniformly in $|\mathcal{A}|$, the Kruskal-Katona theorem gives a lower bound on $|\partial \mathcal{A}|$ for every possible choice of $|\mathcal{A}|$, each one attainable. There are a range of possible proofs of this theorem, we present a particular one based on UV-compressions [5].

Definition 5 (UV-compression). *For $U, V \subseteq X$ with $|U| = |V|$ and $U \cap V = \varnothing$, the UV-compression C_{UV} is defined as follows: For $A \subseteq X$,*

$$C_{UV}(A) = \begin{cases} A \cup U \setminus V & \text{if } V \subseteq A, U \cap A = \varnothing \\ A & \text{otherwise} \end{cases}$$

and if $\mathcal{A} \subseteq X^{(r)}$,

$$C_{UV}(\mathcal{A}) = \{C_{UV}(A) : A \in \mathcal{A}\} \cup \{A \in \mathcal{A} : C_{UV}(A) \in \mathcal{A}\}.$$

Say $\mathcal{A}$ is UV-compressed if $C_{UV}(\mathcal{A}) = \mathcal{A}$.

Note that $|C_{UV}(\mathcal{A})| = |\mathcal{A}|$. Our key proposition describes how compression changes the size of the shadow:

Proposition 1. *Let $\mathcal{A} \subseteq X^{(r)}$ and $U, V \subseteq X$ with $|U| = |V|$ and $U \cap V = \varnothing$. Suppose $\forall x \in U \; \exists y \in V$ such that $\mathcal{A}$ is $(U \setminus \{x\}, V \setminus \{y\})$-compressed. Then $|\partial C_{UV}(\mathcal{A})| \leq |\partial \mathcal{A}|$.*

Proof (Sketch). We will construct an injection from $\partial C_{UV}(\mathcal{A}) \setminus \partial \mathcal{A}$ to $\partial \mathcal{A} \setminus \partial C_{UV}(\mathcal{A})$, which is sufficient for the result. In particular, show that if $B \in \partial C_{UV}(\mathcal{A}) \setminus \partial \mathcal{A}$ then $U \subseteq B$, $V \cap B = \varnothing$ and $(B \cup V) \setminus U \in \partial \mathcal{A} \setminus \partial C_{UV}(\mathcal{A})$. Each of these requires some nontrivial set calculations, and the first and third use the compression hypothesis, but we omit the details here for brevity.

Proof (of Kruskal-Katona theorem). Define

$$\Gamma = \{(U, V) : U, V \subseteq X, |U| = |V| > 0, U \cap V = \emptyset, \max U < \max V\},$$

the set of useful compressions. Iteratively find the compression (U, V) in this set for which the current set is not already compressed by it, and with $|U|$ minimal. At each stage, we remain the same size, but the shadow may only decrease by Proposition 1. The process must terminate, as $\sum_{A \in \mathcal{A}_n} \sum_{i \in A} 2^i$ is decreasing in n. The final system $\mathcal{B}$ satisfies $|\mathcal{B}| = |\mathcal{A}|$ and $|\partial \mathcal{B}| \leq |\partial \mathcal{A}|$, and is (U, V)-compressed for every (U, V) in Γ.

Finally, it is enough to show $\mathcal{B} = \mathcal{C}$. As they have the same size, it suffices to demonstrate $\mathcal{B}$ is an initial segment, so suppose otherwise. Then there are sets A, B with $A < B$ in colex with $A \notin \mathcal{B}$ and $B \in \mathcal{B}$, but then $U := A \setminus B$ and $V := B \setminus A$ have $(U, V) \in \Gamma$ but $C_{UV}(B) = A$, contradicting that $\mathcal{B}$ is (U, V)-compressed. $\square$

This theorem has a number of interesting consequences. First, taking $k \leq n$, $[k]^{(r)}$ is an initial segment of $X^{(r)}$ so if $|\mathcal{A}| = \binom{k}{r}$ then $|\partial\mathcal{A}| \geq \binom{k}{r-1}$. Note that applying local LYM to this case, we are left instead with

$$|\partial\mathcal{A}| \geq \binom{k}{r-1}\frac{k-r+1}{n-r+1},$$

which is a weaker lower bound as $k \leq n$ demonstrating directly that we have a stronger result. Note also that the theorem easily self-strengthens:

Theorem 5. *Let $\mathcal{A} \subseteq X^{(r)}$ for $1 \leq r \leq n$ and $\mathcal{C}$ the initial segment of colex on $X^{(r)}$ with $|\mathcal{A}| \geq |\mathcal{C}|$. Then $|\partial\mathcal{A}| \geq |\partial\mathcal{C}|$.*

In this form, it is natural to iterate the theorem.

Theorem 6. *Let $\mathcal{A} \subseteq X^{(r)}$ for $1 \leq r \leq n$ and $\mathcal{C}$ the initial segment of colex on $X^{(r)}$ with $|\mathcal{A}| \geq |\mathcal{C}|$. Then for any t with $1 \leq t \leq r$, we have $|\partial^t\mathcal{A}| \geq |\partial^t\mathcal{C}|$.*

Proof (Sketch). Iterate the strengthened Kruskal-Katona theorem, observing that the shadow of an initial segment of colex in $X^{(r)}$ is an initial segment of colex in $X^{(r-1)}$. □

In particular, taking $k \leq n$, $[k]^{(r)}$ is an initial segment of $X^{(r)}$ so if $|\mathcal{A}| = \binom{k}{r}$ then $|\partial^t\mathcal{A}| \geq \binom{k}{r-t}$. This final form gives a quick proof of the Erdős-Ko-Rado theorem, for which we introduce the notion of intersecting families.

Definition 6. *A family $\mathcal{A} \subseteq \mathcal{P}(X)$ is* intersecting *if $A \cap B \neq \varnothing$ for every $A, B \in \mathcal{A}$.*

It is immediate to see that 2^{n-1} is the maximum size of an intersecting family in $\mathcal{P}(X)$, as at most one of any set and its complement can be in $\mathcal{A}$, and that this is achievable by taking $\mathcal{A} = \{A \subseteq X : 1 \in A\}$. The Erdős-Ko-Rado theorem gives the maximum size of an intersecting family in $X^{(r)}$ for $r < \frac{n}{2}$.

Theorem 7 (Erdős-Ko-Rado). *Let $r < \frac{n}{2}$ and $\mathcal{A} \subseteq X^{(r)}$ be intersecting. Then $|\mathcal{A}| \leq \binom{n-1}{r-1}$.*

Proof. Write $\overline{\mathcal{A}} = \{A^c : A \in \mathcal{A}\}$. Then as $\mathcal{A}$ is intersecting, $\partial^{n-2r}\overline{\mathcal{A}}$ is disjoint from $\mathcal{A}$. Then $|\overline{\mathcal{A}}| = |\mathcal{A}| > \binom{n-1}{r-1} = \binom{n-1}{n-r}$, so the iterated Kruskal-Katona theorem gives $|\partial^{n-2r}\overline{\mathcal{A}}| \geq \binom{n-1}{r}$, and so $|\partial^{n-2r}\overline{\mathcal{A}}| + |\mathcal{A}| > |X^{(r)}|$, a contradiction. □

4   Formalising the Shadow

In these following sections, we describe the Lean formalisation of these concepts and proofs.

We begin by describing the basic data types and definitions used for these formalisations, then move on to talk about the implementation of the proofs given in Sect. 3. As mentioned previously, some arguments are omitted when the types are clear from the context, and occasionally declarations are given shorter names.

4.1   Definitions

The notion in mathlib of `finset` provides a type of finite sets which can used for finitary operations, such as a sum or a finite union. While all of the theorems are typically stated in mathematics in terms of the ground set $X = \{1, \ldots, n\}$, we formalise them in terms of a finite ground type α with a linear order; these are classically equivalent as every such type is isomorphic to some $\{1, \ldots, n\}$, but it remains convenient to have this adjustment: switching back to the special case is immediate.

We can then define a predicate for when $\mathcal{A} \subseteq X^{(r)}$ which, the observant reader will have noticed, is virtually the only way in which we needed to refer to $X^{(r)}$, so it is not needed to define this set family directly:

```
variables {α : Type*} [decidable_eq {α}]

/-- `all_sized A r` states that every set in A has size r. -/
def all_sized (A : finset (finset α)) (r : ℕ) : Prop :=
   ∀ x ∈ A, card x = r
```

To define the shadow, we first make an auxiliary useful definition.

```
def all_removals (A : finset α) : finset (finset α) :=
  A.image (erase A)

def shadow (A : finset (finset α)) : finset (finset α) :=
  A.bind all_removals

notation ∂A := shadow A
```

Here we have made use of some existing mathlib constructors on finite sets, and given the mathematical prefix notation for the shadow: notably `erase A i` means the finite set A without the element i, i.e. just $A \setminus \{i\}$, and `bind` allows us to construct the union of `all_removals A` across all $A \in \mathcal{A}$. We can confirm the correctness of this definition of the shadow by showing it has precisely the property given in its definition in Sect. 3, as well as verifying Example 1. Due to the computable nature of the definition given, Lean can automatically run the function and produce the output in a convenient form. This would not be possible for Lean's sets, or finsets implemented in a different way - indeed many definitions in mathlib are explicitly marked `noncomputable`, and while they can be proved to have the appropriate properties, they cannot be *ran* to check specific example directly.

```
lemma mem_shadow {A : finset (finset α)} (B : finset α) :
  B ∈ shadow A ↔ ∃ A ∈ A, ∃ i ∈ A, erase A i = B :=
by simp only [shadow, all_removals, mem_bind, mem_image]

#eval ∂({{1, 2, 3}, {1, 2, 4}, {1, 3, 4}, {1, 3, 5}} :
    finset (finset (fin 6)))
-- outputs {{1, 2}, {1, 3}, {1, 4}, {1, 5},
--           {2, 3}, {2, 4}, {3, 4}, {3, 5}}
```

Observe that the proof of `mem_shadow` is just to unfold the definitions, and use the defining properties of `bind` and `image`. This example also introduces another important Lean type, `fin`: for a natural `n`, `fin n` is the type of natural numbers below `n`. Note we are making good use of the fact that Lean is a dependently-typed language to form this definition. It will play the role of $[n]$, except starting from 0 and ending at $n - 1$.

The definition we have given does not require that $\mathcal{A} \subseteq X^{(r)}$, indeed even the mathematical definition does not require this. Instead, we prove as a theorem that $\partial \mathcal{A} \subseteq X^{(r-1)}$.

```
lemma shadow_sized {𝒜 : finset (finset α)} {r : ℕ}
  (a : all_sized 𝒜 r) :
  all_sized (∂𝒜) (r-1) := ...
```

4.2   Local LYM Inequality

The usual mathematical proof is a double counting on a set of pairs: one precise count and one inequality. Here, we instead form two sets and find their exact cardinality, and demonstrate a subset relation. Our first set (`the_pairs`) will be the same as P from Sect. 3, but our second set (`from_below`) can be written as $Q = \{(A, B) : B \in \partial \mathcal{A}, \exists i \notin B : A = B \cup \{i\}\}$. We then immediately establish the key properties of these definitions which, when combined with a cancellation step, prove the inequality. We assume here that α is a finite type (else the inequality does not make sense), and abbreviate n as its cardinality, recalling that the main application is when α = `fin N`, which does indeed have N elements.

```
variables [fintype α] (𝒜 : finset (finset α))
local notation `n` := card α

def the_pairs : finset (finset α × finset α) := ...
def from_below : finset (finset α × finset α) := ...

lemma card_the_pairs {r : ℕ} (a : all_sized 𝒜 r) :
  (the_pairs 𝒜).card = 𝒜.card * r := ...

lemma card_from_below {r : ℕ} (a : all_sized 𝒜 r) :
  (from_below 𝒜).card = (∂𝒜).card * (n - (r - 1)) := ...

lemma above_sub_below : the_pairs 𝒜 ⊆ from_below 𝒜 := ...

lemma multiply_out {A B n r : ℕ} (hr1 : 1 ≤ r) (hr2 : r ≤ n)
  (h : A * r ≤ B * (n - r + 1)) :
  (A : ℚ) / nat.ch₁se n r ≤ B / nat.ch₁se n (r-1) := ...

theorem local_lym {r : ℕ} (hr1 : 1 ≤ r) (H : all_sized 𝒜 r) :
  (𝒜.card : ℚ) / nat.ch₁se n r ≤ (∂𝒜).card / nat.ch₁se n (r-1) := ...
```

4.3 LYM Inequality

To simplify the statement and proof of the LYM inequality, we define the slice of a set family, analogous to the definition $\mathcal{A}_r = \mathcal{A} \cap X^{(r)}$, which we denote using the infix symbol #.

```
def slice (A : finset (finset α)) (r : N) : finset (finset α) :=
  A.filter (λ i, card i = r)
notation A#r := slice A r
```

As the proof given in Sect. 3 was a downward induction, we mimic its structure, first defining the iterated shadow and union construction shown in Fig. 2b. In particular, `falling A k` consists of the sets with cardinality $n - k$ which are a subset of some set in $\mathcal{A}$. For example, `falling A 2` is the union of the two red sets in the lowest displayed layer in Fig. 2b.

```
def falling (A : finset (finset α)) : Π (k : N), finset (finset α)
  | 0 := A#n
  | (k+1) := A#(n - (k+1)) ∪ ∂ (falling k)

lemma disjoint_of_antichain {A : finset (finset α)} {k : N}
  (hk : k + 1 ≤ n) (H : antichain A) :
  disjoint (A#(n - (k + 1))) (∂falling A k) := ...
```

Observe that Lean's equation compiler (the | notation) allows making this definition in the natural inductive fashion. The lemma `disjoint_of_antichain` shows that the union at each point is disjoint, as performed in the original proof in Sect. 3.

These lemmas together with induction let us prove a bound on parts of the main sum, which then directly gives the theorem.

```
lemma card_falling {k : N} (hk : k ≤ card α) (H : antichain A) :
  (range (k+1)).sum (λ r, ((A#(n-r)).card : Q) / nat.ch₁se n (n - r))
  ≤ (falling A k).card / nat.ch₁se n (n - k) := ...

theorem lubell_yamamoto_meshalkin (H : antichain A) :
  (range (n + 1)).sum (λ r, ((A#r).card : Q) / nat.ch₁se n r) ≤ 1 :=
  ...
```

Note that `range n` provides the finite set of naturals below n, so we use `range (n+1)` to sum over the finite set of naturals $\leq n$: this is distinct from `fin n` as the latter is a type, whereas we wish to sum over the set. For the sake of clarity, we write the conclusion of `card_falling` in more familiar notation:

$$\sum_{r=0}^{k} \frac{|\mathcal{A}_{n-r}|}{\binom{n}{n-r}} \leq \frac{|\text{falling } \mathcal{A} \text{ k}|}{\binom{n}{n-k}}$$

and make the observation that this lemma has an easy proof by induction on k given the previous facts, and when applied at $k = n$ immediately gives the main theorem.

We remark additionally that the author followed an informal proof virtually identical to the one given in Sect. 3 in order to generate the formal inductive statement. However, the exact same rigorous version is presented in a standard textbook (of 1986) about combinatorics of set families (observed by the author after the formalisation process had occurred), in particular Theorem 5 of [4] in which $\mathcal{G}_{n-k}$ denotes precisely falling $\mathcal{A}$ k.

With this result formalised, it is now easy (indeed, just 20 lines of Lean) to show Sperner's theorem, noting that the default division operator on naturals in Lean is floor division, so we may write n / 2 to mean $\lfloor \frac{n}{2} \rfloor$.

```
theorem sperner {A : finset (finset α)} (H : antichain A) :
  A.card ≤ nat.ch₁se n (n / 2) := ...
```

5   Kruskal-Katona Theorem

We now move on to work towards the main theorem of this paper. We first define the colex order so we may state the theorem, then define compressions before proving the Kruskal-Katona theorem.

5.1   Colexicographic Ordering

```
def colex_lt [has_lt α] (A B : finset α) : Prop :=
  ∃ (k : α), (∀ {x}, k < x → (x ∈ A ↔ x ∈ B)) lk k ∉ A lk k ∈ B
def colex_le [has_lt α] (A B : finset α) : Prop := colex_lt A B ∨ A = B
infix ' <ᶜ ':50 := colex_lt
infix ' ≤ᶜ ':50 := colex_le

lemma colex_hom [linear_order α] [decidable_eq β] [preorder β]
  {f : α → β} (h₁ : strict_mono f) (A B : finset α) :
  image f A <ᶜ image f B ↔ A <ᶜ B := ...

instance [linear_order α] [decidable_eq α] :
  is_linear_order (finset α) (≤ᶜ) := ...

lemma binary_iff (A B : finset ℕ) :
  A.sum (pow 2) < B.sum (pow 2) ↔ A <ᶜ B := ...

example : ({2,3} : finset (fin 5)) <ᶜ {1,4} := dec_trivial
```

The definition of the colex ordering is given in a markedly more general setting than the one given previously: it requires only a type with some < defined on it, not necessarily finite, and not necessarily linear. While almost all of the applications of this will be in a linear order (such as the instance demonstrating that the colex order is itself linear) the generalisation afforded by not insisting on finiteness means this definition also makes sense for subsets of naturals. Indeed, this is helpful to consider - the colex order on each $[k]^{(r)}$ is an initial segment of the colex order on $\mathbb{N}^{(r)}$, and we may then state the binary definition of colex more cleanly without inserting the function of type fin n → ℕ, as shown in binary_iff. We finally use this characterisation to confirm that $\{2,3\} < \{1,4\}$ in the colex order by using dec_trivial, which performs direct computation to decide the truth of the statement.

5.2   Compressions

We begin by defining the compression of a set $C_{UV}(A)$, which is just a direct translation of the mathematical definition.

```
def compress (U V : finset α) (A : finset α) :=
if disjoint U A ∧ V ⊆ A then (A ∪ U) \ V else A
```

For the compression of a set family $C_{UV}(\mathcal{A})$, we adjust the definition slightly, instead writing

$$C_{UV}(\mathcal{A}) = \{C_{UV}(A) : A \in \mathcal{A},\, C_{UV}(A) \notin \mathcal{A}\} \cup \{A \in \mathcal{A} : C_{UV}(A) \in \mathcal{A}\}.$$

This is the same as our previous definition because the compression operator on sets is idempotent, but has the advantage of being a disjoint union, and hence slightly easier to prove results about: the proposition $A \in C_{UV}(\mathcal{A})$ gives a disjunction where the left hand side has an additional piece of information.

```
variables (U V : finset α) (𝒜 : finset (finset α))

def comp :=
(𝒜.filter (λ A, compress U V A ∉ 𝒜)).image (λ A, compress U V A) ∪
  𝒜.filter (λ A, compress U V A ∈ 𝒜)

lemma comp_all_sized {r : ℕ} (h₁ : U.card = V.card)
  (h₂ : all_sized 𝒜 r) : all_sized (comp U V 𝒜) r := ...
lemma comp_idempotent : comp U V (comp U V 𝒜) = comp U V 𝒜 := ...
lemma comp_card : (comp U V 𝒜).card = 𝒜.card := ...
```

Finally we state and prove the main proposition on compressions: Proposition 1, that sufficiently good compressions decrease the size of the shadow. The proof here is particularly detailed - this single proof takes more lines of code than the entirety of Sect. 4.2, and almost as many as the proof of the LYM inequality.

Part of this can be explained by the fact that the proof on paper here involves a good amount of casework, especially in comparison to the results just mentioned, but another particular difficulty arises from lengthy set calculations. For instance, one part of the proof requires showing $((((B \cup \{x\}) \cup V) \setminus U) \cup (U \setminus \{x\})) \setminus (V \setminus \{y\}) = B \cup \{y\}$, assuming certain assumptions such as $x \notin B$, $V \cap B = \varnothing$, $x \in U$, $y \in V$, $U \subseteq B \cup \{x\}$. A direct proof by extensionality and set lemmas is not quite as effective as one may hope here, as results on equality must additionally be used to deal with the singleton sets, as x and y are fixed elements (in particular, reducing the statement to one about propositional logic and appealing to automation on propositional logic cannot succeed: it needs to additionally handle quantifiers such as those in $\subseteq$ and also automatically solve steps such as $y \in V \wedge t = y \wedge V \cap B = \varnothing \Rightarrow t \notin B$). Instead, we showed a number of elementary lemmas missing from mathlib, such as $(s \cup t) \setminus t = s \setminus t$ and used chains of lemmas like these to establish the equality. While this works, it is a non-trivial contribution to the length of the proof.

```
theorem comp_reduces_shadow {𝒜 : finset (finset α)} {U V : finset α}
  (h₁ : ∀ x ∈ U, ∃ y ∈ V, is_compressed (erase U x) (erase V y) 𝒜)
  (h₂ : U.card = V.card) : (∂ comp U V 𝒜).card ≤ (∂𝒜).card := ...
```

5.3 Main Theorem

```
def family_measure (𝒜 : finset (finset (fin n))) : ℕ := ...

variables (𝒜 : finset (finset (fin n)))
lemma comp_reduces_family_measure {U V : finset (fin n)} {hU : U ≠ ∅}
  {hV : V ≠ ∅} (h : max' U hU < max' V hV) (a : comp U V 𝒜 ≠ 𝒜) :
  family_measure (comp U V 𝒜) < family_measure 𝒜 :=

def useful_compression (U V : finset (fin n)) : Prop :=
  ∃ (HU : U ≠ ∅), ∃ (HV : V ≠ ∅), disjoint U V ∧
    finset.card U = finset.card V ∧ max' U HU < max' V HV

lemma comp_improved (U V : finset (fin n)) (h₁ : useful_compression U V)
  (h₂ : ∀ U₁ V₁, useful_compression U₁ V₁ ∧ U₁.card < U.card →
    is_compressed U₁ V₁ 𝒜) :
  (∂ comp U V 𝒜).card ≤ (∂𝒜).card := ...

lemma kruskal_katona_helper {r : ℕ} (h : all_sized 𝒜 r) :
  ∃ (ℬ : finset (finset (fin n))),
    (∂ℬ).card ≤ (∂𝒜).card ∧ 𝒜.card = ℬ.card ∧ all_sized ℬ r
  ∧ (∀ U V, useful_compression U V → is_compressed U V ℬ) := ...
```

We now assemble the final ingredients for the proof. The natural-valued function `family_measure` is important to show that the iteration process terminates: each compression successfully applied will strictly decrease this quantity, shown in `comp_reduces_family_measure`. It is defined just as in the usual proof, as $\sum_{A \in \mathcal{A}} \sum_{i \in A} 2^i$, as we know that the function $A \mapsto \sum_{i \in A} 2^i$ is an order isomorphism from the colex ordering to the usual ordering on naturals. Note that now we do indeed assume our ground set is `fin n` rather than an arbitrary type, as it becomes convenient to view the family measure as an absolute quantity on set families, rather than merely a relative comparison.

We also define when a pair (U, V) is a useful compression, analogous to the set Γ (defined in Sect. 3.3), and show that applying a useful compression to a family which is already compressed by all smaller useful compressions cannot increase the size of the shadow. Finally, these two combine (together with some simple extra results elided here) to prove the primary workhorse of the proof, `kruskal_katona_helper`. The iteration process is performed by essentially using strong induction on the naturals, where the induction step required `comp_reduces_family_measure`. Observing that this completes the first two paragraphs of the proof sketch given in Sect. 3, it remains only to show that a compressed family is indeed an initial segment of colex, and conclude.

```
def is_init_seg_of_colex (r : ℕ) : Prop :=
all_sized 𝒜 r ∧ (∀ A ∈ 𝒜, ∀ B, B <ᶜ A ∧ B.card = r → B ∈ 𝒜)

lemma init_seg_of_compressed {r : ℕ} (h₁ : all_sized ℬ r)
  (h₂ : ∀ U V, useful_compression U V → is_compressed U V ℬ):
  is_init_seg_of_colex ℬ r := ...

theorem kruskal_katona {r : ℕ} (h₁ : all_sized 𝒜 r)
  (h₂ : 𝒜.card = 𝒞.card) (h₃ : is_init_seg_of_colex 𝒞 r) :
  (∂𝒞).card ≤ (∂𝒜).card := ...

theorem strengthened_kk {r : ℕ} (h₁ : all_sized 𝒜 r)
  (h₂ : 𝒞.card ≤ 𝒜.card) (h₃ : is_init_seg_of_colex 𝒞 r) :
  (∂𝒞).card ≤ (∂𝒜).card := ...
```

To prove the iterated Kruskal-Katona theorem however, we first must show that the shadow of an initial segment is itself an initial segment. This fact is typically given without proof (e.g. [2,4,10]) or with only minor justification (e.g. [14]), and we found it particularly difficult to formalise. In contrast to the previous similar case however, this seems to be merely because the proof breaks down into many different cases, each one needing to be dealt with in an entirely different way, so adding additional basic lemmas do not seem to be the solution. Perhaps here further automation would be of use however, meaning that each case can be swiftly handled; or perhaps the answer is simply that the proof found by the author is not as simple as the one experts have in mind.

For completeness we present a rough proof sketch that was formalised, a clearer description is included in comments in the source code snapshot: in particular in `shadow_of_everything_up_to` in `kruskal_katona.lean`. First, observe that an initial segment of colex can be specified by its largest (under colex) element; the proof sketch in [14] suggests that the shadow of some set $\leq A$ should be exactly the sets B which are $B \leq A' := A \setminus \{\min A\}$, which we will prove directly.

In the forward direction, we have $B \cup \{i\} \leq A$ - if this is an equality we are clearly done if $i = \min A$ as $B = A'$, and if $i \neq \min A$ we can prove explicitly that $B < A'$ by using i as the colex witness (satisfying the existential from the definition). If $B \cup \{i\} < A$ with colex witness k, certainly $i \neq k$. If $k < i$, then i is our colex witness for $B < A'$, while otherwise $(i < k)$ if $\min A < k$ then k witnesses $B < A'$ and if $\min A = k$ then we have $B = A'$.

In the reverse direction, we must choose j from $[n]$ and prove that $B \cup \{j\} \leq A$. We know that $B \leq A'$, if this is equality then we may use $j = \min A$ and we are done. Hence suppose $B < A'$, and say k is the colex witness, and choose j as the smallest element of $[n]$ which is not in B (there must be such an element, as k is not in B). Certainly $j \leq k$, if the inequality is strict then k works as the colex witness, while if $j = k$ then we know that $[k] \subseteq B$, so it suffices to show $A \subseteq B \cup \{j\}$, which we may do by taking $t \in A$ and further splitting into cases $j < t$, $j = t$ and $j > t$ each of which can be resolved directly to show $t \in B \cup \{j\}$.

```
variables {A C : finset (finset (fin n))}

lemma shadow_of_init_seg (r : ℕ) (h₁ : is_init_seg_of_colex A r) :
  is_init_seg_of_colex (∂A) (r - 1) := ...

theorem iterated_kk {r k : ℕ} (h₁ : all_sized A r)
  (h₂ : C.card ≤ A.card) (h₃ : is_init_seg_of_colex C r) :
  (shadow^[k] C).card ≤ (shadow^[k] A).card := ...

theorem special_iterated_kk {r k i : ℕ} (hir : i ≤ r)
  (hrk : r ≤ k) (hkn : k ≤ n) (h₁ : all_sized A r)
  (h₂ : ch₁se k r ≤ A.card) :
  ch₁se k (r-i) ≤ (shadow^[i] A).card := ...
```

The special case `special_iterated_kk` has a relatively straightforward proof, we must simply note that $[k]^{(r)}$ is an initial segment of colex on $X^{(r)}$ provided $r \leq n$, and that its i-iterated shadow is just $[k]^{(r-i)}$. With this case at hand, we can now state and prove the Erdős-Ko-Rado theorem.

```
def intersecting : Prop :=
∀ A ∈ 𝒜, ∀ B ∈ 𝒜, ¬ disjoint A B

theorem intersecting_all (h : intersecting 𝒜) : 𝒜.card ≤ 2^(n-1) := ...

theorem EKR {r : ℕ} (h₁ : intersecting 𝒜) (h₂ : all_sized 𝒜 r)
  (h₃ : r ≤ n/2) : 𝒜.card ≤ ch₁se (n-1) (r-1) := ...
```

We define intersecting families, show the upper bound on intersecting families in general, then show the bound on intersecting families which lie in $X^{(r)}$: each of these proofs are straightforward direct translations of the versions in Sect. 3. The eagle-eyed reader will have noticed that our condition on r here appears weaker than that in Theorem 7. In the cases when n is odd, this is not strictly true since $r \leq n/2$ should be interpreted as $r \leq \lfloor \frac{n}{2} \rfloor < \frac{n}{2}$. However if n is even, we do have a technically stronger statement. Nonetheless, this is not particularly notable, as the $r = n/2$ case can be proved directly in the same way as `intersecting_all`.

6 Future Work and Conclusion

Despite the formalisation here being implemented in early versions of Lean, the code has since been updated by the author as well as Alena Gusakov and Yaël Dillies, and is in the process of being merged into mathlib itself: indeed the majority of the results presented here have already been merged.

The choice of the theorems formalised here was in part simply following a lecture series - about a quarter of a Masters' level course on combinatorics has been verified, and the majority with only trivial modifications. However, the compressions proof of the Kruskal-Katona theorem is particularly appealing because the technique of compressions for extremal combinatorics is particularly powerful, and can be used to show results such as Harper's theorem, the edge-isoperimetric inequality in the cube, the vertex-isoperimetric inequality in the grid, and the Sauer-Shelah lemma (see [4,14]).

'The important ideas of combinatorics do not usually appear in the form of precisely stated theorems, but more often as general principles of wide applicability,' writes Timothy Gowers in his 1997 essay [11]. We aim to have demonstrated a formalisation of one use of one such principle, and sincerely hope that others can follow this work as a guide for using the principle of compressions further.

Acknowledgement. We would like to give particular thanks to Imre Leader for his inspiring lecture series with demonstrations of these proofs, Yaël Dillies for their continuous and determined efforts to migrate the code here to mathlib, and the anonymous reviewers for their helpful feedback.

References

1. Alon, N.: Problems and results in extremal combinatorics-III. J. Comb. **7**(2), 233–256 (2016)
2. Anderson, I.: Combinatorics of Finite Sets. Courier Corporation (2002)
3. Bollobás, B.: On generalized graphs. Acta Math. Hungar. **16**(3–4), 447–452 (1965)
4. Bollobás, B., Béla, B.: Combinatorics: Set Systems, Hypergraphs, Families of Vectors, and Combinatorial Probability. Cambridge University Press, Cambridge (1986)
5. Bollobás, B., Leader, I.: Compressions and isoperimetric inequalities. J. Comb. Theory Ser. A **56**(1), 47–62 (1991)
6. Dahmen, S.R., Hölzl, J., Lewis, R.Y.: Formalizing the solution to the cap set problem. In: 10th International Conference on Interactive Theorem Proving, ITP 2019, pp. 1–19. Schloss Dagstuhl-Leibniz-Zentrum für Informatik (2019)
7. Dillies, Y., Mehta, B.: Formalizing Szemerédi's regularity lemma in lean. In: 13th International Conference on Interactive Theorem Proving (ITP 2022). Schloss Dagstuhl-Leibniz-Zentrum für Informatik (2022, to appear)
8. Edmonds, C., Paulson, L.C.: Formalising fisher's inequality: formal linear algebraic techniques in combinatorics. In: 13th International Conference on Interactive Theorem Proving (ITP 2022). Schloss Dagstuhl-Leibniz-Zentrum für Informatik (2022, to appear)
9. Erdős, P., Ko, C., Rado, R.: Intersection theorems for systems of finite sets. Quart. J. Math. Oxford Ser. **2**(12), 313–320 (1961)
10. Frankl, P., Tokushige, N.: Extremal Problems for Finite Sets, vol. 86. American Mathematical Soc. (2018)
11. Gowers, W.T.: The two cultures of mathematics. In: Mathematics: Frontiers and Perspectives vol. 65, p. 65 (1997)
12. Gusakov, A., Mehta, B., Miller, K.A.: Formalizing Hall's Marriage Theorem in lean. arXiv preprint arXiv:2101.00127 (2021)
13. Katona, G.: A theorem of finite sets. In: Classic Papers in Combinatorics, pp. 381–401. Springer, Boston (1968). https://doi.org/10.1007/978-0-8176-4842-8_27
14. Leader, I.: Part III Combinatorics, December 2018. https://github.com/b-mehta/maths-notes/blob/master/iii/mich/combinatorics.pdf. Lecture notes transcribed by Mehta, B. Accessed May 2022
15. Lubell, D.: A short proof of Sperner's lemma. J. Comb. Theory **1**(2), 299 (1966)
16. Marić, F., Živković, M., Vučković, B.: Formalizing Frankl's conjecture: FC-families. In: Jeuring, J., et al. (eds.) CICM 2012. LNCS (LNAI), vol. 7362, pp. 248–263. Springer, Heidelberg (2012). https://doi.org/10.1007/978-3-642-31374-5_17

17. Meshalkin, L.D.: Generalization of Sperner's theorem on the number of subsets of a finite set. Theory Probab. Appl. **8**(2), 203–204 (1963)
18. de Moura, L., Kong, S., Avigad, J., van Doorn, F., von Raumer, J.: The lean theorem prover (system description). In: Felty, A.P., Middeldorp, A. (eds.) CADE 2015. LNCS (LNAI), vol. 9195, pp. 378–388. Springer, Cham (2015). https://doi.org/10.1007/978-3-319-21401-6_26
19. Sperner, E.: Ein Satz über Untermengen einer endlichen Menge. Math. Z. **27**(1), 544–548 (1928)
20. The mathlib Community: The Lean Mathematical Library. In: Proceedings of the 9th ACM SIGPLAN International Conference on Certified Programs and Proofs, CPP 2020, pp. 367–381. Association for Computing Machinery, New York (2020)
21. Yamamoto, K.: Logarithmic order of free distributive lattice. J. Math. Soc. Japan **6**(3–4), 343–353 (1954)

Wetzel: Formalisation of an Undecidable Problem Linked to the Continuum Hypothesis

Lawrence C. Paulson[(✉)]

Computer Laboratory, University of Cambridge, Cambridge, UK
lp15@cam.ac.uk
https://www.cl.cam.ac.uk/~lp15/

Abstract. In 1964, Paul Erdős published a paper [5] settling a question about function spaces that he had seen in a problem book. Erdős proved that the answer was yes if and only if the continuum hypothesis was false: an innocent-looking question turned out to be undecidable in the axioms of ZFC. The formalisation of these proofs in Isabelle/HOL demonstrate the combined use of complex analysis and set theory, and in particular how the Isabelle/HOL library for ZFC [17] integrates set theory with higher-order logic.

Keywords: Isabelle · Erdős · continuum hypothesis · set theory · complex analysis · formalisation of mathematics

1 Introduction

This story [6] expresses the richness of mathematics. It seems that Paul Erdős found the following question in a problem book belonging to the mathematics department of Ann Arbor University:

> Suppose that F is a family of analytic functions on $\mathbb{C}$ such that for each z the set $\{f(z) : f \in F\}$ is countable. (Call this *property P_0*.) Then is the family F itself countable?

This question apparently arose in the PhD work of John E. Wetzel, in connection with spaces of harmonic functions. Erdős was able to show that if the continuum hypothesis failed, every family satisfying P_0 had to be countable; but if the hypothesis held, there existed (by a transfinite construction) an uncountable family satisfying P_0. His proof appears in *Proofs from THE BOOK* [2],[1] Aigner and Ziegler's collection of "perfect proofs" inspired by Erdős.

Cantor's celebrated *continuum hypothesis* (CH), number one on Hilbert's list of fundamental questions, asks whether there exists a cardinality between that of the natural numbers, namely $\aleph_0$, and that of the real numbers, namely $\mathfrak{c}$.

---

[1] End of Chap. 19. Also https://doi.org/10.1007/978-3-662-57265-8_19.

© The Author(s), under exclusive license to Springer Nature Switzerland AG 2022
K. Buzzard and T. Kutsia (Eds.): CICM 2022, LNAI 13467, pp. 92–106, 2022.
https://doi.org/10.1007/978-3-031-16681-5_6

Since the next cardinal after $\aleph_0$ is denoted $\aleph_1$ and the cardinality of the continuum is known to equal $2^{\aleph_0}$, CH can be written symbolically as $\aleph_1 = 2^{\aleph_0}$. It's fundamental because the notion of a countable set is straightforward, as is the proof by diagonalisation that the real numbers cannot be enumerated (Cantor's theorem). CH asserts that no set exists of intermediate size between the natural numbers and the reals. It was shown to be consistent with the axioms of set theory by Gödel and to be independent from them by Cohen. The details can be found in any set theory text [11].

These results have been machine verified by a variety of authors. Han and van Doorn [10] were the first to mechanise Cohen's proof of the independence of CH, using Lean.[2] I have formalised Gödel's model, the *constructible sets* [16], and some of its properties. Gunther et al. [9] have formalised model constructions both to confirm and refute CH using forcing. These latter formalisations were done using Isabelle/ZF [20], the instance of Isabelle for first-order logic and set theory. However, Isabelle/HOL is much better developed than Isabelle/ZF; in particular, the Wetzel problem requires its complex analysis library. So this paper demonstrates how to tackle a problem that combines the worlds of analysis and set theory, including such mysteries as holomorphic functions, transfinite cardinals and recursion up to uncountable ordinals, on the common basis of higher-order logic.

The paper outlines the development of formal treatments of ZF set theory in Isabelle/HOL (§2), continuing to the Wetzel problem in the non-CH case (§3) and the more difficult CH case (§4). Next comes some discussion of the work in the context of the formalisation of mathematics ganerally (§5), and finally conclusions.

The formal development itself is available online [19] in Isabelle's Archive of Formal Proofs (AFP), as is the ZFC formalisation [17] on which it is built.

2   Isabelle and Set Theory

Isabelle is a generic theorem prover, ultimately based on a uniform representation of logical syntax in the typed lambda calculus, with inference rules expressed syntactically and combined using higher-order unification [14]. Although Isabelle/HOL [12]—the version for higher-order logic—is by far the best known and most developed instance of Isabelle, other instances exist. These include Isabelle/FOL (classical first-order logic) and Isabelle/ZF. The latter is a faithful development of set theory from the Zermelo-Fraenkel axioms within first-order logic. Since the axiom of choice (AC) is kept separate from the other axioms, one can investigate weaker forms of AC and equivalents of AC [20].

During the 1990s, I made some significant formal developments using Isabelle/ZF [15], culminating in a proof of the relative consistency of the axiom of choice using Gödel's constructible universe [16]. Recently, several highly impressive formalisations of forcing were done within Isabelle/ZF [8,9].

---

[2] https://leanprover.github.io.

On the other hand, Isabelle/ZF lacks much of the automation found in Isabelle/HOL and has no theory even of the real numbers. That makes it unsuitable for the Wetzel problem. However, it is also possible to formalise ZF set theory within higher-order logic.

2.1   Formalising ZF in Higher-Order Logic

During the 1990s, Michael JC Gordon conducted several experiments [7] involving the formalisation of ZF set theory in his HOL proof assistant. His HOL-ST simply introduced a type $\mathbf{V}$ of all sets and a relation $\in : \mathbf{V} \times \mathbf{V} \rightarrow$ bool, then asserted all the Zermelo-Fraenkel axioms. Sten Agerholm [1] formalised Dana Scott's inverse limit construction of the set D_∞, satisfying $D_\infty \cong [D_\infty \rightarrow D_\infty]$ and yielding a model of the untyped λ-calculus.

The use of higher-order logic as opposed to first-order logic as the basis makes this version of set theory somewhat stronger than standard ZF. In HOL-ST it is possible to define the syntax of first-order logic and the semantics of set-theoretic formulas in terms of $\mathbf{V}$, verifying the ZF axioms and thus proving their consistency. On the other hand, a model for HOL-ST can be constructed in ZF plus one inaccessible cardinal. These remarks (which Gordon credited to Kenneth Kunen) together imply that the strength of HOL-ST is somewhere between ZF and ZF plus one inaccessible cardinal. This is a weak assumption, much weaker than the dependent type theories used in Coq and Lean, which are stronger than any finite number of inaccessible cardinals [23].

Some time later, Steven Obua performed a similar experiment [13], using Isabelle/HOL. He adopted the same axioms and overall approach as Gordon, and demonstrated his system by formalising John H Conway's *partizan games* [21]. Obua obtained some interoperability with the existing infrastructure of Isabelle/HOL, such as its recursive function definitions.

2.2   Axiomatic Type Classes

Deeper integration requires the use of axiomatic type classes, introduced by Wenzel in 1997. An *axiomatic type class* [22] defines an open-ended collection of types on the basis of a signature and a possibly empty list of axioms. The signature specifies certain operations and their types, which will be polymorphic with type variables referring to that type class. The operations may also be equipped with concrete syntax, such as infix declarations. Any type, whether already existing or defined in the future, can be shown to be an instance of the class if it provides definitions of all the operations in the signature that satisfy the associated axioms.

Isabelle/HOL defines a series of type classes for orderings:

- *ord* introduces the relations $\leq$ and $<$, of type $'a \Rightarrow 'a \Rightarrow bool$ but no axioms
- *preorder* extends *ord* with axioms for reflexivity and transitivity
- *order* extends *preorder* with antisymmetry
- *linorder* extends *order* with the axiom $x \leq y \lor y \leq x$, for linear orderings

The arithmetic types *nat, int, real* belong to each of these classes; the type operator × preserves membership in each of these classes.

2.3   The Development ZFC-in-HOL

My own framework [17] for set theory within Isabelle/HOL is mathematically equivalent to Gordon's and Obua's. Its aims were practical:

1. To reproduce as much of Isabelle/ZF as possible,
2. while achieving maximum integration with the underlying higher-order logic.

To achieve these aims I relied on axiomatic type classes, whenever possible overloading existing symbols for set theory rather than introducing new vocabulary.

It introduces the type V of sets and the function *elts* of type $V \Rightarrow V \text{ set}$ mapping a set to its elements. Thus it uses Isabelle/HOL's existing typed sets to represent classes, but not all elements of $V \text{ set}$ correspond to sets. The predicate *small* identifies those that are small enough, but here comes the first stage of integration: *small* needs to be polymorphic, accepting a set of any type and with the more general meaning that a set is small if its elements can be put into one-to-one correspondence with the elements of a ZF set. A set that is small in this more general sense does not in itself denote a ZF set, but this condition can be necessary if it is used in constructions that ultimately lead to a ZF set.

The fundamentals of set theory are built up as usual, starting with the ZF axioms. Set membership is expressed using the existing (typed) set membership operator: $x \in \text{elts } y$. Union, intersection and the subset relation are expressed using the type class *distrib_lattice* (distributive lattices), which already provides the symbols ⊔, ⊓, ≤, etc. Unions and intersections of families rely on the type class *conditionally_complete_lattice*, which provides the symbols ⨆ and ⨅. Type classes also allow the overloading of 0 to denote the empty set (which is also the natural number zero) and 1 for the natural number one.

On this foundation, it was easy to import significant chunks of Isabelle/ZF: above all, cardinal arithmetic and the ℵ operator. It was also possible to reuse Isabelle/HOL's existing theory of recursion to obtain transfinite recursion on ordinals, and hence order types, Cantor normal form and much else. So the first aim was met, but more needed to be done to achieve the second.

2.4   The Integration of ZFC-in-HOL with Isabelle/HOL

In theory, ZFC suffices for the formalisation of all the mathematics in this problem, and much more. In practice, we absolutely do not want to be forced to develop complex analysis from first principles in set theory when it already exists in the Isabelle/HOL libraries. And while transfinite cardinalities and other constructions are typically understood from the framework of ZFC, they are perfectly intelligible in a broader context.

The simplest generalisation is for cardinality. The *cardinality* $|x|$ of a ZF set x is simply the minimal ordinal that is equipollent to[3] x. This definition generalises

---

[3] in one-to-one correspondence with.

naturally to sets of any type. We can now refer to the cardinality of sets of real and complex numbers.

Next comes *transfinite recursion*, also known as ϵ-recursion. It is a sort of fixedpoint operator allowing the definition of functions over the whole of V. If a function H is given, then transfinite recursion yields F such that for any a,

$$F(a) = H(F \restriction a), \tag{1}$$

where $F \restriction a$ denotes F itself, restricted to the elements of a. It is just an instance of well-founded recursion on the membership relation. The definition is simple:

definition *transrec* :: "$((V \Rightarrow \text{'}a) \Rightarrow V \Rightarrow \text{'}a) \Rightarrow V \Rightarrow \text{'}a$"
   where "*transrec H a* $\equiv$ *wfrec* $\{(x,y). \ x \in elts \ y\} \ H \ a$"

This version differs from the original [17] only in its type, which is now polymorphic as shown, allowing the recursively defined function to return anything. Here, *wfrec* is Isabelle/HOL's built-in operator for well-founded recursion. The recursion equation (1) easily follows.

lemma *transrec*: "*transrec H a* = *H* ($\lambda x \in elts \ a$. *transrec H x*) *a*"

Although transfinite recursion is typically used to define operations within the set-theoretic universe **V**, we can now use it to create an uncountable set of analytic functions.

2.5   Embedding Isabelle/HOL Types into V

The ZFC-in-HOL library defined the class of types that can be *embedded* into the set theoretic universe, V, by some injective map, V_of:

class *embeddable* =
   assumes *ex_inj*: "$\exists V\_of :: \text{'}a \Rightarrow V$. *inj V_of*"

If a type is embeddable then each of its elements corresponds to some ZF set.

As it happens, Isabelle/HOL already provides the class *countable* of all types that can be embedded into the natural numbers. The latter are trivially embedded into V as finite ordinals, so it is easy to show that all countable types are *embeddable*. Examples include *nat, int, rat, bool*. Trivially, V can be embedded into itself, and the type constructors $\times$, $+$ and *list* are straightforwardly shown to preserve the embeddable property.

The library also defines the class of types that are small, which means that the type itself corresponds to some ZF set. It is defined in terms of the predicate *small*:

class *small* =
   assumes *small*: "*small* (*UNIV*::'a *set*)"

Every *countable* type (and in particular those listed above) is *small*, because ω is a set. And it's obvious that every *small* type is *embeddable*. By proving types to be small, we further extend the embeddable class. The type constructors $\times$, $+$ and *list* preserve smallness. The situation for the function type constructor ($\rightarrow$) is a little subtle:

instance *"fun" :: (small,embeddable) embeddable*
instance *"fun" :: (small,small) small*

The (straightforward) proofs are omitted. It should be obvious that we cannot expect to embed $\mathbf{V} \rightarrow \mathbf{V}$ into $\mathbf{V}$, but any function $f \in A \rightarrow \mathbf{V}$ must be a set provided A is a set.

Types *real* and *complex* are also *small*. How do we know? The reals are obtained by quotienting type *nat* $\Rightarrow$ *rat*. The details of their construction do not concern us. This is the full text of the proof.

instance *real :: small*
proof -
  have *"small (range (Rep_ real))"*
    by *simp*
  then show *"OFCLASS(real, small_ class)"*
    by *intro_ classes*
      (metis Rep_ real_ inverse image_ inv_f_f inj_ on_ def replacement)
qed

And type *complex* is essentially the same thing as $\mathbb{R}^2$.

The type classes *embeddable* and *small* aren't especially visible in the formalisation below. However, they provide the crucial link between type V and other Isabelle/HOL types. They are the key to combining set theory and complex analysis. For example, the check that some entity is small (representable by a ZF set) is usually automatic, thanks to this type class setup. Because *real* and *complex* are small, we can reason about their cardinality.

3 Wetzel's Problem: The ¬CH Case

Now it is time to formalise the proof itself, following Aigner and Ziegler's presentation [2]. We begin by defining the predicate *Wetzel*, corresponding to P_0 above, on sets of complex-valued functions. It holds if every element of the given set F is analytic on the complex plane and if, for all z, the set $\{f(z) : f \in F\}$ is countable:

definition *Wetzel :: "(complex $\Rightarrow$ complex) set $\Rightarrow$ bool"*
  where *"Wetzel $\equiv$ λF. ($\forall$ f$\in$F. f analytic_ on UNIV) $\wedge$ ($\forall$ z. countable((λf. f z) ' F))"*

Remarkably, the Isabelle/HOL proof is barely 50 lines. First, here is the theorem statement:

proposition *Erdos_ Wetzel_ nonCH:*
  assumes *W:* *"Wetzel F"* **and** *NCH:* *"C_ continuum >* $\aleph 1$*"*
  shows *"countable F"*

We set out to prove the contrapositive, negating the conclusion:

have *"*$\exists$ *z0. gcard ((*λ*f. f z0) ' F)* $\geq \aleph 1$*"* **if** *"uncountable F"*

Given the uncountable family F, we find a subset $F' \subseteq F$ of cardinality $\aleph_1$ and thus a bijection $\phi : \aleph_1 \to F'$ between the ordinals below $\aleph_1$ and F'.

  have *"gcard F* $\geq \aleph 1$*"*
    using *that uncountable_ gcard_ ge* **by** *force*
  then obtain *F'* **where** *"F'* $\subseteq$ *F"* **and** *F': "gcard F'* $= \aleph 1$*"*
    by *(meson Card_ Aleph subset_ smaller_ gcard)*
  then obtain φ **where** φ*: "bij_ betw* φ *(elts* $\omega 1$*) F'"*
    by *(metis TC_ small eqpoll_ def gcard_ eqpoll)*

We next define the family of sets $S(\alpha, \beta)$ as $\{z. \phi_\alpha(z) = \phi_\beta(z)\}$. Here α and β range over ordinals, and ω_1 is the first uncountable ordinal (thus the same set as $\aleph_1$, but regarded as an ordinal). The ϕ_α for $\alpha < \omega_1$ are the given analytic functions.

  define *S* **where** *"S* $\equiv \lambda\alpha$ β*. {z.* φ α *z =* φ β *z}"*

It takes 10 lines to prove that $S(\alpha, \beta)$ is countable for $\alpha < \beta < \omega_1$, since two distinct holomorphic functions on $\mathbb{C}$ can agree at only countably many arguments (a fact proved by Aigner and Ziegler).

  have *"gcard (S* α β*)* $\leq \aleph 0$*"* **if** *"*$\alpha \in$ *elts* β*"* *"*$\beta \in$ *elts* $\omega 1$*"* **for** α β

The next step is to define SS as the union of all $S(\alpha, \beta)$ for $\alpha < \beta < \omega_1$.

  define *SS* **where** *"SS* $\equiv \bigsqcup \beta \in$ *elts* $\omega 1$*.* $\bigsqcup \alpha \in$ *elts* β*. S* α β*"*

A 14 line calculation shows that $|SS| \leq \aleph_1$, but we are assuming the negation of CH, so SS cannot be the entire complex plane: there exists some $z_0 \notin SS$.

  finally have *"gcard SS* $\leq \aleph 1$*"* *.*
  with *NCH* **obtain** *z0* **where** *"z0* $\notin$ *SS"*
    by *(metis Complex_ gcard UNIV_ eq_ I less_ le_ not_ le)*

That z_0 satisfies our requirements follows straightforwardly by the definitions of S and SS.

4 Wetzel's Problem: The CH Case

Assuming CH, it is possible to construct an uncountable family of analytic functions that makes the Wetzel property P_0 fail. It's a rather delicate transfinite recursion, so let's review the argument before examining the formal proof.

4.1   The Transfinite Construction

CH implies that $|\mathbb{C}| = \aleph_1$ and we can write $\mathbb{C} = \{\zeta_\alpha : \alpha < \omega_1\}$. Now consider the set $D \subseteq \mathbb{C}$ of *rational* complex numbers:

$$D = \{p + iq : p, q \in \mathbb{Q}\}.$$

Suppose that we had a family of functions $\{f_\beta : \beta < \omega_1\}$ such that

$$f_\beta(\zeta_\alpha) \in D \quad \text{if} \quad \alpha < \beta. \tag{2}$$

Since the set D is countable and the ζ_α for $\alpha < \omega_1$ include all the complex numbers, the desired result would follow.

Erdős showed how to construct this family by transfinite induction. For an arbitrary $\gamma < \omega_1$, assume that a family of distinct analytic functions $\{f_\beta : \beta < \gamma\}$ is defined below γ. To conclude the inductive argument and therefore prove the theorem, we must extend it with a new function, f_γ.

Since γ is countable, the set $\{f_\beta : \beta < \gamma\}$ can be enumerated as $\{g_0, g_1, \ldots\}$ and $\{\zeta_\alpha : \alpha < \gamma\}$ can be enumerated as $\{w_0, w_1, \ldots\}$; both sets are finite or infinite according to whether γ itself is finite or infinite. The sought-for analytic function f_γ should satisfy, for all n,

$$f_\gamma(w_n) \in D \quad \text{and} \quad f_\gamma(w_n) \neq g_n(w_n). \tag{3}$$

The second condition above ensures that f_γ is new by diagonalisation, while the first is simply (2). We construct f_γ by putting

$$
\begin{aligned}
f_\gamma(z) := {} & \epsilon_0 + \epsilon_1(z - w_0) + \epsilon_2(z - w_0)(z - w_1) \\
& + \epsilon_3(z - w_0)(z - w_1)(z - w_2) + \cdots .
\end{aligned}
\tag{4}
$$

Again, this sum is finite or infinite according to γ. The $\{\epsilon_m\}$ are complex numbers chosen one at a time to satisfy conditions (3) above; ensuring that $f_\gamma(w_n)$ avoids $g_n(w_n)$ is possible because D is dense in $\mathbb{C}$. In the finite case the sum is a polynomial, so trivially analytic. In the infinite case, the choice of the $\{\epsilon_m\}$ needs to be carefully calibrated in order to satisfy the conditions while converging to zero sufficiently rapidly. We have a lot of leeway and I chose

$$|\epsilon_m| < \left[m! \cdot \prod_{i < m} (1 + |w_i|) \right]^{-1}. \tag{5}$$

The summation (4) converges to an analytic function because the uniform limit of holomorphic functions is holomorphic, and the limit can be shown to be uniform by the Weierstrass M-test.[4]

A clever aspect of the construction is that the conditions (3) constrain only $f_\gamma(w_n)$, whose value depends only on ϵ_m for $m \leq n$. That is, $f_\gamma(w_0) = \epsilon_0$, $f_\gamma(w_1) = \epsilon_0 + \epsilon_1(w_1 - w_0)$, etc. The desired values of ϵ_m can be calculated sequentially.

---

[4] Thanks to Manuel Eberl for suggesting this argument.

4.2   The Isabelle/HOL Formalisation

The full proof is some 280 lines of Isar text. Let's examine some highlights. First, let's state the theorem formally:

proposition *Erdos_ Wetzel_ CH:*
  assumes *CH: "C_ continuum = $\aleph 1$"*
  obtains *F* **where** *"Wetzel F"* **and** *"uncountable F"*

The proof begins with a self-evident definition of *D*. It's then shown to be countably infinite.

define *D* **where** *"D ≡ {z. Re z ∈ ℚ ∧ Im z ∈ ℚ}"*
have *Deq: "D = ($\bigcup x ∈ ℚ$. $\bigcup y ∈ ℚ$. {Complex x y})"*
  using *complex.collapse* **by** *(force simp: D_ def)*
with *countable_ rat* **have** *"countable D"*
  by *blast*
have *"infinite D"*
  ⟨proof⟩
have *"∃ w. Re w ∈ ℚ ∧ Im w ∈ ℚ ∧ norm (w - z) < e" if "e > 0"* **for** *z e*
  ⟨proof⟩
then have *cloD: "closure D = UNIV"*
  by *(auto simp: D_ def closure_ approachable dist_ complex_ def)*

The closure of *D* equals the universal set *UNIV* of type *complex set*. We obtain the transfinite enumeration $\{\zeta_\alpha : \alpha < \omega_1\}$ of the complex plane.

obtain *ζ* **where** *ζ: "bij_ betw ζ (elts ω1) (UNIV::complex set)"*
  by *(metis Complex_ gcard TC_ small assms eqpoll_ def gcard_ eqpoll)*

Next come some technical definitions: *inD* for functions whose range for certain arguments lies within *D* and *Φ* to express that *f* is a family of analytic functions indexed by the ordinals up to *β*.

define *inD* **where** *"inD ≡ λβ f. (∀ α ∈ elts β. f (ζ α) ∈ D)"*
define *"Φ ≡ λβ f. f β analytic_ on UNIV ∧ inD β (f β) ∧ inj_ on f (elts(succ β))"*

4.3   The Transfinite Construction

The following lemma is the step of the transfinite induction. A family *f* defined below the ordinal *γ* is extended with a new function, f_γ (the variable *h* below).

have **: "∃ h. Φ γ ((restrict f (elts γ))(γ:=h))"*
  if *γ: "γ ∈ elts ω1"* **and** *"∀ β ∈ elts γ. Φ β f"* **for** *γ f*

The construction of *h* depends on whether or not *γ* is finite:

obtain *h* **where** *"h analytic_ on UNIV" "inD γ h" "∀ β ∈ elts γ. h ≠ f β"*
  proof *(cases "finite (elts γ)")*

The finite case is easier, since the function we have to construct will simply be a polynomial, and trivially analytic. The finite ordinal γ is simply some natural number n, and the construction of f_γ is by induction on n.

Let's examine the infinite case, in which similar ideas are taken to the max. We formalise the step of writing $\{f_\beta : \beta < \gamma\}$ as $\{g_0, g_1, \ldots\}$ and $\{\zeta_\alpha : \alpha < \gamma\}$ as $\{w_0, w_1, \ldots\}$ by picking some bijection η between $\mathbb{N}$ and γ, a countable ordinal which of course equals $\{\alpha : \alpha < \gamma\}$.

> **case** *False*
> **then obtain** η **where** η: *"bij_ betw η (UNIV::nat set) (elts γ)"*
>   **by** *(meson γ countable_ infiniteE' less_ω1_ imp_ countable)*
> **define** g **where** *"g $\equiv$ f o η"*
> **define** w **where** *"w $\equiv$ ζ o η"*

The following three definitions set up the finite approximants of our new analytic function, h. Note that $p(n)$ is the product of $z - w_i$ for $i < n$, while $q(n)$ is $\prod_{i<n}(1 + |w_i|)$.

> **define** p **where** *"p $\equiv$ λn z. $\prod$ i<n. z - w i"*
> **define** q **where** *"q $\equiv$ λn. $\prod$ i<n. 1 + norm (w i)"*
> **define** h **where** *"h $\equiv$ λn c z. $\sum$ i<n. ε i * p i z"*

The following three definitions constrain the next value ϵ_n, a calculation left unspecified by previous authors [2,5]. To ensure convergence, each approximant needs to be close to the previous one, hence a neighbourhood (open ball) whose radius has $n!q(n)$ for its denominator. DD is the intersection of this neighbourhood with the set D, excluding the one point that must be avoided. Finally, the function dd picks an arbitrary element of DD. In each of these functions, the argument ϵ is an integer-valued function representing the $\{\epsilon_i\}$ for $i < n$.

> **define** *BALL* **where**
>   *"BALL $\equiv$ λn ε. ball (h n ε (w n)) (norm (p n (w n)) / (fact n * q n))"*
> **define** *DD* **where** *"DD $\equiv$ λn ε. D $\cap$ BALL n ε - {g n (w n)}"*
> **define** *dd* **where** *"dd $\equiv$ λn ε. SOME x. x $\in$ DD n ε"*

Because D is dense in $\mathbb{C}$, the set DD is always nonempty. Therefore, dd always chooses an element satisfying the constraints described above.

> **have** *"DD n ε $\neq$ {}"* **for** n ε
> $\langle proof \rangle$
> **then have** *dd_ in_ DD:* *"dd n ε $\in$ DD n ε"* **for** n ε
>   **by** *(simp add: dd_ def some_ in_ eq)*

Each ϵ_n (written *coeff n* in Isabelle) is defined in terms of the previous epsilons by course of values recursion. The ugly details of the definition are omitted but the following line shows the recursion that it satisfies; note that h n *coeff (w n)* refers recursively to all the ϵ_i for $i < n$. The last line verifies the bound (5).

define *coeff* **where** *"coeff ≡ ... "*
have *coeff_ eq:* *"coeff n = (dd n coeff - h n coeff (w n)) / p n (w n)"* **for** *n*
  by *(simp add: def_wfrec [OF coeff_def])*
have *norm_ coeff:* *"norm (coeff n) < 1 / (fact n * q n)"* **for** *n*
  ⟨*proof*⟩

To conclude, define the function f_γ (calling it *hh*) and prove that it is analytic.

define *hh* **where** *"hh ≡ λz. suminf (λi. coeff i * p i z)"*
have *"hh holomorphic_ on UNIV"*
  ⟨*proof*⟩
then show *"hh analytic_ on UNIV"*
  by *(simp add: analytic_ on_ open)*

Here are a few details of the proof that was skipped above. For a fixed n and complex number z, we show uniform convergence within a circle of radius 1. First, we need a bound on $\prod_{i<n} |z' - w_i|$ for $|z - z'| < 1$:

have *norm_ p_ bound:* *"norm (p n z') ≤ q n * (1 + norm z) ^ n"*
  if *"dist z z' ≤ 1 "* **for** *n z z'*
  ⟨*proof*⟩

To apply the Weierstrass M-test to *hh*, exhibit a summable series M whose terms bound the finite approximants of the summation, which are given by *h*:

have *"uniform_ limit (cball z 1) (λn. h n coeff) hh sequentially"*
  unfolding *hh_ def h_ def*
proof *(rule Weierstrass_ m_ test)*
  let *?M = "λn. (1 + norm z) ^ n / fact n"*
    show *"summable ?M"*
      ⟨*proof*⟩
  fix *n z'*
  assume *"z' ∈ cball z 1"*
  show *"norm (coeff n * p n z') ≤ ?M n"*
      ⟨*proof*⟩
qed

The hard part is over. Checking that the image of $\{\zeta_\alpha : \alpha < \gamma\}$ under *hh* lies within D is easy, because for every complex number of the form w_n, the summation is actually finite.

have *hh_ eq_ dd:* *"hh (w n) = dd n coeff"* **for** *n*
  ⟨*proof*⟩
then have *"hh (w n) ∈ D"* **for** *n*
  using *DD_ def dd_in_ DD* **by** *fastforce*

Last comes the diagonalisation argument, showing that the function just constructed is indeed new:

have *"hh (w n) ≠ f (η n) (w n)"* **for** *n*
  using *DD_ def dd_in_ DD g_ def hh_ eq_ dd* **by** *auto*
then show *"∀ β∈elts γ. hh ≠ f β "*
  by *(metis η bij_ betw_ imp_ surj_ on imageE)*

4.4   Concluding the Proof

About 20 lines of boilerplate are needed to code up the formal application of transfinite recursion to the construction formalised above. Skipping over this, let's look at the conclusion of the proof. We finally have a family of analytic functions $\{f_\beta : \beta < \omega_1\}$ satisfying (2) above.

We now show that for all z, the set $\{f_\beta(z) : \beta < \omega_1\}$ is countable by writing $z = \zeta_\alpha$ for some $\alpha < \gamma$. The point is that

$$\{f_\beta(\zeta_\alpha) : \beta < \omega_1\} = \{f_\beta(\zeta_\alpha) : \beta \leq \alpha\} \cup \{f_\beta(\zeta_\alpha) : \alpha < \beta\}$$

and that both parts of the union are countable: the first is a collection below α, a countable ordinal, and the second is a subset of D, a countable set. In the proof below, $?B$ is the set $\{\beta : \alpha < \beta < \gamma\}$, while *elts (succ α)* is the set of all ordinals up to α.

```
show ?thesis
proof
  let ?F = "f ' elts ω1"
  have "countable ((λf. f z) ' f ' elts ω1)" for z
  proof -
    obtain α where α: "ζ α = z" "α ∈ elts ω1" "Ord α"
      by (meson Ord_ω1 Ord_in_Ord UNIV_I ζ bij_betw_iff_bijections)
    let ?B = "elts ω1 - elts (succ α)"
    have eq: "elts ω1 = elts (succ α) ∪ ?B"
  using α by (metis Diff_partition Ord_ω1 OrdmemD less_eq_V_def succ_le_iff)
    have "(λf. f z) ' f ' ?B ⊆ D"
      using α inD by clarsimp (meson Ord_ω1 Ord_in_Ord Ord_linear)
    then have "countable ((λf. f z) ' f ' ?B)"
      by (meson ‹countable D› countable_subset)
    moreover have "countable ((λf. f z) ' f ' elts (succ α))"
      by (simp add: α less_ω1_imp_countable)
    ultimately show ?thesis
      using eq by (metis countable_Un_iff image_Un)
  qed
  then show "Wetzel ?F"
    unfolding Wetzel_def by (blast intro: anf)
  show "uncountable ?F"
    using Ord_ω1 countable_iff_less_ω1 countable_image_inj_eq injf by blast
  qed
qed
```

The following corollary to the two cases summarises the equivalence between the Wetzel property and the negation of CH:

theorem *Erdos_Wetzel:* "C_continuum = ℵ1 ⟷ (∃ F. Wetzel F ∧ uncountable F)"
  by *(metis C_continuum_ge Erdos_Wetzel_CH Erdos_Wetzel_nonCH less_V_def)*

5   Discussion

The formalisation of mathematics within proof assistants is being pursued actively. There are too many contributions to list, but notable recent ones include the Buzzard–Commelin–Massot formalisation of perfectoid spaces [3] and the striking progress on the Liquid Tensor Experiment, which is led by Johan Commelin.[5] Both of these involve formalising the sophisticated work of Fields Medallist Peter Scholze, using Lean.

The small example reported here cannot be compared with such accomplishments, but shines a light on an issue that makes mathematics difficult: its interconnectedness. The connections between number theory and complex analysis are well known. If your problem concerns itself with sets of permutations, you might suddenly find yourself needing group theory. If you are trying to count the elements of the group, you might find yourself requiring sophisticated combinatorial arguments. If you find yourself talking about countable sets (and countability is an everyday notion), you could find yourself wandering into the world of cardinals, then into the world of ordinals and then perhaps into the undecidable. To quote Wetzel himself, "once again a natural analysis question has grown horns" [6, p.244].

I created the ZFC-in-HOL library in October 2019 with no specific application in mind but merely thinking that it might be useful occasionally to talk about really big sets. Then I decided to apply it to a project that had been proposed by Mirna Džamonja and Angeliki Koutsoukou-Argyraki: to formalise some ordinal partition theory. This concerns advanced generalisations of Ramsey's theorem; the formalisation [18] was complete by August 2020 and a paper is now available [4]. However, this work essentially belongs to straight set theory, borrowing little from Isabelle/HOL apart from natural numbers and lists.

The Wetzel example showed that there were a few gaps limiting the integration between that theory and Isabelle/HOL, above all, a more general definition of cardinality. It's easy to define the cardinality of an arbitrary Isabelle/HOL set (i.e. having a type of the form T set as the least ordinal whose elements can be put into bijection with that set. The equivalent notion of cardinality for ZF sets is a trivial consequence.

The original AFP entry [19], published online in February 2022, represents about three weeks' work. It begins with a variety of extensions to the Isabelle/HOL libraries, showing precisely how the original ZFC library needed to be extended. It includes an early and rather cumbersome definition of cardinality for HOL sets. In updated versions of the development,[6] the library material has been moved elsewhere. The latest version is 360 lines, 2833 tokens, which is not bad compared with the original text by Erdős [5], which is one and a half pages long. The exposition by Aigner and Ziegler [2] is two full pages long (67 lines, 1026 tokens). A crude comparison based on compressing the texts by gzip

---

[5] https://tinyurl.com/5n8rh297.

[6] At https://devel.isa-afp.org/entries/Wetzels_Problem.html. Future readers should be able to locate the original Wetzel entry via "Older releases" on the download page.

suggests a de Bruijn factor of under 2.3. For the original version, including the material later moved to libraries, the de Bruijn factor is 4.2.

6   Conclusions

The intriguing solution to Wetzel's problem by Erdős was easily formalised in Isabelle/HOL, using a library designed to integrate set theory with higher-order logic. The ease with which we can intermix analytic functions with ordinals gives reason to hope that harder mixed-domain problems will be amenable to formalisation without particular difficulties. It would be interesting to see this example tackled using other proof assistants.

Acknowledgements. This work was supported by the ERC Advanced Grant ALEXANDRIA (Project GA 742178). Dmitriy Traytel provided a particularly slick formal axiomatisation of type V. Manuel Eberl provided a crucial tip for the CH case. Angeliki Koutsoukou-Argyraki and the referees made insightful suggestions.

References

1. Agerholm, S.: A comparison of HOL-ST and Isabelle/ZF. In: Paulson, L.C. (ed.) Proceedings of the First Isabelle Users Workshop, pp. 53–70. Technical Report 379, Computer Laboratory, University of Cambridge, September 1995
2. Aigner, M., Ziegler, G.M.: Proofs from THE BOOK. Springer, 6th edn. (2018). https://doi.org/10.1007/978-3-662-57265-8
3. Buzzard, K., Commelin, J., Massot, P.: Formalising perfectoid spaces. In: Certified Programs and Proofs, pp. 299–312. Association for Computing Machinery (2020). https://doi.org/10.1145/3372885.3373830
4. Džamonja, M., Koutsoukou-Argyraki, A., Paulson, L.C.: Formalising ordinal partition relations using Isabelle/HOL. Experimental Mathematics (in press). https://doi.org/10.1080/10586458.2021.1980464
5. Erdős, P.: An interpolation problem associated with the continuum hypothesis. Mich. Math. J. **11**(1), 9–10 (1964)
6. Garcia, S.R., Shoemaker, A.L.: Wetzel's problem, Paul Erdős, and the continuum hypothesis: a mathematical mystery. Notices AMS **62**(3), 245–247 (2015)
7. Gordon, M.: Set theory, higher order logic or both? In: Goos, G., Hartmanis, J., van Leeuwen, J., von Wright, J., Grundy, J., Harrison, J. (eds.) TPHOLs 1996. LNCS, vol. 1125, pp. 191–201. Springer, Heidelberg (1996). https://doi.org/10.1007/BFb0105405
8. Gunther, E., Pagano, M., Sánchez Terraf, P.: Formalization of forcing in Isabelle/ZF. In: Peltier, N., Sofronie-Stokkermans, V. (eds.) IJCAR 2020. LNCS (LNAI), vol. 12167, pp. 221–235. Springer, Cham (2020). https://doi.org/10.1007/978-3-030-51054-1_13
9. Gunther, E., Pagano, M., Terraf, P.S., Steinberg, M.: The independence of the continuum hypothesis in Isabelle/ZF. Archive of Formal Proofs, March 2022. https://isa-afp.org/entries/Independence_CH.html, Formal proof development

10. Han, J.M., van Doorn, F.: A formal proof of the independence of the continuum hypothesis. In: 9th ACM SIGPLAN Conference on Certified Programs and Proofs, pp. 353–366. CPP 2020, Association for Computing Machinery (2020). https://doi.org/10.1145/3372885.3373826

11. Kunen, K.: Set Theory: An Introduction to Independence Proofs. Elsevier, North-Holland (1980)

12. Nipkow, T., Paulson, L.C., Wenzel, M.: Isabelle/HOL: a Proof Assistant for Higher-order Logic. Springer, Berlin (2002). https://doi.org/10.1007/3-540-45949-9, http://isabelle.in.tum.de/dist/Isabelle/doc/tutorial.pdf

13. Obua, S.: Partizan games in Isabelle/HOLZF. In: Barkaoui, K., Cavalcanti, A., Cerone, A. (eds.) ICTAC 2006. LNCS, vol. 4281, pp. 272–286. Springer, Heidelberg (2006). https://doi.org/10.1007/11921240_19

14. Paulson, L.C.: The foundation of a generic theorem prover. J. Autom. Reasoning **5**(3), 363–397 (1989). https://doi.org/10.1007/BF00248324, https://arxiv.org/abs/cs/9301105

15. Paulson, L.C.: The reflection theorem: a study in meta-theoretic reasoning. In: Voronkov, A. (ed.) CADE 2002. LNCS (LNAI), vol. 2392, pp. 377–391. Springer, Heidelberg (2002). https://doi.org/10.1007/3-540-45620-1_31

16. Paulson, L.C.: The relative consistency of the axiom of choice–mechanized using Isabelle/ZF. LMS J. Comput. Math. **6**, 198–248 (2003). https://doi.org/10.1112/S1461157000000449

17. Paulson, L.C.: Zermelo Fraenkel set theory in higher-order logic. Arch. Formal Proofs, October 2019. http://isa-afp.org/entries/ZFC_in_HOL.html, formal proof development

18. Paulson, L.C.: Ordinal partitions. Archive of Formal Proofs, August 2020. http://isa-afp.org/entries/Ordinal_Partitions.html, Formal proof development

19. Paulson, L.C.: Wetzel's problem and the continuum hypothesis. Archive of Formal Proofs, February 2022. https://isa-afp.org/entries/Wetzels_Problem.html, Formal proof development

20. Paulson, L.C., Grabczewski, K.: Mechanizing set theory: cardinal arithmetic and the axiom of choice. J. Autom. Reasoning **17**(3), 291–323 (1996). https://doi.org/10.1007/BF00283132

21. Schleicher, D., Stoll, M.: An introduction to Conway's games and numbers(2004). https://doi.org/10.48550/ARXIV.MATH/0410026, https://arxiv.org/abs/math/0410026

22. Wenzel, M.: Type classes and overloading in higher-order logic. In: Gunter, E.L., Felty, A. (eds.) TPHOLs 1997. LNCS, vol. 1275, pp. 307–322. Springer, Heidelberg (1997). https://doi.org/10.1007/BFb0028402

23. Werner, B.: Sets in types, types in sets. In: Abadi, M., Ito, T. (eds.) TACS 1997. LNCS, vol. 1281, pp. 530–546. Springer, Heidelberg (1997). https://doi.org/10.1007/BFb0014566

Hall's Theorem for Enumerable Families of Finite Sets

Fabián Fernando Serrano Suárez[1], Mauricio Ayala-Rincón[2]([⊠]) [iD],
and Thaynara Arielly de Lima[3] [iD]

[1] Universidad Nacional de Colombia - Sede Manizales, Manizales, Colombia
ffserranos@unal.edu.co
[2] Universidade de Brasília, Brasília, Brazil
ayala@unb.br
[3] Universidade Federal de Goiás, Goiânia, Brazil
thaynaradelima@ufg.br

Abstract. This work discusses the mechanisation in Isabelle/HOL of a general version of Hall's Theorem. It states that an enumerable family of finite sets has a system of distinct representatives (SDR) if it satisfies the "marriage condition". The marriage condition states that every finite subfamily of the possible infinite family of sets contains at least as many distinct members as the number of sets in the subfamily. The proof applies a formalisation of the Compactness Theorem for propositional logic. It checks the marriage condition for finite subfamilies of sets using Jiang and Nipkow's formalisation of the finite version of Hall's Theorem.

Keywords: Hall's Theorem · Compactness Theorem · Formalisation · Isabelle/HOL

1 Hall's Theorem

Let $\mathcal{A}$ be a finite family of arbitrary subsets of a set S such that sets in the family may repeat. Hall's theorem (also known as the "marriage theorem") was proved initially by Philip Hall in 1935 [11]. It establishes a necessary and sufficient condition to select a distinct element for each set in the collection. This theorem is equivalent to other significant results applied in the study of combinatory and graph theory problems (cf. [2,4,20]): Menger's theorem (1929), König's minimax theorem (1931), König-Egerváry theorem (1931), Birkhoff-von Neumann's theorem (1946), Dilworth's theorem (1950), Max Flow-Min Cut theorem (Ford-Fulkerson algorithm) (1956), and also to probability theory results as Strassen's theorem (1965). For instance, the König-Egerváry theorem states that the number of lines (rows or columns) that cover all ones in a binary matrix is precisely

Second and third authors supported by FAP-DF DE 00193.00001175/2021-11 and CNPq Universal 409003/2021-2 grants. Second author partially funded by CNPq grant 313290/2021-0.

the cardinality of a set of ones in different lines of the matrix. Taking the sets of ones in the matrix lines as the family of finite sets and selecting the ones that do not share lines as the system of distinct representatives, the equivalence between both problems is evident.

Hall's theorem is established using the notion of a system of distinct representatives (SDR) for a family of sets.

Definition 1 (SDR). *Let S be an arbitrary set and $\{S_i\}_{i \in I}$ a collection of not necessarily distinct subsets of S with indices in the set I.*

A sequence $(x_i)_{i \in I}$ is a system of distinct representatives for $\{S_i\}_{i \in I}$ if:

1. *for all $i \in I$, $x_i \in S_i$, and;*
2. *for all $i, j \in I$, $x_i \neq x_j$, whenever $i \neq j$.*

Alternatively, one can define SDR as follows.
A function $f : I \to \bigcup_{i \in I} S_i$ is a SDR for $\{S_i\}_{i \in I}$ if:

1. for all $i \in I$, $f(i) \in S_i$, and;
2. f is an injective function.

Theorem 1 (Hall's Theorem | finite case). *Consider an arbitrary set S and a positive integer n. A finite collection $\{S_1, S_2, \ldots, S_n\}$ of finite subsets of S has a SDR if and only if the so called* marriage condition *(M) below is satisfied.*

> *For every $1 \leq k \leq n$ and an arbitrary set of k distinct indices $1 \leq i_1, \ldots, i_k \leq n$, one has that $|S_{i_1} \cup \ldots \cup S_{i_k}| \geq k$.*   (M)

Hall's Theorem also holds for an infinite enumerable collection $\{S_i\}_{i \in I}$ of finite subsets of S (Theorem 2). Indeed, other versions of such a theorem are considered and proved in [19].

Theorem 2 (Hall's Theorem | enumerable case). *Let S be an arbitrary set and I an enumerable set of indices of finite subsets of S. The family $\{S_i\}_{i \in I}$ has a SDR if and only if the condition (M*) below holds.*

> *For every finite subset of indices $J \subseteq I$, one has that $|\bigcup_{j \in J} S_j| \geq |J|$.*   (M*)

Jiang and Nipkow formalized the finite case of Hall's theorem in Isabelle/HOL [14,15]. The distinguishing feature of their formalisation was the use of functional indexations of collections of subsets of S instead of a representation of such collections as sequences. Indeed, using such indexation structure, they formalized this theorem applying both the Halmos and Vaughan's and the Rado's approaches (see [12], and [19], respectively). The former proof is nicely presented by Aigner and Ziegler using sequences in [1].

This work discusses a formalisation in Isabelle/HOL of the enumerable version of Hall's Theorem (Theorem 2). The demonstration consists in proving the sufficiency of the marriage condition for the existence of SDR: $M^* \Rightarrow$ SDR. The proof applies the Compactness Theorem for propositional logic (as formalised in [23]), where the marriage condition for finite families is verified by using Jiang and Nipkow's formalisation. As in Jiang and Nipkow's approach, we use functional (infinite) indexations of families of sets. Such indexations representation allows us to apply their formalisation straightforwardly, allowing elegant and simple specifications.

The formalisation approach follows the logical constructive-model lines of reasoning of Cameron's informal proof in [4].

As far as we know, there is only another formalisation of the enumerable version of Hall's theorem, which follows an approach different from the one used in this paper. Indeed, Gusakov, Mehta and Miller [10] formalised the theorem in Lean following a purely combinatorial approach. Based on the inverse limit, it depends on a formalisation of König's lemma. This proof is discussed in more detail in Sect. 4. Our formalisation of Hall's theorem infers the enumerable from the finite version of the theorem by applying the Compactness Theorem for propositional logic. Therefore, our proof technique also depends on combinatorial elements since Jiang and Nipkow's formalisation of the finite version relies on a combinatorial approach. Other equivalent theorems to Hall's theorem were also formalised using the Compactness Theorem. The Isabelle/HOL distribution that accompanies this paper includes formalisations of the De Bruijn-Erdös's graph colouring theorem [5] and the König's lemma (cf., exercise in Chapter I.6 in [17]). Both formalisations follow the constructive-model approach and are the subject of another work.

Using the enumerable version of Hall's Theorem, it is possible to formalise equivalences with enumerable versions of other combinatorial theorems. For instance, the enumerable version of the König-Egerváry Theorem may be stated as follows. First, observe that binary matrices correspond to adjacency matrices of bipartite graphs, where the nodes are given by the row and column indices. Thus, this theorem relates maximal matchings and minimal cover sets in bipartite graphs. Now, let $G = \langle X, Y, E \rangle$ be a bipartite graph, where the vertices are given by the enumerable set $X \uplus Y$, and all edges in E connect a vertex in X with another in Y. Additionally, suppose that all vertices have a non-zero finite degree, and the maximal degree of the vertices in the graph is also finite. Thus, the graph theoretical enumerable (infinite) version of the König-Egerváry Theorem guarantees the existence of a X-perfect matching in G if and only if the vertices X are a minimal cover of the graph. A X-perfect matching is a subset of edges, say $E' \subseteq E$, such that there is exactly one edge in E' incident to each vertex in X, and for any pair of different edges in E' the incident vertices in Y are different. This enumerable version of the theorem is proved by constructing a translation from Hall's Theorem. Indeed, the family S of subsets, S_x, for $x \in X$, is given by the finite set of vertices in the neighbourhood of x. Thus, the set X corresponds to the indices of the family of finite subsets S. Finally, the set of vertices in Y incident to a X-perfect matching $E' \subseteq E$ corresponds to a system of distinct representatives for the family S.

Organisation. The paper is organised as follows. Section 2 presents Cameron's informal proof that is the one followed in our formalisation approach. Section 3 briefly describes the formalisation of the Compactness Theorem (Subsect. 3.1) for propositional logic, and the formalisation of the enumerable infinite version of Hall's theorem (Subsect. 3.2). Section 4 discusses related work, considering other formalisations of the two key ingredients, namely, the finite version of Hall's theorem and the Compactness Theorem. Finally, Sect. 5 concludes and proposes future work. The formalisation is available through hyperlinks ($\mathcal{C}$) in the body of the paper.

2   Cameron's Informal Proof

The formalisation approach follows the lines of reasoning of Cameron's informal proof given in [4], page 318.

Assume that the marriage condition (M^*) holds.

Consider the propositional language with constant symbols given by the set below

$$\mathcal{P} = \{C_{n,x} \mid n \in I, x \in S_n\}.$$

For each $n \in I$, the constant $C_{n,x}$ is interpreted as "select the element x from the set S_n".

The following three sets of propositional formulas describe the existence of a SDR for $\{S_n\}_{n\in I}$.

1. Select at least an element from each S_n:

$$\mathcal{F} = \{\vee_{x\in S_n} C_{n,x} \mid n \in I\}.$$

   The disjunction $\vee_{x\in S_n} C_{n,x}$ of atomic formulas is well-defined, since each constant corresponds to an element of the set S_n, that by hypothesis is finte.
2. Select at most an element from each S_n:

$$\mathcal{G} = \{\neg(C_{n,x} \wedge C_{n,y}) \mid x,y \in S_n, x \neq y, n \in I\}.$$

3. Do not select more than once the same element from $\bigcup_{n\in I} S_n$:

$$\mathcal{H} = \{\neg(C_{n,x} \wedge C_{m,x}) \mid x \in S_n \cap S_m, n \neq m, n, m \in I\}.$$

   Let $\mathcal{T} = \mathcal{F} \cup \mathcal{G} \cup \mathcal{H}$. We apply the Compactness Theorem to prove that $\mathcal{T}$ is satisfiable.

   Let $\mathcal{T}_0$ be any finite subset of formulas in $\mathcal{T}$ and let $J = \{i_1, \ldots, i_m\}$ be the corresponding finite subset of indices in I that are "referred" in $\mathcal{T}_0$, i.e., the set of all indices i such that $C_{i,x}$ for some $x \in S_i$, occurs in some formula of $\mathcal{T}_0$.

   Let us consider the family of sets $\{S_{i_1}, \ldots, S_{i_m}\}$. Then, $\mathcal{T}_0$ is contained in the set $\mathcal{T}_1 = \mathcal{F}_0 \cup \mathcal{G}_0 \cup \mathcal{H}_0$, where

1. $\mathcal{F}_0 = \{\vee_{x\in S_n} C_{n,x} \mid n \in J\}$,

2. $\mathcal{G}_0 = \{\neg(C_{n,x} \wedge C_{n,y}) \mid x, y \in S_n, x \neq y, n \in J\}$,
3. $\mathcal{H}_0 = \{\neg(C_{n,x} \wedge C_{m,x}) \mid x \in S_n \cap S_m, n \neq m, n, m \in J\}$.

By hypothesis, $\{S_{i_1}, \ldots, S_{i_m}\}$ satisfies the condition (M^*) and, in particular, the condition (M). Therefore, by the finite version of Hall's Theorem there exists a function $f : J \to \bigcup_{i \in J} S_i$ such that f is a SDR for $\{S_{i_1}, \ldots, S_{i_m}\}$.

Consequently, a model for $\mathcal{T}_1$ is given by the interpretation $v : \mathcal{P} \to \{V, F\}$ defined by,

$$v(C_{n,x}) = \begin{cases} V, & \text{if } n \in J \text{ and } f(n) = x, \\ F, & \text{otherwise.} \end{cases}$$

Thus, $\mathcal{T}_1$ is satisfiable and so is $\mathcal{T}_0$. In this manner, $\mathcal{T}$ is finitely satisfiable and consequently, by the Compactness Theorem, it is satisfiable.

Let $v' : \mathcal{P} \to \{V, F\}$ be a model of $\mathcal{T}$. We define the function $f : I \to \bigcup_{n \in I} S_n$ as

$$f(m) = x \text{ if and only if } v'(C_{m,x}) = V.$$

Then, f is a SDR for $\{S_n\}_{n \in I}$:

Since $\mathcal{F}$ and $\mathcal{G}$ are satisfiable, for each $m \in I$ there is exactly an element in S_m, thus f is a function. Also, since $\mathcal{H}$ is satisfiable, one has that f is injective fuction.

3   Formalisation

In this section, we discuss the formalisation of Hall's Theorem. This paper does not focus on the formalisation of the Compactness Theorem, but it is briefly explained for completeness.

The formalisation of the enumerable version of Hall's Theorem consists of less than 6.000 words in ca. 900 lines of code. It includes seven definitions and 46 lemmas and theorems.

Pertinently, we include links ($\mathbb{C}$) to the specific parts of the formalisation under analysis.

3.1   Notes on the Formalisation of the Propositional Compactness Theorem

For completeness, this subsection sketches the formalisation of the Propositional Compactness Theorem, which is used here but is not part of this work. The formalisation was first given in [23] and follows closely Fitting's textbook presentation in [9].

We present the most important definitions and proofs used in the formalisation.

The language of propositional formulas is specified by the following datatype.

Datatype $'b$ *formula* $\mathbb{C}$ =

⊥
| ⊤
| *atom* '*b*
| *negation*   '*b formula*   (¬.(-) [*110*] *110*)
| *conjunction* '*b formula*   '*b formula*   (**infixl** ∧. *109*)
| *disjunction* '*b formula*   '*b formula*   (**infixl** ∨. *108*)
| *implication* '*b formula*   '*b formula*   (**infixl** →. *100*)

To evaluate the *truth-value* of propositional formulas over an interpretation we specify the operator *t-v-evaluation*.

Primrec *t-v-evaluation* ⬚ :: ('*b* ⇒   *truth-value*) ⇒ '*b formula*   ⇒ *truth-value*
   where
     t-v-evaluation I ⊥ = *Ffalse*
  | *t-v-evaluation I* ⊤ = *Ttrue*
  | *t-v-evaluation I* (*Atom P*) = *I P*
  | *t-v-evaluation I* (¬. *F*) = (*v-negation* (*t-v-evaluation I F*))
  | *t-v-evaluation I* (*F* ∧. *G*) = (*v-conjunction* (*t-v-evaluation I F*) (*t-v-evaluation I G*))
  | *t-v-evaluation I* (*F* ∨. *G*) = (*v-disjunction* (*t-v-evaluation I F*) (*t-v-evaluation I G*))
  | *t-v-evaluation I* (*F* →. *G*) = (*v-implication* (*t-v-evaluation I F*) (*t-v-evaluation I G*))

The operator *t-v-evaluation* uses the definitions below.

Definition *v-negation* ⬚ :: *truth-value* ⇒ *truth-value* **where**
   v-negation x ≡ (*if x* = *Ttrue then Ffalse else Ttrue*)

Definition *v-conjunction* ⬚ ::   *truth-value* ⇒ *truth-value* ⇒ *truth-value*
where
   v-conjunction x y ≡ (*if x* = *Ffalse then Ffalse else y*)

Definition *v-disjunction* ⬚ ::   *truth-value* ⇒ *truth-value* ⇒ *truth-value*
where
   v-disjunction x y ≡ (*if x* = *Ttrue then Ttrue else y*)

Definition *v-implication* ⬚ :: *truth-value* ⇒ *truth-value* ⇒ *truth-value*
where
   v-implication x y ≡ (*if x* = *Ffalse then Ttrue else y*)

The notion of satisfiability is specified through the existence of *models*.

Definition *model* ⬚ :: ('*b* ⇒ *truth-value*) ⇒ '*b formula set* ⇒ *bool*
where *I model S* ≡ (∀ *F* ∈ *S*. *t-v-evaluation I F* = *Ttrue*)

Definition *satisfiable* ⬚ :: '*b formula set* ⇒ *bool* **where**

$$satisfiable\ S \equiv (\exists\, v.\ v\ model\ S)$$

The notion of compactness is specified using the Isabelle specification for finite sets and a specification for enumerable sets.

The next lemma, from Isabelle, formalised the fact that a *finite* set A is finite if and only if there exists a surjective function f from I_n onto A, where $I_n = \{m \in \mathbb{N} \mid m < n\}$, for some $n \in \mathbb{N}$.

Lemma *finite* $A \longleftrightarrow (\exists\, n\, f.\ A = f\ ` \{i::nat.\ i < n\})$

We specify enumerable sets using the notion of *enumeration*, i.e., the existence of a surjective function with domain $\mathbb{N}$, given below.

Definition *enumeration* ⚙ :: $(nat \Rightarrow\, 'b) \Rightarrow bool$ **where** *enumeration* $f = (\forall\, y.\exists\, n.\ y = (f\ n))$

König's lemma is used in classic textbooks to prove the Compactness Theorem. In the formalisation, we follow Fitting's textbook approach in [9] that instead applies the propositional model existence theorem.

Theorem 3 (Propositional model existence (Th. 3.6.2 in [9]). *If $\mathcal{C}$ is a propositional consistency property, and $S \in \mathcal{C}$, then S is satisfiable.*

Theorem 4 (Propositional Compactness (Th. 3.6.3 in [9]). *Let S be a set of propositional formulas. If every finite subset of S is satisfiable, so is S.*

Both these theorems require the definition of propositional consistency. Let $\mathcal{C}$ be a collection of sets of propositional formulas. We call $\mathcal{C}$ a propositional consistency property if it meets the conditions for each $S \in \mathcal{C}$, given in the definition *consistenceP*, as specified below. In this definition *FormulaAlpha* and *FormulaBeta* correspond respectively to conjunctive and disjunctive propositional formulas as defined in [9].

Definition *consistenceP* ⚙ :: $'b\ formula\ set\ set \Rightarrow bool$ **where**
   consistenceP $\mathcal{C}$ =
   $(\forall\, S.\ S \in \mathcal{C} \longrightarrow (\forall\, P.\ \neg\ (atom\ P \in S \wedge (\neg.atom\ P\ ) \in S)) \wedge$
   $\bot\ \notin S \wedge (\neg.\top) \notin S \wedge$
   $(\forall\, F.\ (\neg.\neg.F) \in S \longrightarrow S \cup \{F\} \in\ \mathcal{C}) \wedge$
   $(\forall\, F.\ ((FormulaAlpha\ F) \wedge F{\in}S) \longrightarrow (\ S \cup \{Comp1\ F,\ Comp2\ F\}) \in$
$\mathcal{C})\ \wedge$
   $(\forall\, F.\ ((FormulaBeta\ F) \wedge F{\in}S) \longrightarrow (\ S \cup \{Comp1\ F\} \in \mathcal{C}) \vee$
   $(\ S \cup \{Comp2\ F\} \in \mathcal{C})))$

The formalisations of the model existence and the Compactness Theorems are given below.

Theorem *TheoremExistenceModels* ⚙:
   assumes $h1:\ \exists\, g.\ enumeration\ (g::\ nat \Rightarrow\, 'b\ formula)$
   and $h2:\ consistenceP\ \mathcal{C}$

and *h3*: $(S:: \text{'}b \text{ formula set}) \in \mathcal{C}$
shows *satisfiable S*

The following auxiliary lemma is required to apply *TheoremExistenceModels* to obtain the Compactness Theorem. This lemma states that the collection of sets of propositional formulas given by $\mathcal{C}$ below is a propositional consistency property

$$\mathcal{C} = \{W \mid \forall A \, (A \subseteq W \wedge A \, finite \rightarrow A \, satisfiable)\}.$$

Lemma *ConsistenceCompactness* 🖝:
    shows *consistenceP*$\{W:: \text{'}b \text{ formula set. } \forall A. \ (A \subseteq W \wedge finite \ A) \longrightarrow \ satisfiable \ A\}$

Finally, the Compactness Theorem is specified as below.

Theorem *Compactness-Theorem* 🖝:
    assumes $\exists g. \ enumeration \ (g:: nat \Rightarrow \text{'}b \ formula)$
    and $\forall A. \ (A \subseteq (S:: \text{'}b \ formula \ set) \wedge finite \ A) \longrightarrow satisfiable \ A$
    shows *satisfiable S*

3.2   Formalisation of Hall's Theorem - the Enumerable Version

As in [15], we represent the collection of enumerable sets $\{S_n\}_{n \in I}$ in Isabelle as a function $S :: a \Rightarrow b \ set$ together with a set of indices $I :: a \ set$, where a and b are variable sets, and such that for all $i \in I$, the set $(S \ i)$ is finite. Unlike Jian and Nipkow's formalisation, for the enumerable version of the Hall's theorem, a and b are constrained to be arbitrary enumerable types.

A SDR for S and I is any function $R :: a \Rightarrow b$, which satisfies the predicate below.

Definition *system-representatives* 🖝 :: $(\text{'}a \Rightarrow \text{'}b \ set) \Rightarrow \text{'}a \ set \Rightarrow (\text{'}a \Rightarrow \text{'}b) \Rightarrow bool$ **where**
system-representatives $S \ I \ R \ \equiv (\forall i \in I. \ (R \ i) \in (S \ i)) \wedge (inj\text{-}on \ R \ I)$

Above, *(inj-on R I)* means that the function R is injective on I.
The marriage condition for S and I is formalized by the proposition,

$$\forall J \subseteq I. \ finite \ J \longrightarrow card \ J \leq card \left(\bigcup (S \ \text{'} \ J) \right)$$

where $S \ \text{'} \ J = \{S \ j \mid j \in J\}$.
    Using the previous notions, Hall's Theorem is specified as:

Theorem *Hall* 🖝:
    fixes $S \ :: \ \text{'}a \Rightarrow \text{'}b \ set$ **and** $I \ :: \ \text{'}a \ set$
    assumes $\exists g. \ enumeration \ (g:: nat \Rightarrow \text{'}a)$ **and** $\exists h. \ enumeration \ (h:: nat \Rightarrow \text{'}b)$

and *Finite*: $\forall\, i{\in}I.\ finite\ (S\ i)$
and *Marriage*: $\forall\, J{\subseteq}I.\ finite\ J \longrightarrow card\ J \le card\ (\bigcup\ (S\ `\ J))$
shows $\exists\, R.\ system\text{-}representatives\ S\ I\ R$

The following four definitions in Isabelle correspond to the formalisation of the sets $\mathcal{F}, \mathcal{G}, \mathcal{H}$, and $\mathcal{T}$ used in the informal proof. The definition of $\mathcal{F}$ uses *disjunction-atomic* to build the disjunction associated with each finite set in the collection.

Primrec *disjunction-atomic* 🔗 :: $`b\ list \Rightarrow `a \Rightarrow (`a \times\ `b)formula$ **where**
  disjunction-atomic $[]\ i = \bot$
| *disjunction-atomic* $(x\#D)\ i = (atom\ (i,\ x))\ \vee.\ (disjunction\text{-}atomic\ D\ i)$

Definition $\mathcal{F}$ 🔗 :: $(`a \Rightarrow `b\ set) \Rightarrow `a\ set \Rightarrow ((`a \times `b)formula)\ set$ **where**
  $\mathcal{F}\ S\ I \equiv (\bigcup i{\in}I.\ \{\ disjunction\text{-}atomic\ (set\text{-}to\text{-}list\ (S\ i))\ i\ \})$

Definition $\mathcal{G}$ 🔗 :: $(`a \Rightarrow `b\ set) \Rightarrow `a\ set \Rightarrow (`a \times `b)formula\ set$ **where**
  $\mathcal{G}\ S\ I \equiv \{\neg.(atom\ (i,x)\ \wedge.\ atom(i,y))$
    $|x\ y\ i\ .\ x{\in}(S\ i)\ \wedge\ y{\in}(S\ i)\ \wedge\ x{\neq}y\ \wedge\ i{\in}I\}$

Definition $\mathcal{H}$ 🔗 :: $(`a \Rightarrow `b\ set) \Rightarrow `a\ set \Rightarrow (`a \times `b)formula\ set$ **where**
  $\mathcal{H}\ S\ I \equiv \{\neg.(atom\ (i,x)\ \wedge.\ atom(j,x))$
      $|\ x\ i\ j.\ x \in (S\ i) \cap (S\ j)\ \wedge\ (i{\in}I\ \wedge\ j{\in}I\ \wedge\ i{\neq}j)\}$

Definition $\mathcal{T}$ 🔗 :: $(`a \Rightarrow `b\ set) \Rightarrow `a\ set \Rightarrow (`a \times `b)formula\ set$ **where**
  $\mathcal{T}\ S\ I \equiv (\mathcal{F}\ S\ I) \cup (\mathcal{G}\ S\ I) \cup (\mathcal{H}\ S\ I)$

The above definitions illustrate the benefit of using sets of indices in our specification. The set of indices occurring in a set of formulas (*indices-set-formulas* 🔗) is the union of set of indices occurring in each formula (*indices-formula*) that are defined recursively.

Let $To \subseteq (\mathcal{T}\ S\ I)$ any finite subset of formulas, (T_0 in the informal proofs), such that the collection of finite sets of formulas with indices used by the formulas in To, (*indices-set-formulas   To*), satisfies the marriage condition, then there is a SDR. The proof uses Jiang and Nipkow's finite version of Hall's Theorem given in [15]. Indeed, the proof can apply either Halmos and Vaughan's or Rado's formalisations in [15] without any modification. This is possible since our specification of predicates, as SDR, are independent of any definition in Jiang and Nipkow's formalisation.

Lemma *system-distinct-representatives-finite* 🔗:
  assumes
    $\forall\, i{\in}I.\ (S\ i){\neq}\{\}$ **and** $\forall\, i{\in}I.\ finite\ (S\ i)$ **and** $To \subseteq (\mathcal{T}\ S\ I)$ **and** *finite* To
    and $\forall\, J{\subseteq}(indices\text{-}set\text{-}formulas\ To).\ card\ J \le card\ (\bigcup\ (S\ `\ J))$
  shows $\exists\, R.\ system\text{-}representatives\ S\ (indices\text{-}set\text{-}formulas\ To)\ R$

The following lemma states that if there exists a SDR R for a collection of finite sets given by A and $\mathcal{I}$, then any subset of formulas $X \subseteq (\mathcal{T} A \mathcal{I})$ is satisfiable. A model for X is given by the next interpretation of formulas.

Fun *Hall-interpretation* $\mathbb{C}$ :: $('a \Rightarrow 'b\ set) \Rightarrow 'a\ set \Rightarrow ('a \Rightarrow 'b) \Rightarrow (('a \times 'b) \Rightarrow truth\text{-}value)$ **where**
Hall-interpretation $A\ I\ R = (\lambda(i,x).(if\ i \in I \wedge x \in (A\ i) \wedge (R\ i) = x\ then\ Ttrue\ else\ Ffalse))$

Lemma *SDR-satisfiable* $\mathbb{C}$:
    assumes $\forall i \in I.\ (A\ i) \neq \{\}$ **and** $\forall i \in I.\ finite\ (A\ i)$ **and** $X \subseteq (\mathcal{T}\ A\ I)$
    and *system-representatives* $A\ I\ R$
  shows *satisfiable* X

Lemma *SDR-satisfiable* above is the kernel of the formalisation. It proves that the set of formulas $(\mathcal{T} A I)$ built from A and I is satisfiable building and evaluating the model given by the function *Hall-interpretation*.

Previous results allow us to prove the following lemma. It states that any finite subset of formulas $To \subseteq (\mathcal{T} S I)$, such that the collection of finite sets of formulas with indices used by the formulas in To hold the marriage condition, is satisfiable.

Lemma *finite-is-satisfiable* $\mathbb{C}$:
    assumes
    $\forall i \in I.\ (S\ i) \neq \{\}$ **and** $\forall i \in I.\ finite\ (S\ i)$ **and** $To \subseteq (\mathcal{T}\ S\ I)$ **and** *finite* To
    and $\forall J \subseteq (indices\text{-}set\text{-}formulas\ To).\ card\ J \leq card\ (\bigcup\ (S\ {}^{\backprime}\ J))$
  shows *satisfiable* To

The lemma *finite-is-satisfiable* and the Compactness Theorem are then used to prove that the set of formulas $(\mathcal{T} S I)$ is satisfiable.

Lemma *all-formulas-satisfiable* $\mathbb{C}$:
    fixes S :: $'a \Rightarrow 'b\ set$ **and** I :: $'a\ set$
    assumes $\exists g.\ enumeration\ (g:: nat \Rightarrow 'a)$ **and** $\exists h.\ enumeration\ (h:: nat \Rightarrow 'b)$
    and $\forall i \in I.\ finite\ (S\ i)$
    and $\forall J \subseteq I.\ finite\ J \longrightarrow card\ J \leq card\ (\bigcup\ (S\ {}^{\backprime}\ J))$
  shows *satisfiable* $(\mathcal{T}\ S\ I)$

The lemma below, *satisfiable-representant*, states that if $(\mathcal{T} S I)$ is satisfiable then the corresponding (enumerable) collection of finite sets $\{S_i\}_{i \in I}$, given by S and I, has a SDR. For its proof we use the function SDR and lemma *function-SDR*.

Fun *SDR* ☐ :: $((\,'a \times \, 'b) \Rightarrow$ *truth-value*$) \Rightarrow (\,'a \Rightarrow \, 'b$ *set*$) \Rightarrow \, 'a$ *set* $\Rightarrow$
$(\,'a \Rightarrow 'b\,)$
where
SDR M S I = $(\lambda i.\ (THE\ x.\ (value\ M\ (atom\ (i,x)) = Ttrue) \wedge x \in (S\ i)))$

Soundness of the function *SDR* is proved by the lemma *function-SDR* below.

Lemma *function-SDR* ☐:
  assumes $i \in I$ **and** M *model* $(\mathcal{F}\ S\ I)$ **and** M *model* $(\mathcal{G}\ S\ I)$ **and** *finite*(*S*
i)
  shows $\exists\ !x.\ (value\ M\ (atom\ (i,x)) = Ttrue) \wedge x \in (S\ i) \wedge SDR\ M\ S\ I\ i =$
x

Lemma *satisfiable-representant* ☐:
  assumes *satisfiable* $(\mathcal{T}\ S\ I)$ **and** $\forall\ i \in I.\ finite\ (S\ i)$
  shows $\exists\ R.\ system\text{-}representatives\ S\ I\ R$
Finally, we obtain the formalisation of Hall's Theorem.

Theorem *Hall* ☐:
  fixes $S :: \, 'a \Rightarrow \, 'b$ *set* **and** $I :: \, 'a$ *set*
  assumes $\exists\ g.\ enumeration\ (g:: nat \Rightarrow 'a)$ **and** $\exists\ h.\ enumeration\ (h:: nat$
$\Rightarrow 'b)$
    and *Finite*: $\forall\ i \in I.\ finite\ (S\ i)$
    and *Marriage*: $\forall\ J \subseteq I.\ finite\ J \longrightarrow card\ J \leq card\ (\bigcup\ (S\ {}^{\backprime}\ J))$
  shows $\exists\ R.\ system\text{-}representatives\ S\ I\ R$
  proof-
    have *satisfiable* $(\mathcal{T}\ S\ I)$ **using** *assms all-formulas-satisfiable*[*of I*] **by** *auto*
    thus *?thesis* **using** *Finite Marriage satisfiable-representant*[*of S I*] **by** *auto*
  qed

4 Related Work

There are two preliminary Isabelle formalisations over which this one is developed. The first is the formalisation of the finite case of Hall's Theorem by Jiang and Nipkow [18] and the second is the formal verification of the Compactness Theorem given by Serrano in [23]. Also, there exists another formalisation in Isabelle of the Compactness Theorem for propositional logic developed by Michaelis and Nipkow [16]. Nevertheless, we prefer to use the above mentioned developed by Serrano.

Concerning other proof assistants and extensions of the Compactness Theorem to other logics, Harrison [13] provided a formalisation of the Compactness Theorem for propositional and first-order logic in HOL. Also, Braselmann and Koepke [3] formally verified such a Theorem in Mizar for first-order logic.

Regarding the finite case of Hall's Theorem, the first formalisation of such theorem was developed by Romanowicz and Grabowski [21] in Mizar following

Rado's analytical proof [19]. Also, there is a formalisation in Coq of the finite version of Hall's theorem that uses formalisations of combinatorial arguments as Dilworth's decomposition theorem and existence of bi-partitions in graphs [24]. Indeed, there are earlier combinatorial formalisations of Dilworth's theorem in Mizar as the one presented in [22]. This theorem states that in a finite partially ordered set, the size of minimal chains and maximal anti-chains are the same. Recently, Gusakov, Mehta and Miller [10] presented three different proofs of the finite version of Hall's theorem formalised in Lean in terms of indexed families of finite subsets, of existence of *matchings* (injections) that saturate binary relations over finite sets, and of matchings in bipartite graphs.

There are a myriad of formalisations related to Hall's theorem, which are based on purely combinatorial approaches and not on the compactness approach followed in this paper. Among them, we could mention recent works by Doczkal et al. in their graph theory Coq library (e.g., [6,8], and [7]). Finally, Singh and Natarajan formalized in Coq other combinatorial resuls as the perfect graph theorem and a weak version of this theorem (e.g., [25,26]).

As far as we know, the unique formalisation of the enumerable version of Hall's theorem is the one by Gusakov, Mehta and Miller cited in the introduction [10]. As above mentioned, the authors formalised three versions of the finite case of Hall's theorem in Lean. Also, they apply an *inverse limit* version of the König's lemma to conclude the enumerable case as specified in this paper. The inverse limit version of the König's lemma states that if $\{X_i\}, i \in \mathbb{N}$ is an indexed family of sets with functions $f_i : X_{i+1} \to X_i$, for each i, then if each X_i is a nonempty finite set, then there exists a family of elements $x \in \prod_i X_i$ such that $x_i = f_i(x_{i+1})$, for all i. The usual version of the König's lemma follows form this one, by choosing as set X_i the paths of length i from the root vertex v_0 in a tree. So, the function f_i maps paths in X_{i+1} into the paths without their last edge in X_i. The inverse limit consist of the infinite chain of functions $f_1, f_2, \ldots$. König's lemma is applied to prove the enumerable version of Hall's theorem by taking M_n as the set of all *matchings* on the first n indices of I (i.e., the set of all possible SDRs for the sets $S_1, \ldots, S_n$), and $f_n : M_{n+1} \to M_n$ as the restriction of a matching to a smaller index set. Since the marriage condition holds for the finite indexed families, each M_n is nonempty and by König's lemma an element of the inverse limit gives a matching on I. Differently for our formalisation, Gusakov, Mehta and Miller proof does not follow a constructive approach as the one given in our development in which a model is built to guarantee the hypotheses of the Compactness Theorem for propositional logic.

5   Conclusion

This paper presented a formalisation of Hall's theorem for infinite enumerable collections of finite sets in Isabelle. The proof uses a formalisation of the Compactness Theorem for propositional logic and, in addition, Jiang and Nipkow's formalisation of Hall's theorem for considering the case of finite collections of finite sets.

The distinctive characteristics of our formalisation are:

- it inherits the advantages of the representation of collections of sets through set-indexations from Jiang and Nipkow's formalisation of the finite version of Hall's theorem [15];
- it profits from the Isabelle/HOL deductive features to follow a line of reasoning that remains close to the analytical proofs, and;
- in contrast with pure combinatorial proofs, it follows the logical constructive-model approach that applies the Compactness Theorem.

Interesting applications include the formalisation of the extension of Hall's theorem to non-enumerable families of finite sets, and the formalisation of other related combinatorial theorems (applying Hall's theorem) as those mentioned in the introduction.

Other applications of the Compactness Theorem, which are not discussed in this paper, were formalised similarly and are also available in the distribution. For instance, the De Bruijn-Erdös's graph colouring theorem ☑ ([5]), and König's lemma ☑ (cf., exercise in Chapter I.6 in [17]). These formalisations follow the logical constructive-model approach described in this paper. Of course, a variety of consequences of the propositional Compactness Theorem would also be welcome as those presented in textbooks (e.g., [4,9], and [17]). Finally, formalising enumerable combinatorial consequences of Hall's theorem (or of De Bruijn-Erdös's graph colouring theorem, or König's lemma), as the ones discussed in the introduction, is a subject of work in progress.

References

1. Aigner, M., Ziegler, G.M.: Proofs from THE BOOK. Springer Berlin, Heidelberg, 6th edn. (2018), chapter 30: Three famous theorems on finite sets. https://doi.org/10.1007/978-3-662-57265-8
2. Borgersen, R.D.: Equivalence of seven major theorems in combinatorics (2004), talk available at Department of Mathematics, University of Manitoba, Canada. https://home.cc.umanitoba.ca/~borgerse/Presentations/GS-05R-1.pdf
3. Braselmann, P., Koepke, P.: Gödel's completeness theorem. Formalized Math. (University of Białystok) **13**(1), 49–53 (2005). https://fm.mizar.org/2005-13/pdf13-1/goedelcp.pdf
4. Cameron, P.J.: Combinatorics: Topics, Techniques, Algorithms. Cambridge University Press, Cambridge (1994)
5. De Bruijn, N.G., Erdös, P.: A colour problem for infinite graphs and a problem in the theory of relations. Indagationes Mathematicae (Proceedings) **54**, 371–373 (1951). https://doi.org/10.1016/S1385-7258(51)50053-7
6. Doczkal, C., Combette, G., Pous, D.: A formal proof of the minor-exclusion property for treewidth-two graphs. In: Avigad, J., Mahboubi, A. (eds.) ITP 2018. LNCS, vol. 10895, pp. 178–195. Springer, Cham (2018). https://doi.org/10.1007/978-3-319-94821-8_11
7. Doczkal, C., Pous, D.: Completeness of an axiomatization of graph isomorphism via graph rewriting in Coq. In: Proceedings of the 9th ACM SIGPLAN International Conference on Certified Programs and Proofs - CPP, pp. 325–337. ACM (2020). https://doi.org/10.1145/3372885.3373831

8. Doczkal, C., Pous, D.: Graph theory in coq: minors, treewidth, and isomorphisms. J. Autom. Reasoning **64**(5), 795–825 (2020). https://doi.org/10.1007/s10817-020-09543-2

9. Fitting, M.: First-Order Logic and Automated Theorem Proving. 2nd edn. Springer, New York (1996). https://doi.org/10.1007/978-1-4612-2360-3

10. Gusakov, A., Mehta, B., Miller, K.A.: Formalizing hall's marriage theorem in lean. arXiv abs/2101.00127[math.CO] (2021). https://doi.org/10.48550/arxiv.2101.00127

11. Hall, P.: On representatives of subsets. London Math. Soc. **10**, 26–30 (1935). https://doi.org/10.1112/jlms/s1-10.37.26

12. Halmos, P.R., Vaughan, H.E.: The marriage problem. Am. J. Math. **72**(1), 214–215 (1950). https://doi.org/10.2307/2372148

13. Harrison, J.: Formalizing basic first order model theory. In: Grundy, J., Newey, M. (eds.) TPHOLs 1998. LNCS, vol. 1479, pp. 153–170. Springer, Heidelberg (1998). https://doi.org/10.1007/BFb0055135

14. Jiang, D., Nipkow, T.: Hall's marriage theorem. Archive of Formal Proofs, December 2010. https://www.isa-afp.org/browser_info/current/AFP/Marriage/document.pdf

15. Jiang, D., Nipkow, T.: Proof pearl: the marriage theorem. In: Jouannaud, J.-P., Shao, Z. (eds.) CPP 2011. LNCS, vol. 7086, pp. 394–399. Springer, Heidelberg (2011). https://doi.org/10.1007/978-3-642-25379-9_28

16. Michaelis, J., Nipkow, T.: Formalized proof systems for propositional logic. In: Proceedings 23rd International Conference on Types for Proofs and Programs - TYPES 2017. Leibniz International Proceedings in Informatics (LIPIcs), vol. 104, pp. 5:1–5:16 (2018). https://doi.org/10.4230/LIPIcs.TYPES.2017.5

17. Nerode, A., Shore, R.A.: Logic for Applications. Graduate Text in Computer Science, Springer, New York, 2nd edn. (2012). https://doi.org/10.1007/978-1-4612-0649-1

18. Nipkow, T.: Linear quantifier elimination. J. Autom. Reasoning **45**, 189–212 (2010). https://doi.org/10.1007/s10817-010-9183-0

19. Rado, R.: Note on the transfinite case of Hall's theorem on representatives. London Math. Soc. **S1–42**(1), 321–324 (1967). https://doi.org/10.1112/jlms/s1-42.1.321

20. Reichmeider, P.F.: The Equivalence of some Combinatorial Matching Theorems. Polygonal Publishing House, Washington (1985)

21. Romanowicz, E., Grabowski, A.: The hall marriage theorem. Formalized Math. (University of Białystok) **12**(3), 315–320 (2004). https://fm.mizar.org/2004-12/pdf12-3/hallmar1.pdf

22. Rudnicki, P.: Dilworth's decomposition theorem for posets. Formalized Math. **17**(4), 223–232 (2009). https://doi.org/10.2478/v10037-009-0028-4

23. Serrano, F.F.: Formalización en Isar de la Meta-Lógica de Primer Orden. Ph.D. thesis, Departamento de Ciencias de la Computación e Inteligencia Artificial, Universidad de Sevilla, Spain (2012). https://idus.us.es/handle/11441/57780. (in Spanish)

24. Singh, A.K.: Formalization of some central theorems in combinatorics of finite sets. arXiv abs/1703.10977[cs.Lo] (2017). https://doi.org/10.48550/arxiv.1703.10977, short presentation at the 21st International Conference on Logic for Programming, Artificial Intelligence and Reasoning - LPAR-21

25. Singh, A.K., Natarajan, R.: Towards a constructive formalization of perfect graph theorems. In: Khan, M.A., Manuel, A. (eds.) ICLA 2019. LNCS, vol. 11600, pp. 183–194. Springer, Heidelberg (2019). https://doi.org/10.1007/978-3-662-58771-3_17

26. Singh, A.K., Natarajan, R.: A constructive formalization of the weak perfect graph theorem. In: Proceedings of the 9th ACM SIGPLAN International Conference on Certified Programs and Proofs - CPP, pp. 313–324. ACM (2020). https://doi.org/10.1145/3372885.3373819

Graded Rings in Lean's Dependent Type Theory

Eric Wieser[1]([⊠])[iD] and Jujian Zhang[2][iD]

[1] Cambridge University Engineering Department, Cambridge, UK
efw27@cam.ac.uk
[2] Imperial College London, London, UK
jujian.zhang19@imperial.ac.uk

Abstract. In principle, dependent type theory should provide an ideal foundation for formalizing graded rings, where each grade can be of a different type. However, the power of these foundations leaves a plethora of choices for how to proceed with such a formalization. This paper explores various different approaches to how formalization could proceed, and then demonstrates precisely how the authors formalized graded algebras in Lean's mathlib. Notably, we show how this formalization was used as an API; allowing us to formalize various graded structures such as those on tuples, free monoids, tensor algebras, and Clifford algebras.

Keywords: Graded rings · Dependent types · Formalization · mathlib

1 Introduction

One way to introduce graded rings and algebras is by noting that they generalize an early staple in mathematics education; that of single-variate polynomials in X. Any polynomial can be written as a (finite) weighted sum of powers of X, and multiplication only requires the knowledge that $X^m X^n = X^{m+n}$.

If we define the $\mathbb{N}$-indexed family of homogeneous polynomials $A = (i \mapsto \{aX^i \mid a : R\})$, then we can say "the polynomials in a ring R over X, $R[X]$ are an algebra graded by A"[1]; by which we mean:

1. Each of the elements of the family A_i are closed under addition and scalar multiplication by elements of R.
2. There is a $1 \in A_0$.
3. For any $p \in A_i$ and $q \in A_j$, we have $pq \in A_{i+j}$. Equivalently, as sets $A_i A_j \subseteq A_{i+j}$.
4. Every element p can be expressed uniquely as $p = \sum_i p_i$ where $p_i \in A_i$.

---

[1] Or "a graded algebra of type $\mathbb{N}$ over the ring R with graduation A" in the language of [6, III, §3, 1.].

K. Buzzard and T. Kutsia (Eds.): CICM 2022, LNAI 13467, pp. 122–137, 2022.
https://doi.org/10.1007/978-3-031-16681-5_8

The above acts as a general definition of an algebra graded by some arbitrary family of submodules A, which can in general be indexed by any additive monoid ι, not just the natural numbers.

To build some intuition for this generalization, it is worth enumerating some other examples:

Multivariate polynomials, $R[X, Y, \ldots]$. Over two variables we can grade either by the $\mathbb{N}$-indexed family of elements of homogeneous degree $A = (i \mapsto \{aX^jY^{i-j} \mid a : R, j \leq i\})$ where X^3 and XY^2 e.t.c. have the same grade, or by a $\mathbb{N} \oplus \mathbb{N}$-indexed family on the individual variables, $A = ((i, j) \mapsto \{aX^iY^j \mid a : R\})$.

The tensor algebra, $\mathcal{T}(V)$. Conventionally we grade this by the $\mathbb{N}$-indexed family where A_i spans the ith tensor powers $V^{\otimes i}$.

The exterior algebra, $\bigwedge(V)$. The exterior algebra is graded in exactly the same way, but when V is of dimension n we find that A_i for $n < i$ is the trivial submodule.

The Clifford algebra, $\mathcal{C}\ell(V, Q)^2$. We cannot[3] use exactly the same approach for the Clifford algebra, as for a vector v, we have $v^2 = Q(v)$, where the LHS would be of grade 2 and the RHS would be of grade 0. This can be resolved by having just two grades; one corresponding to sums of "even" monomials (those which are a product of an even number of elements of V), and one corresponding to sums of odd monomials. Phrased another way, the family is indexed by[4] $\mathbb{Z}/2\mathbb{Z}$, the integers modulo two.

Any ring α. Any ring can be equipped with the trivial grading structure, where the index type contains only one element 0 corresponding to the entire ring.

As this is a paper about formalization, we will predictably proceed by finding all the different ways to "pull legs off"[5] this definition. By relaxing items 2 and 3, we can talk about about gradings of additive monoids, additive groups ([6, II, §11, 1.]), and R-modules[6]. By relaxing item 1, we can additionally talk about gradings by families of additive subgroups, additive submonoids, or even just sets; which we refer to as graded rings, graded semirings, and graded monoids respectively. For graded monoids (such as the n-tuples α^n or tensor powers $M^{\otimes n}$) there is no summation, so item 4 is interpreted as the statement that p must belong to exactly one A_i. Figure 1 outlines the connection between these various generalizations.

While the existence of many examples following the same pattern is already a good reason to formalize that pattern, it is only half the picture; just as impor-

---

[2] Where $\mathcal{C}\ell(V, Q)$ is notation to specify the quadratic form Q and vector space V.

[3] At least, when $Q \neq 0$. If $Q = 0$ then $\mathcal{C}\ell(V, Q) = \bigwedge(V)$ and we can proceed as above.

[4] Note that literature referring to an $\mathbb{N}$-grading is referring to the grading on $\bigwedge(V)$ via the canonical module equivalence.

[5] https://en.wikipedia.org/wiki/Centipede_mathematics.

[6] Some sources use a more general definition of graded R-modules and R-algebras, where R is itself a graded ring such that $R_iM_j \subseteq M_{i+j}$. For brevity we will not discuss these here (in essence considering only the special case when R has the trivial graduation), but our approach would extends to this straightforwardly .

tant is to have situations where the generalization itself is necessary. For instance, without a formalization of commutative additive monoids, we can't even define what it means to take the sum of a finite set, and would instead be forced to repeat this definition for ℕ, ℤ, etc. A particularly motivating example for need a formalization of graded rings is that of the Proj S construction in algebraic geometry [12, Tag 01M3], a definition which requires a notion of homogeneous ideals, which in turn requires precisely the notion of graded rings this paper is about. Another recent motivation appeared in the "Liquid Tensor Experiment"[8], which required a proof of Gordan's lemma; one proof of which goes via graded rings[7].

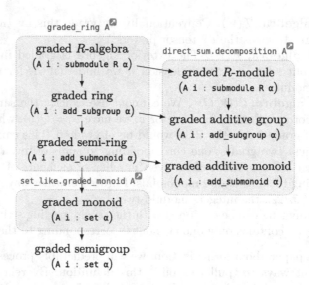

Fig. 1. The algebraic hierarchy of graded objects discussed in this paper. The meanings of the typeclasses introduced in Sect. 3 for grading α internally by A are shown as labelled gray regions, with the objects representing each internal grade (that is, the type of A_i) shown in parentheses.

1.1   Prior Formalizations

In Coq, [9, 3.] refers to graded modules as `nat -> FreeModule R` in the context of homology. No mention of graded multiplicative structures appears. A similar formalization of graded modules exists for Lean 2 in [2, `algebra/graded.hlean`], although the language has changed substantially since then, in particular dropping the experimental HoTT mode which [2] builds upon.

In Agda, [7] shows that the axioms of a commutative graded ring are satisfied by a particular construction, the graded cohomology groups. No attempt seems

---

[7] In the absence of our work a proof via convexity was used instead.

to be made to provide a general definition of what it means for an object to satisfy those axioms.

Extensive discussion about formalizing "Commutative Differential Graded algebras", algebraic objects with some additional axioms on top of the graded algebras discussed in this paper, has taken place on the Lean Zulip Chat [5]. While these discussions refer to a current version of Lean 3 [10], the ideas explored in [5] never resolved into a contribution to Lean's monolithic mathematics library mathlib [11]. It is by this metric that [2,9] are substantially different in scope to this work; in those formalizations, a definition was chosen for the particular use case of interest to the authors, with little regard for interoperability with large amounts of existing code. Conversely, one of the reasons the work in [5] never made it into mathlib is likely the lack of applications to verify that the design is the right one. Since the closure of that thread, mathlib has grown by a factor of five in terms of total lines, and has gained formalizations of many new objects of interest to us: tensor algebras, Clifford algebras, and tensors powers.

1.2   Additive Vs Multiplicative Grading

Note that some work uses the convention that multiplication in item 3 behaves as $A_i \rightarrow A_j \rightarrow A_{ij}$ instead of $A_i \rightarrow A_j \rightarrow A_{i+j}$. For simplicity, this paper and mathlib only develop the additive version, as this has more pre-existing applications in mathlib. Using additive ι as the index type where ι is a multiplicative monoid allows the former to be expressed in the language of the latter, so if we were to have both versions it would only be for convenience.

2   External Gradings

There are two ways to think about a grading in dependent type theory; either as a family of sets of a single type (internal), or as an indexed family of distinct types. There are merits to both approaches; which is most useful depends on whether it is more natural to define the single type then break it into pieces (as with the monoid under concatenation of lists graded by their length), or to define the family of types then glue them together (as with the monoid under concatenation of the tuples fin n → α graded by n). A crucial factor in the coherence of mathlib as a unified library is its ability to translate between multiple ways of stating the same thing, so we do not want to have to choose between these approaches in an exclusive manner. Thankfully, the former approach can be represented via the latter; an internal grading can be written as an external grading over the family of subtypes corresponding to each grade, shown in parentheses in Fig. 1. We will revisit this equivalence in Sect. 3.

It is worth remembering that when building externally-graded objects in this way, that the grades are disjoint by definition. If for example our indexed family of types is λ i : N, R (a family indexed by the naturals, all equal to the same ring), then this is viewed as a countable sequence of *copies* of R, which makes this construction exactly analogous to the single-variate polynomials.

2.1   Graded Semigroups

Let us now try to develop the framework for talking about externally-graded semigroups[8] over a family of types A indexed by an additive semigroup ι. We would like to be able to express these via Lean's "typeclasses", as this matches how the usual non-graded algebraic structure is expressed. This means that to talk about a graded monoid, a user might write:

```
def sq {ι : Type*} {A : ι → Type*} [add_semigroup ι] [g_semigroup A]
  (i : ι) (x : A i) : A (i + i) :=
g_semigroup.mul x x
```

To explain this syntax briefly; sq names the definition, {name : type} and (name : type) introduce implicit and explicit variables, [type] introduces a typeclass variable, the trailing : prefixes the result type, and := prefixes the value of the definition. Typeclass variables are special; the [add_semigroup ι] variable is used to define the meaning of i + i via mathlib's algebra framework, while the g_semigroup.mul would be defined by the [g_semigroup A] variable. A user calling this sq function might write sq _ x, where _ acts as a wildcard which Lean works out automatically by looking at x. The same mechanism is used to infer the implicit ι and A arguments, but typeclass search is used to populate the two arguments in square brackets; for instance, if ι := N then Lean finds nat.add_semigroup : add_semigroup N . More thorough introductions to typeclasses in Lean and mathlib can be found in [4, §2] and [13, §1.1].

Attempting to define a new g_semigroup A typeclass by directly writing down item 3 and a suitable associativity axiom as

```
variables {ι : Type*} (A : ι → Type*)

class g_semigroup [add_semigroup ι] :=
(mul {i j} : A i → A j → A (i + j))
(mul_assoc {i j k} (x : A i) (y : A j) (z : A k) :
    mul (mul x y) z = mul x (mul y z))
```

leads to the error "term mul x (mul y z) has type A (i + (j + k)) but is expected to have type A ((i + j) + k) ". While these are "obviously" equal, that's not enough for Lean; for the statement to type-check, we need the types to be *definitionally* equal. We would have similar problems with A (i + 0) and A i if we were trying to prove x one = x for a graded monoid. To escape this problem, we could:

1. Use heterogenous equality (denoted ==), which allows us to express equality between distinct types:

```
class g_semigroup [add_semigroup ι] :=
(mul {i j : ι} : A i → A j → A (i + j))
(mul_assoc {i j k : ι} (x : A i) (y : A j) (z : A k) :
  mul (mul x y) z == mul x (mul y z))
```

2. Express the equality in terms of sigma types or dependent pairs, denoted Σ i, A i: :

---

[8] Chosen for brevity due to having the fewest axioms, not because they are interesting.

```
class g_semigroup [add_semigroup ι] :=
(mul {i j : ι} : A i → A j → A (i + j))
(mul_assoc {i j k : ι} (x : A i) (y : A j) (z : A k) :
  ((_, mul (mul x y) z) : Σ i, A i) = (_, mul x (mul y z)))
```

3. Express the grading constraint as an equality on sigma types:

```
class g_semigroup [add_semigroup ι] extends semigroup (Σ i, A i) :=
(fst_mul {i j : ι} (x : A i) (y : A j) :
  ((_, x) * (_, y) : Σ i, A i).fst = i + j)
```

4. Provide an explicit proof that the equality is type correct using the recursor for equality, eq.rec:

```
class g_semigroup [add_semigroup ι] :=
(mul {i j : ι} : A i → A j → A (i + j))
(mul_assoc {i j k : ι} (x : A i) (y : A j) (z : A k) :
  (add_assoc i j k).rec (mul (mul x y) z) = mul x (mul y z))
```

5. Store a canonical map between objects of the "same" grade to use instead of using eq.rec, to allow better definitional control:

```
class g_semigroup [add_semigroup ι] :=
(cast {i j : ι} (h : i = j) : A i → A j)
(cast_rfl {i} (x : A i) : cast rfl x = x)
(mul {i j : ι} : A i → A j → A (i + j))
(mul_assoc {i j k : ι} (x : A i) (y : A j) (z : A k) :
  cast (add_assoc i j k) (mul (mul x y) z) = mul x (mul y z))
```

6. Take an additional index into mul and a proof that it is equal to i + j: :

```
class g_semigroup [add_semigroup ι] :=
(mul {i j k : ι} (h : i + j = k) : A i → A j → A k)
(mul_assoc {i j k ij jk ijk : ι}
  (hij : i + j = ij) (hjk : j + k = jk)
  (hi_jk : i + jk = ijk) (hij_k : ij + k = ijk)
  (x : A i) (y : A j) (z : A k) :
    (mul hij_k (mul hij x y) z) = mul hi_jk x (mul hjk y z))
```

Many of these options are derived from the discussions in [5]. When deciding between these options, we need to consider both the ease of providing an instance with **instance : g_semigroup A**, and the ease of consumer working with one using [g_semigroup A]. Note that it is straightforward to expose different interfaces to the consumer and producer, and provide a layer of translation in between. This is especially the case when the interfaces differ only in their statement of the propositional fields. Table 1 outlines a rough comparison between these approaches.

Taking a step back from this problem, we also need to decide on a spelling for the consumer, as writing mul instead of a multiplication symbol is hardly pleasant. There are essentially two options here: either introduce new notation for our graded mul, or hook into the existing * notation. The latter is a far more appealing option, as it means we can reuse all the lemmas we have about * by providing the appropriate algebraic typeclasses. The only catch is that the existing * notation requires the operation to be homogeneous; acting on a single type, rather than three elements of a family[9].

Table 1. Merits of the various approaches to defining g_semigroup. "Producer" refers to the code providing the `instance : g_semigroup A`, while "consumer" refers to the code with a `[g_semigroup A]` argument.

Approach	item 1 `==`	item 2 `Σ i, A i`	item 3 `extends`	item 4 `eq.rec`	item 5 `cast`	item 6 `h : i+j=k`
h : (i+j)+k=i+(j+k) needed by	producer	producer	—	—	—	consumer
Estimated difficulty for the producer	medium	harder	easier	harder	easier	medium
Directions of consumer rw tactic use	0	2	2	1	1	2

To achieve this homogenization, we can use the builtin `sigma` type of dependent pairs, storing the grade of the monoid alongside the value at that grade such that `x : A i` is represented by `sigma.mk i x : Σ i, A i`.

```
instance g_semigroup.to_semigroup [add_semigroup ι] [g_semigroup A] :
  semigroup (Σ i, A i) :=
{ mul := λ (x y : Σ i, A i), (x.fst + y.fst, g_semigroup.mul x.snd y.snd,
  mul_assoc := λ (x y : Σ i, A i), sorry }
```

If we choose item 3 from Table 1, then this code is generated for us automatically! However, the fact that the grade of the multiplication is known only propositionally and not definitionally can make things harder in Sect. 2.3 so we avoid this choice. As the next most appealing option, choosing item 2 makes the `sorry` above[10] fall out immediately, and so this is what `mathlib` does. This decision is far from final, but the best way to compare the option of Table 1 is to thoroughly implement one of them, and then come back and see whether changing the definition to something different makes the existing proofs better or worse.

2.2   Graded Monoids

In reality, we do not define `gsemigroup` at all in `mathlib`, and jump straight to graded monoids due to lack of need for the former. We also don't actually put the instance on the `sigma` type, as this would not be a sufficiently canonical choice to be worthy of a global instance. Instead, we define `graded_monoid A` as an alias for `sigma A`, and place the instances on that. By splitting apart the typeclasses a little and interleaving the instances for `graded_monoid A`:

---

[9] Lean 4 lifts this notation restriction, but the algebraic typeclasses provided by mathlib would need reworking.

[10] The syntax in Lean for an incomplete proof.

```
class ghas_one [has_zero ι] := (one : A 0)
class ghas_mul [has_add ι] := (mul {i j} : A i → A j → A (i + j))

instance ghas_one.to_has_one [has_zero ι] [ghas_one A] :
   has_one (graded_monoid A) := { one := (_, ghas_one.one) }
instance ghas_mul.to_has_mul [has_add ι] [ghas_mul A] :
   has_mul (graded_monoid A) := { mul := λ x y, (_, ghas_mul.mul x.snd y.snd) }
```

we make the notation for the instance much more pleasant

```
class gmonoid [add_semigroup ι] extends ghas_one A, ghas_mul A :=
(one_mul (a : graded_monoid A) : 1 * a = a)
(mul_one (a : graded_monoid A) : a * 1 = a)
(mul_assoc (a b c : graded_monoid A) : a * b * c = a * (b * c))
```

It is worth remembering that while this may look identical to the definition of a regular monoid, it is constraining the grade-preserving behavior of the multiplication by construction. As is always the case with formalization, it is never quite as simple as you would hope it would be. In fact, the definition[2] of gmonoid in mathlib contains three additional fields!

```
(gnpow : Π (n : N) {i}, A i → A (n • i))
(gnpow_zero' : Π (a : graded_monoid A), graded_monoid.mk _ (gnpow 0 a.snd) = 1)
(gnpow_succ' : Π (n : N) (a : graded_monoid A),
   (graded_monoid.mk _ $ gnpow n.succ a.snd) = a * (_, gnpow n a.snd))
```

These describe the power operator by the natural numbers, and ensure that its grade too is known definitionally following the "forgetful inheritance"[1] pattern, the relevance of which is explored in [13, §5].

Example 1 (the n-tuples). This typeclass allows us to express the graded monoid structure of the *n*-tuples under concatenation, as[2]

```
instance : gmonoid (λ n : N, fin n → α) :=
{ one := ![], -- the empty tuple
  mul := λ i j a b, fin.add_cases a b,
  ..sorry /- boring proofs -/ }
```

2.3 Graded (semi)rings

For graded rings, we do not run into any new equality problems, as addition remains within a grade. To state the requirement for a family of types to represent a graded ring, we can simply extend the monoid structure from earlier:[2]

```
class gsemiring [add_monoid ι] [Π i, add_comm_monoid (A i)]
   extends gmonoid A :=
(mul_zero : ∀ {i j} (a : A i), mul a (0 : A j) = 0)
(zero_mul : ∀ {i j} (b : A j), mul (0 : A i) b = 0)
(mul_add : ∀ {i j} (a : A i) (b c : A j), mul a (b + c) = mul a b + mul a c)
(add_mul : ∀ {i j} (a b : A i) (c : A j), mul (a + b) c = mul a c + mul b c)
-- For "forgetful inheritance" like the previous `gnpow` field
(nat_cast : N → A 0) (nat_cast_zero : sorry) (nat_cast_succ : sorry)
```

We *almost* do not need any axioms about negation; to work with a graded ring as opposed to a graded semiring, the user could write [Π i, add_comm_group (A i)] [gsemiring A]. Unfortunately, our hand is forced by "forgetful inheritance" to define gring[✎] anyway in order to add an int_cast operation and associated axioms.

Just as we used graded_monoid A in Sect. 2.1 to bundle our graded monoid with its grade to enable reuse of the monoid API, we'd like to be able to enable reuse of the semiring API on graded semirings. We cannot use graded_monoid A here, as in a graded ring an element can consist of distinct grades added together; instead we use direct_sum ι A, with notation ⊕ i, A i. Here, the element x : A i is represented by direct_sum.of A i x. This comes with all the additive structure we need already; all we have to do is extend our multiplicative structure onto it linearly, with our end goal being to produce a (⊕ i, A i) instance.

To do this, we first promote our mul to a bundled homomorphism [11, §4.1.2] that is additive in each argument[✎]

```
def gmul_hom [gsemiring R] {i j} : A i →+ A j →+ A (i + j) :=
{ to_fun := λ a,
  { to_fun := λ b, gsemiring.mul a b,
    map_zero' := gsemiring.mul_zero _,
    map_add' := gsemiring.mul_add _ },
  map_zero' := add_monoid_hom.ext $ λ a, gsemiring.zero_mul a,
  map_add' := λ a₁ a₂, add_monoid_hom.ext $ λ b, gsemiring.add_mul _ _ _}
```

as this allows us to lift this map to consume and produce elements of the direct sum as:[✎]

```
def mul_hom : (⊕ i, A i) →+ (⊕ i, A i) →+ ⊕ i, A i :=
direct_sum.to_add_monoid $ λ i,
  add_monoid_hom.flip $ direct_sum.to_add_monoid $ λ j, add_monoid_hom.flip $
    (direct_sum.of A _).comp_hom.comp $ gmul_hom A
```

Unfortunately working with bundled maps in mathlib forces you to write things in the rather unreadable point-free style as above. The benefit of working in a theorem prover is that we can at least verify that something unreadable still behaves as we want:[✎]

```
lemma mul_hom_of_of {i j} (a : A i) (b : A j) :
  mul_hom A (of _ i a) (of _ j b) = of _ (i + j) (gsemiring.mul a b) := sorry
```

It might feel like we're on the home stretch to providing the semiring instance; indeed, the proofs relating multiplication and the additive structure follow trivially from mul_hom. Unfortunately, we're now faced with actually using the API we built in Sect. 2.2 to prove the multiplicative properties! The key result we need to do this is that our two notions of equality are equivalent; that is,[✎]

```
lemma of_eq_of_graded_monoid_eq {i j : ι} {a : A i} {b : A j}
  (h : graded_monoid.mk i a = graded_monoid.mk j b) :
  direct_sum.of A i a = direct_sum.of A j b := sorry
```

After once again fighting against point-free nonsense to turn associativity into equality of two tri-additive maps, we can use the ext tactic to reduce our problem to associativity of three terms of the form of A i x:[✎]

```
private lemma mul_assoc (a b c : ⊕ i, A i) : a * b * c = a * (b * c) :=
-- `λ a b c, a * b * c = λ a b c, a * (b * c)` as bundled homomorphisms
suffices (mul_hom A).comp_hom.comp (mul_hom A)
      = (add_monoid_hom.comp_hom flip_hom $
          (mul_hom A).flip.comp_hom.comp (mul_hom A)).flip,
  from fun_like.congr_fun (fun_like.congr_fun (fun_like.congr_fun this a) b) c,
begin
  ext ai ax bi bx ci cx : 6,
  show mul_hom A (mul_hom A (of A ai ax) (of A bi bx)) (of A ci cx) =
      mul_hom A (of A ai ax) (mul_hom A (of A bi bx) (of A ci cx)),
```

from which the rest follows using our previous results:

```
  rw [mul_hom_of_of, mul_hom_of_of, mul_hom_of_of, mul_hom_of_of],
  exact of_eq_of_graded_monoid_eq
    (mul_assoc (graded_monoid.mk ai ax) (bi, bx) (ci, cx)),
end
```

A similar approach can be used to prove the mul_one and one_mul fields in order to finish the construction of semiring (⊕ i, A i). The "point-free nonsense" approach is not the only path available to us; but the alternative of using induction creates annoying side-goals to prove that the map is additive.

There is another important construction we will want for working with this direct sum representation of a graded ring; a way to construct a ring homomorphism out of the graded ring given a suitable family of homomorphisms on each piece. This can be written as:[ ]

```
def direct_sum.to_semiring
  (f : Π {i}, A i →+ R)
  (hone : f gsemiring.one = 1)
  (hmul : ∀ {i j} (ai : A i) (aj : A j), f (gsemiring.mul ai aj) = f ai * f aj) :
  (⊕ i, A i) →+* R :=
{ map_one' := sorry, map_mul' := sorry, .. direct_sum.to_add_monoid f }
```

We will find this instrumental for relating external and internal direct sums in Sect. 3.4.

Example 2 (the nth tensor powers). This typeclass allows us to express the graded ring structure of the nth tensor power over an R-module V as[ ]

```
instance : gsemiring (λ n : N, ⊗[R]^n V) := sorry
```

We can then show with the aid of direct_sum.to_semiring that $\mathcal{T}(V)$ is isomorphic as a ring (and algebra) to $\bigoplus_n V^{\otimes n}$; that is tensor_algebra R V ≃ₐ[R] (⊕ n, ⊗[R]^n V)[ ] .

3   Internal Gradings

3.1   Decompositions of Sets

For an external decomposition with no algebraic structure, the task is simple; a unique decomposition of a type α into its pieces A : ι → Type* can be spelled as the equivalence to a sigma type, decompose : α ≃ Σ i : ι, A i. For an internal decomposition where `A : ι → set α`, we have some other options. If we don't care about a constructive decomposition and are happy with a classical one, we can just state that the components span the entire type and are disjoint, as:

```
(⋃ i, A i) = set.univ ∧ pairwise (disjoint on A)
```

If we do care about constructiveness, we can instead have a function `grade : α → ι` that respects `∀ (a : α) (i : ι), a ∈ A i → grade a = i`. Whichever approach we pick, it is straightforward to recover the externally-graded viewpoint via the map `decompose : α ≃ Σ i : ι, ↥(A i)`. Here, `↥` is the operator that lets us view a set as a subtype of the type of its elements.

3.2 Graded Monoids

A preliminary attempt at formalizing an internal multiplicative grading structure might look like

```
class set.graded_monoid [monoid M] [add_monoid ι] (A : ι → set M) : Prop :=
(one_mem : 1 ∈ A 0)
(mul_mem : ∀ {i j : ι} {gi gj : M}, gi ∈ A i → gj ∈ A j → gi * gj ∈ A (i + j))
```

This works fine for graded monoids; but for graded semirings, rings, and algebras we need to apply the additional constraints that each `A i` is closed under the appropriate operations.

   To avoid having to write separate typeclasses for each case and ending up with our API in triplicate, we instead generalize over `A : ι → set M`; with the goal being able to talk about `A : ι → add_submonoid R`, `A : ι → add_subgroup R`, and `A : ι → submodule S R` in a unified way. As well as avoiding the need for three different typeclasses, this also means we can reuse all the theory we already have about `add_submonoid`, `add_subgroup`, and `submodule`. To do that, we introduce a `set_like` class, which was one of the inspirations for the `fun_like` generalization described in [4, §6.3]. This class lets us express that elements `s` of a type `S` has a canonical intepretation as a set `↑s : set α`, and is defined as:[6]

```
class set_like (S : Type*) (α : out_param Type*) :=
(coe : S → set α)    -- the function that is used for `↑` coercion notation
(coe_injective' : function.injective coe)
```

This equips `s : S` with membership notation `a ∈ s : Prop` and a coercion to type `↑s : Type*`, and permits us to write[7]

```
class set_like.graded_monoid {ι M S : Type*}
  [set_like S M] [monoid M] [add_monoid ι] (A : ι → S) : Prop :=
(one_mem : 1 ∈ A 0)
(mul_mem : ∀ {i j : ι} {gi gj : M}, gi ∈ A i → gj ∈ A j → gi * gj ∈ A (i + j))
```

At this point, we can deliver on the earlier claim that the internal viewpoint can be expressed via the external viewpoint. To do this, we show that the family of subtypes `λ i, ↥(A i)` has a graded multiplicative structure, as:[8]

```
-- this implies `monoid (graded_monoid (λ i, ↥(A i)))` via `gmonoid.to_monoid`
instance set_like.gmonoid [set_like S M] [monoid M] [add_monoid ι]
  (A : ι → S) [set_like.graded_monoid A] :
  gmonoid (λ i, ↥(A i)) :=
{ one := ⟨1, set_like.graded_monoid.one_mem,
  mul := λ i j a b, ⟨(a * b : R), set_like.graded_monoid.mul_mem a.prop b.prop⟩,
  mul_assoc := λ ⟨i, a, ha⟩ ⟨j, b, hb⟩ ⟨k, c, hc⟩,
    sigma.subtype_ext (add_assoc _ _ _) (mul_assoc _ _ _),
  ..sorry /- etc -/ }
```

Note here that we are paying the cost outlined in the first row of Table 1 of having to reprove associativity of addition of the grades, which is a mark against our choice of item 2.

Example 3 (the free monoid over α). This typeclass allows us to express[11] the internal graded monoid structure of the free monoid, with elements graded by the number of generators

```
instance :
  set_like.graded_monoid
    (λ i : N, (set.range (free_monoid.of : α → free_monoid α)) ^ i) :=
{ one_mem := by rw [pow_zero, set.mem_one],
  mul_mem := λ i j x y hx hy, by { rw pow_add, exact set.mul_mem_mul hx hy} }
```

We are not quite done yet; we have shown that the subtypes can be glued together to form an object with graded multiplication, but the `set_like.graded_monoid` typeclass above does nothing to ensure that the glued-together type `graded_monoid (λ i, ↥(A i))` is in bijection with M. We can reuse either the classical or constructive approach from Sect. 3.1, but with our new `monoid (graded_monoid (λ i, ↥(A i)))` instance we can promote the equivalence `decompose : α ≃ Σ i : ι, ↥(A i)` to a multiplicative isomorphism, stated as `decompose : α ≃* graded_monoid (λ i, ↥(A i))`.

3.3 Decompositions of Additive Monoids and *R*-Modules

For a unique decomposition with an additive structure, we cannot use the same approach as Sect. 3.1 but instead need to decompose into a direct sum, as `decompose : α ≃+ ⊕ i : ι, ↥(A i)`. Let us consider the internal decomposition of an additive group α into the family `A : ι → add_subgroup α`. We need to be a little more careful when describing the disjointness condition, as what we actually require is that every component is disjoint (in the sense of having trivial intersection {0}) from the span of all the others. We can spell that as

```
(⨆ i, A i) = ⊤ ∧ complete_lattice.independent A
```

but for additive submonoids this condition, while necessary, is still not sufficient; consider when $A_+ = \{z : \mathbb{Z} \mid 0 \leq z\}$ and $A_- = \{z : \mathbb{Z} \mid 0 \geq z\}$, which are disjoint and span all of $\mathbb{Z}$, but clearly do not permit a decomposition . As such, we

---

[11] After enabling the appropriate by-default-disabled instances.

cannot use this as our definition. Instead, we require that the canonical map from $\oplus$ i : ι, A i to α (defined as roughly λ x, Σ i, ι(x i)) is bijective.

Once again we're on the precipice of stating things in triplicate, as we want to state this condition (and the consequences of it) for add_monoid α and submodule R α as well to cover the cases on the right of Fig. 1. Until very recently, stating this condition in triplicate was exactly what mathlib did; but thanks to [3] which transfers the success in [4] from fun_like to set_like, we can now generalize over add_subgroup α as S where (S : Type*) [set_like S α] [add_submonoid_class S α]. This lets us define a single canonical map coe_add_monoid_hom[■] that works for all three cases as

```
protected def coe_add_monoid_hom [add_comm_monoid α]
  [set_like S α] [add_submonoid_class S α] (A : ι → S) :
  (⊕ i, A i) →+ M :=
direct_sum.to_add_monoid (λ i, add_submonoid_class.subtype (A i))
```

which in turn allows us to state our condition just once to cover all three cases. We can state it either constructively, by carrying around an explicit inverse[■]:

```
class decomposition (A : ι → S) :=
(decompose' : α → ⊕ i, A i) -- split elements into their grades
(left_inv : function.left_inverse (coe_add_monoid_hom A) decompose')
(right_inv : function.right_inverse (coe_add_monoid_hom A) decompose')
```

or classically by simply proving bijectivity[■] . We provide a proof in mathlib that on submodules over additive groups the complete_lattice.independent formulation is equivalent[■] to these definitions.

3.4   Graded (semi)rings

To talk about a semiring with an internally grade-compatible multiplication, we thankfully need to define no further typeclasses; we can write

```
variables {ι R S : Type*} [add_monoid ι] [semiring R] [set_like S R]
variables [add_submonoid_class S R] (A : ι → S) [set_like.graded_monoid A]
```

where set_like.graded_monoid handles our conditions on the multiplicative structure, add_submonoid_class handles our conditions on the additive structure, and semiring R ensures the compatibility between the two. We'd like to end up with a gsemiring (λ i, ι(A i)) instance as a result of these hypotheses so as to also have a semiring (⊕ i, ι(A i)) instance, which we achieve as[■]

```
instance set_like.gsemiring : direct_sum.gsemiring (λ i, ι(A i)) :=
{ mul_zero := λ i j _, subtype.ext (mul_zero _),
  zero_mul := λ i j _, subtype.ext (zero_mul _),
  mul_add := λ i j _ _ _, subtype.ext (mul_add _ _ _),
  add_mul := λ i j _ _ _, subtype.ext (add_mul _ _ _),
  ..set_like.gmonoid A }
```

Once again, the introduction of add_submonoid_class excused us from needing three copies of this definition.

Now that we have this instance, we can build the canonical ring morphism from $\oplus$ i, ι(A i) to R that amounts to summing the elements from each grade (after passing them through the canonical additive morphism from the subtype), building upon the `direct_sum.to_semiring` definition at the end of Sect. 2.3:

```
def direct_sum.coe_ring_hom [add_monoid ι] [semiring R] [set_like S R]
  [add_submonoid_class S R] (A : ι → S) [set_like.graded_monoid A] :
  (⊕ i, ↥(A i)) →+* R :=
direct_sum.to_semiring
  (λ i, add_submonoid_class.subtype (A i)) rfl (λ _ _ _ _, rfl)
```

The `direct_sum.coe_add_monoid_hom` we mention in Sect. 3.3 has a definitionally equal underlying function to this, meaning we can reuse the `decomposition` A from that section defined in terms of the former to obtain the canonical ring isomorphism `decompose_ring_equiv` : R ≃+* ⊕ i, ↥(A i) between an internally-graded ring R and the direct sum of its grades ⊕ i, ↥(A i). For convenience, we provide a single typeclass that provides access to this operation:

```
class graded_ring (A : ι → σ) extends graded_monoid A, decomposition A.
```

Among other functions and lemmas which build on this convenience, we provide projection maps as additive maps `A i : R →+ R` so that any $x : R$ can be written as $x = \sum_i x_i$, with x_i being the i-th projection of x with respect to grade A, without having to explicitly go via the direct sum of subtypes ⊕ i, ↥(A i).

4   Graded R-Algebras

This paper would not be complete without building external and internal graded R-algebras on top of the graded rings; indeed, we have added definitions of these to mathlib, as the typeclass `galgebra` , the instance `submodule.galgebra` , and the shorthand `graded_algebra` ; but the process of defining these presented no new challenges over those already faced when defining graded rings. Just as we were able to recover a ring isomorphism in Sect. 3.4, these definitions let us recover the analogous R-algebra isomorphism `decompose_alg_equiv` : X ≃ₐ[R] ⊕ i, ↥(A i) .

Example 4 (the multivariate polynomials). Returning to the examples in Sect. 1, we can define the homogeneous graduation :

```
instance : graded_algebra (λ i : ℕ, homogeneous_submodule σ R i) := sorry
```

where σ represents the variables in the multivariate polynomial ring and `homogeneous_submodule σ R i` the homogeneous polynomials of degree i.

Example 5 (the tensor algebra $\mathcal{T}(V)$). Given ι R as the canonical map $V \to \mathcal{T}(V)$, we write the typical internally grading as:

```
instance : graded_algebra
  ((^) (ι R : M →ₗ[R] tensor_algebra R M).range : ℕ → submodule R _) := sorry
```

Example 6 (the Clifford algebra $\mathcal{C}\ell(V,Q)$). Similarly, we can write[12]

---

[12] Thus resolving the further work in [14, §8.1].

```
def even_odd (i : zmod 2) : submodule R (clifford_algebra Q) :=
⨆ (j : {n : N // ↑n = i}), (ι Q).range ^ (j : N)
instance : graded_algebra (clifford_algebra.even_odd Q) := sorry
```

where ι Q is a similar canonical map, and even_odd 0 and even_odd 1 are the even and odd submodules respectively.

5   Conclusions

This paper outlined the substantial amount of busy-work required to define the various types of graded structure in mathlib, and the various design choices made along the way. Instead of choosing between internal and external grading, we opted to use the latter to implement the former. When it comes to classical vs constructive decompositions, we opted to have both, just like mathlib has classical vs constructive finite sets. Finally, for stating equality between non-definitionally-equal grades, we opted to use sigma types.

To verify that our design decisions are sensible, we demonstrated a variety of results stated using our new typeclasses. While not discussed in detail in this paper, our supplemental repository includes an extended example by the second author of a preliminary development of the Proj construction in algebraic geometry⬚ , which using the result from example 4 culminates in defining the projective n-space⬚ . The success of this formalization indicates that the design of graded objects is coherent with other theories in mathlib.

A snapshot of the unabridged code from this work, permalinked throughout via "⬚ ", is available at https://github.com/eric-wieser/lean-graded-rings. It comprises around 1150 sloc[13] of API development, 1250 sloc of applications, and 5300 sloc of the extended Proj example. Permalinks resembling "⬚ " refer to declarations in mathlib itself that were infeasible to extract into our isolated repository. Much of this code has already been integrated into mathlib over the course of over 30 pull requests[14].

The monolithic nature of mathlib development means that our design decisions are easy to revisit at a later date, as the assumption is already that code written against one version of mathlib is not guaranteed to work without modification on a later version. Meanwhile, the act of having made the decisions enables downstream work to progress; as well as unblocking diverging formalization projects by the two authors[15], our foundations have allowed mathlib to gain formalizations by other contributors about various internal decompositions in inner product spaces and torsion modules.

Acknowledgments. The authors would like to thank Kevin Buzzard for his justified insistence on needing an interface to talk about internal gradings, Anne Baanen for picking up the mantle dropped by the first author on the rest of the set_like refactor,

---

[13] source lines of code.

[14] Tracked in https://github.com/leanprover-community/mathlib/projects/12.

[15] In geometric algebra and algebraic geometry, respectively!.

and the rest of the mathlib community for the extraordinarily collaborative work providing the context for this paper. The first author is funded by a scholarship from the Cambridge Trust. The second author is funded by the Schrödinger Scholarship Scheme from Imperial College London.

References

1. Affeldt, R., Cohen, C., Kerjean, M., Mahboubi, A., Rouhling, D., Sakaguchi, K.: Competing inheritance paths in dependent type theory: a case study in functional analysis. In: Peltier, N., Sofronie-Stokkermans, V. (eds.) IJCAR 2020. LNCS (LNAI), vol. 12167, pp. 3–20. Springer, Cham (2020). https://doi.org/10.1007/978-3-030-51054-1_1
2. Avigad, J., et al.: Spectral Sequences in Homotopy Type Theory, November 2015. https://github.com/cmu-phil/Spectral
3. Baanen, A.: Leanprover-community/mathlib#11750: define subobject classes from submonoid up to subfield, April 2022. https://github.com/leanprover-community/mathlib/pull/11750
4. Baanen, A.: Use and abuse of instance parameters in the Lean mathematical library. In: ITP 2022, Haifa, Israel, May 2022. http://arxiv.org/abs/2202.01629
5. Barton, R., Commelin, J., Buzzard, K., Lau, K., Carneiro, M.: #maths > CDGAs, June 2019. https://leanprover-community.github.io/archive/stream/116395-maths/topic/CDGAs.html
6. Bourbaki, N.: Algebra I, Chapters 1–3. Elements of Mathematics. Springer, Heidelberg (1989). https://doi.org/10.1007/978-3-642-59312-3
7. Brunerie, G., Ljungström, A., Mörtberg, A.: Synthetic integral cohomology in cubical agda. In: Manea, F., Simpson, A. (eds.) 30th EACSL Annual Conference on Computer Science Logic (CSL 2022). Leibniz International Proceedings in Informatics (LIPIcs), vol. 216, pp. 11:1–11:19. Schloss Dagstuhl - Leibniz-Zentrum für Informatik, Dagstuhl, Germany (2022). https://doi.org/10.4230/LIPIcs.CSL.2022.11. ISSN: 1868-8969
8. Castelvecchi, D.: Mathematicians welcome computer-assisted proof in 'grand unification' theory. Nature **595**(7865), 18–19 (2021). https://doi.org/10.1038/d41586-021-01627-2
9. Domínguez, C., Rubio, J.: Effective homology of bicomplexes, formalized in Coq. Theoret. Comput. Sci. **412**(11), 962–970 (2011). https://doi.org/10.1016/j.tcs.2010.11.016
10. de Moura, L., Kong, S., Avigad, J., van Doorn, F., von Raumer, J.: The lean theorem prover (system description). In: Felty, A.P., Middeldorp, A. (eds.) CADE 2015. LNCS (LNAI), vol. 9195, pp. 378–388. Springer, Cham (2015). https://doi.org/10.1007/978-3-319-21401-6_26
11. The mathlib Community: The lean mathematical library. In: Proceedings of the 9th ACM SIGPLAN International Conference on Certified Programs and Proofs, pp. 367–381. ACM, New Orleans, January 2020. https://doi.org/10.1145/3372885.3373824
12. The Stacks project authors: The Stacks project (2022). https://stacks.math.columbia.edu
13. Wieser, E.: Scalar actions in Lean's mathlib. In: CICM 2021. Timisoara, Romania, August 2021. http://arxiv.org/abs/2108.10700
14. Wieser, E., Song, U.: Formalizing geometric algebra in lean. Adv. Appl. Clifford Algebras **32**(3), 28 (2022). https://doi.org/10.1007/s00006-021-01164-1

Digital Libraries and Mathematical Knowledge Management

An Integrated Web Platform for the Mizar Mathematical Library

Hideharu Furushima[1], Daichi Yamamichi[1], Seigo Shigenaka[1],
Kazuhisa Nakasho[1(✉)] ⓘ, and Katsumi Wasaki[2]

[1] Yamaguchi University, Yamaguchi, Japan
{b089vgv,a102vgu,a085vgu,nakasho}@yamaguchi-u.ac.jp
[2] Shinshu University, Nagano, Japan
wasaki@cs.shinshu-u.ac.jp

Abstract. This paper reports on the development of a Web platform to host the Mizar Mathematical Library (MML). In recent years, the size of formalized mathematical libraries has been drastically increasing, and this has led to a growing demand for tools that support efficient and comprehensive browsing, searching, and annotation of these libraries. This platform implements a Wiki function to add comments to the HTML-ized MML, three types of search function (article, symbol, and theorem), and a function to show the dependency graph of the MML. This platform is designed with consistency, scalability, and interoperability as top priorities for long-term use.

Keywords: mathematical knowledge management · Mizar Mathematical Library · QED manifesto · Web service

1 Introduction

Many systems have been proposed to improve the browsability and searchability of formalized mathematical libraries. The MathWiki Project[1] aims to improve the readability of formalized mathematical libraries and make them accessible to wider communities. A Wiki for Mizar [17], Large Formal Wikis for Coq/CoRN [2], and Agora System for Flyspeck Project [16] were proposed in the MathWiki project. However, since the MathWiki Project was terminated in 2014, its contents cannot follow the libraries' updates. ProofWiki [19] and the Lean Mathematical Library [8] accumulate mathematical libraries and convert them into highly readable HTML documents. However, these two systems do not collaborate with advanced search engines or graphical tools to visualize library dependencies.

We developed the emwiki system, a Wiki service that hosts the Mizar Mathematics Library (MML) [4] while addressing issues with existing systems. The emwiki system is a Web platform based on the Django framework[2], featuring

---

[1] https://www.nwo.nl/en/projects/612066825.
[2] https://www.djangoproject.com/.

K. Buzzard and T. Kutsia (Eds.): CICM 2022, LNAI 13467, pp. 141–146, 2022.
https://doi.org/10.1007/978-3-031-16681-5_9

extensibility and interoperability. Vue.js[3] is used as the front-end framework, making it a partial single-page application. This eliminates extra rendering to reduce user stress. Vuetify[4] is used as the user interface library, which allows for intuitive operation. Currently, this service is available at https://em1.cs.shinshu-u.ac.jp/emwiki/release/.

2 Wiki Function

The Wiki feature is implemented to embed additional comments to read and understand articles in the MML. Although users cannot edit mathematical statements written in the formal language itself, they can add comments to theorems and definitions. We reused the HTMLized MML [18] to improve the convenience of the MML. The HTMLized MML is a document in which the MML is converted to HTML format, and the reference relationships are expressed as hyperlinks. We adopted the LaTeX syntax for mathematical expressions in the comments, which is familiar to mathematicians. MathJax [7] is used as a mathematical expression rendering engine. A real-time preview feature is also implemented to check rendering results while editing.

The version tracking feature manages the history of comments. This function is achieved by using Git, a distributed version control system, as its backend. Moreover, it is necessary to maintain the linkage between theorems/definitions and comments during library updates. Our system exploits the Git merge function for this purpose. As theorems/definitions do not have persistent identifiers in the Mizar language, the most effective way to identify theorems/definitions before and after a library update is to track text differences. Our system embeds comments directly into the MML. Since the comments are written just before the theorems/definitions, administrators can maintain the consistency of the linkage between the theorems/definitions and comments using the three-way merge function during library updates.

The user management function allows administrators to track and block malicious users. The Wiki function stores a history of editors and their revisions, and when a comment is accidentally rewritten, it is possible to contact the user who wrote it and roll back it. Also, if a malicious comment is found, it will be deleted, and the user who wrote it will be blocked. The user management function is available in any component on our platform.

The screenshot of the Wiki feature is shown in Fig. 1.

3 Search Function

The emwiki system has three types of search components. The two incremental search components accept the name of articles and symbols as input. As of 2018, the MML contains 1,290 articles and 8,852 symbols [4], and being able to search

---

[3] https://vuejs.org/.
[4] https://vuetifyjs.com/.

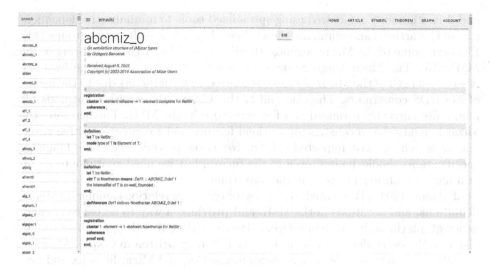

Fig. 1. emwiki

for these articles and symbols efficiently is critical for users. MML Reference [13] generates HTML documents from the MML to help users understand its symbols. Each HTML page contains symbol definitions, referrers, and references. It also has an incremental search function that enables users to search with a fast and intuitive operation. We integrated MML Reference into emwiki and linked this search function with the Wiki function.

We also developed a flexible theorem search component. For a long time, grep has been used for full-text search in the MML [12]. MML Query [5], developed by G. Bancerek et al. in 2001, has dramatically improved the efficiency of the search of the MML. However, MML Query is not easy for beginners, because it has its own query language. MML Query is a pattern-matching-based search system. Therefore, it is not good at searching modified (but logically equivalent) theorems. The Alcor system [6], developed by P. Carins et al., provides an Latent Semantic Indexing (LSI)-based search function for the Mizar Mathematical Library. We have implemented a similar LSI-based search component into the emwiki system. The search component does not require its own query language, and the user can perform searches by inputting the desired theorems in the Mizar language. We have also implemented a function to collect user evaluations by displaying a "good" button on the search results. In the future, we plan to use the evaluations as training data for machine learning.

4 Visualization of Library Dependencies

The dependency graph component is designed to encourage library maintainers to refactor their libraries by visualizing the dependencies of files in the MML. The MML has been maintained for the past 30 years by the University of Bialystok [9] and is continuously refactored using specialized tools to minimize the dependencies of the articles and eliminate the cyclical dependencies between groups [14]. However, some of the Mizar language specifications hinder the refactoring of the MML [15]. The Mizar language does not have namespace or package features. Also, the length of the Mizar file name is limited to 8 characters due to the legacy of MS-DOS constraints. Therefore, all of the 1,290 articles with ambiguous file names are currently arranged in a flat structure in the MML. Furthermore, the Mizar language has a constructor overload feature and a context-sensitive grammar in which the last imported constructor takes precedence. These language specifications should be improved in the future, but we must continue to maintain our vast library of assets in the meantime.

J. Alama (2011) [1] created a database of reference relations among theorems, definitions, and notations in the MML and provided a way to access the dependency graph through a Web interface. J. Heras et al. (2014) [10] provides a tool to visualize the dependency graph of the HoTT library written in Coq. J. Alama et al. (2012) [3] extracted the dependencies in the Coq and Mizar libraries and performed a quantitative comparative analysis of the library features. R. Marcus et al. (2020) [11] proposed the TGView3D system, which renders the dependencies of formalized mathematical libraries as 3D graphs in a hybrid of force-directed and hierarchical layouts.

In this study, we adopted the hierarchical graph. The hierarchical graph is intended to clarify the flow of dependencies and utilize refactoring, such as minimizing dependencies and eliminating circular references among groups. In the Mizar language, reference relationships between articles are described in the environment part at the beginning of an article. Since Mizar does not recursively load external files written in environment sections, a transitive reduction is performed to cut redundant edges before constructing the dependency graph. The data structure is saved in dot and sfdp formats of Graphviz and is drawn in a Web browser using Cytoscape.js library. The component also provides functions such as highlighting nearby nodes, searching nodes, moving nodes, hyperlinking to articles, and zooming in/out. A detailed usage of the dependency graph component can be found in the Help menu in the upper right corner.

5   Conclusions and Future Work

In this paper, we have developed the emwiki system as a Web platform for hosting the MML. This system is differentiated from existing services hosting mathematical libraries in consistency, extendability, and interoperability. One of the most severe difficulties in the long-term use of a service that hosts formalized mathematical libraries is keeping up with library updates. In this study, we attempted to solve this problem using Git and its merge function to maintain reference relationships during version updates. Since this platform is built using Django, a new component development is stylized as adding Django applications. Also, all components are allowed to access their version-controlled library, user management function, and database. The components implemented on the platform share Web pages and data with each other. For example, MML Reference, now integrated into the system, shares a Wiki function and Web pages. Ease of operation is also improved through automated testing during development and deployment using containers.

For the dependency graph component, several issues remain. While our hierarchical graph could properly place the dependencies of the articles, it did not classify the articles into meaningful groups. Mizar articles have metadata such as the Mathematics Subject Classification (MSC)[5]. We consider taking this information into account for graph layout and including it as additional information for article search. We also think it will be effective to analyze the vocabulary contained in the articles and reflect the semantic classification in the layout algorithm. Another beneficial enhancement is drawing theorem graphs. However, handling as many nodes as the theorems in the MML requires a more powerful graph database and drawing library than the conservative library we currently use.

This project aims to improve the browsability, searchability, and comprehensive understandability of formalized mathematical libraries. Although the system currently supports only the Mizar language, it could be extended to other languages. It is also significant to incorporate other components, including a Web IDE.

It is also important to establish a cooperative framework for the long-term operation of the system. In particular, we need to seek the cooperation of the core team that maintains the MML and the Mizar system to create a flow to ensure consistency between the MML versions. Also, we would like to invite a wide range of developers from the Mizar community to work with us in defining requirements and developing new functions.

Acknowledgments. This work was supported by JSPS KAKENHI Grant Number JP20K19863.

References

1. Alama, J.: mizar-items: exploring fine-grained dependencies in the Mizar mathematical library. In: Davenport, J.H., Farmer, W.M., Urban, J., Rabe, F. (eds.) CICM 2011. LNCS (LNAI), vol. 6824, pp. 276–277. Springer, Heidelberg (2011). https://doi.org/10.1007/978-3-642-22673-1_19
2. Alama, J., Brink, K., Mamane, L., Urban, J.: Large formal wikis: issues and solutions. In: Davenport, J.H., Farmer, W.M., Urban, J., Rabe, F. (eds.) CICM 2011. LNCS (LNAI), vol. 6824, pp. 133–148. Springer, Heidelberg (2011). https://doi.org/10.1007/978-3-642-22673-1_10
3. Alama, J., Mamane, L., Urban, J.: Dependencies in formal mathematics: applications and extraction for coq and Mizar. In: Jeuring, J., et al. (eds.) CICM 2012. LNCS (LNAI), vol. 7362, pp. 1–16. Springer, Heidelberg (2012). https://doi.org/10.1007/978-3-642-31374-5_1
4. Bancerek, G., et al.: The role of the Mizar mathematical library for interactive proof development in Mizar. J. Autom. Reasoning **61**(1), 9–32 (2018)

---

[5] https://msc2020.org/.

5. Bancerek, G., Rudnicki, P.: Information retrieval in MML. In: Asperti, A., Buchberger, B., Davenport, J.H. (eds.) MKM 2003. LNCS, vol. 2594, pp. 119–132. Springer, Heidelberg (2003). https://doi.org/10.1007/3-540-36469-2_10

6. Cairns, P., Gow, J.: Integrating searching and authoring in Mizar. J. Autom. Reasoning **39**(2), 141–160 (2007)

7. Cervone, D.: MathJax: a platform for mathematics on the Web. Not. AMS **59**(2), 312–316 (2012)

8. van Doorn, F., Ebner, G., Lewis, R.Y.: Maintaining a library of formal mathematics. In: Benzmüller, C., Miller, B. (eds.) CICM 2020. LNCS (LNAI), vol. 12236, pp. 251–267. Springer, Cham (2020). https://doi.org/10.1007/978-3-030-53518-6_16

9. Grabowski, A., Korniłowicz, A., Naumowicz, A.: Four decades of Mizar. J. Autom. Reasoning **55**(3), 191–198 (2015)

10. Heras, J., Komendantskaya, E.: HoTT formalisation in coq: dependency graphs & ML4PG. arXiv preprint arXiv:1403.2531 (2014)

11. Marcus, R., Kohlhase, M., Rabe, F.: TGView3D: a system for 3-dimensional visualization of theory graphs. In: Benzmüller, C., Miller, B. (eds.) CICM 2020. LNCS (LNAI), vol. 12236, pp. 290–296. Springer, Cham (2020). https://doi.org/10.1007/978-3-030-53518-6_20

12. Matuszewski, R., Rudnicki, P.: Mizar: the first 30 years. Mechanized Math. Appl. **4**(1), 3–24 (2005)

13. Nakasho, K., Shidama, Y.: Documentation generator focusing on symbols for the HTML-ized Mizar library. In: Kerber, M., Carette, J., Kaliszyk, C., Rabe, F., Sorge, V. (eds.) CICM 2015. LNCS (LNAI), vol. 9150, pp. 343–347. Springer, Cham (2015). https://doi.org/10.1007/978-3-319-20615-8_25

14. Naumowicz, A.: Tools for MML environment analysis. In: Kerber, M., Carette, J., Kaliszyk, C., Rabe, F., Sorge, V. (eds.) CICM 2015. LNCS (LNAI), vol. 9150, pp. 348–352. Springer, Cham (2015). https://doi.org/10.1007/978-3-319-20615-8_26

15. Rudnicki, P., Trybulec, A.: On the integrity of a repository of formalized mathematics. In: Asperti, A., Buchberger, B., Davenport, J.H. (eds.) MKM 2003. LNCS, vol. 2594, pp. 162–174. Springer, Heidelberg (2003). https://doi.org/10.1007/3-540-36469-2_13

16. Tankink, C., Kaliszyk, C., Urban, J., Geuvers, H.: Formal mathematics on display: a wiki for flyspeck. In: Carette, J., Aspinall, D., Lange, C., Sojka, P., Windsteiger, W. (eds.) CICM 2013. LNCS (LNAI), vol. 7961, pp. 152–167. Springer, Heidelberg (2013). https://doi.org/10.1007/978-3-642-39320-4_10

17. Urban, J., Alama, J., Rudnicki, P., Geuvers, H.: A wiki for Mizar: motivation, considerations, and initial prototype. In: Autexier, S., et al. (eds.) CICM 2010. LNCS (LNAI), vol. 6167, pp. 455–469. Springer, Heidelberg (2010). https://doi.org/10.1007/978-3-642-14128-7_38

18. Urban, J., Rudnicki, P., Sutcliffe, G.: ATP and presentation service for Mizar formalizations. J. Autom. Reasoning **50**(2), 229–241 (2013)

19. Vyskocil, J., Urban, J.: Disambiguating ProofWiki into Mizar: First steps. AITP 2018 (2018)

Formal Entity Graphs as Complex Networks: Assessing Centrality Metrics of the Archive of Formal Proofs

Fabian Huch[✉]

Technische Universität München, Boltzmannstraße 3, 85748 Garching, Germany
huch@in.tum.de

Abstract. Formalization libraries for interactive theorem provers are rapidly growing in size, but only little is understood structurally about those developments. We aim to address the arising challenges by utilizing dependency graphs of the underlying formal entities. For the Isabelle Archive of Formal Proofs, the individual entry graphs consist of 1.8 million nodes and 2.8 million edges in total, and exhibit certain complex network characteristics: Node in-degrees weakly follow scale-free distributions with an average exponent of $\alpha = 1.81$, and the high clustering coefficient (avg. 0.33) together with the short average path length (3.64) indicate small-world effects. We did not find network centrality metrics to be good indicators of theory quality (measured by lint frequency): The Spearman correlation of our six different centrality metrics was weaker than in similar experiments from software systems, and with a coefficient of $s = 0.38$, the source lines of code metric exhibited a stronger correlation than all centrality metrics considered. In contrast, network centrality metrics worked well in predicting the most important concepts within AFP entries: Of the definitions deemed most important by entry authors, 51.7% could be identified at a precision of 27.7% (optimal F_1-score), using in-degree centrality. At the cost of a few percentage points of precision, a second maximum of 68.8% recall can be achieved.

Keywords: Isabelle · Archive of Formal Proofs · Complex networks · Dependency graph · Formal entity network · Centrality metrics · Formalization quality

1 Introduction

Interactive proof assistants (ITPs) have become quite mature, to a point were even novel research can be performed in a theorem prover, such as the recently completed proof of the Clausen–Scholze theorem in Lean [15]. At the same time, not much is known empirically about the structure of formalizations.

Due to long-term community efforts, large-scale collections of formalized material exist in some (earlier) systems such as *Mizar* [14] (*Mizar Mathematical Library*) and *Isabelle* [13] (*Archive of Formal Proofs*). At the time of writing, the

© The Author(s), under exclusive license to Springer Nature Switzerland AG 2022
K. Buzzard and T. Kutsia (Eds.): CICM 2022, LNAI 13467, pp. 147–161, 2022.
https://doi.org/10.1007/978-3-031-16681-5_10

Archive of Formal Proofs (AFP) consists of over three million lines of code, and more than 195 000 lemmas have been proven in its 675 different articles (called *entries*) [2]. Entries are characterized by their metadata (abstract, authors, topics, etc.) and Isabelle *sessions* (usually a single one per entry), i.e., collections of coherent Isabelle theories.

At the mark of one million lines of code, the size and authorship distribution in the AFP, and how those evolved over time, was analyzed [3], as well as some (mostly) syntactic properties. It was found that proofs made up 58% of the code, and the total proof size has a roughly quadratic relationship with the number of constants in the goal. However, except for some insights into the entry import-graph (where it became apparent that only very few entries were re-used), the syntactical analysis tells us very little about the underlying formalization structure.

Problem. As libraries and archives expand and ITP use becomes more widespread, understanding the formalization structure becomes critical. Due to the vastness of formalization archives, it is important to aid users in grasping material faster and more easily. Moreover, building high-quality formalizations that are re-usable and extensible in the first place is key.

Solution. The related problems in the field of software engineering are widely studied. It is found that methodology from *complex network science*— which is concerned with the structural properties of networks from all kinds of empirical research— is applicable to a wide range of problems and often superior to analysis of software-specific characteristics [22,27]. This gives rise to analyzing the dependency graphs of formalizations to determine whether complex network methodology is applicable and which concrete questions can be answered.

Contribution. In this work, we investigate the formalization network of the Isabelle AFP for patterns commonly found in complex networks. We also examine two concrete problems, namely, assessing formalization quality, and identifying the most important parts of formalizations. Our source code and data is publicly available[1].

Organization. In Sect. 2, we give an overview about concepts of complex networks, and discuss in Sect. 3 how they are used in related work in software engineering. We report our analysis and findings in Sect. 4, and discuss our conclusions in Sect. 5.

2   Complex Network Concepts

Complex systems can frequently be modeled as graphs. Of those, many real-world networks from biology, social science, engineering, and information systems, are found to exhibit certain structural properties that do not appear in graphs generated from simple models, and thus called *complex networks*. In general, those graphs are directed, though some complex network research only considers undirected graphs. We will consider the general case unless stated otherwise.

---

[1] https://github.com/dacit/afp-complex-networks.

In the following, we explain three of the most important complex network properties, as well as graph metrics found useful in complex networks. We express graphs by their adjacency matrix A, i.e., a_{ij} denotes an edge between i and j (we do not consider multiplicities). Thus, in-degree k_{in} of a node i can be defined as follows (out-degree k_{out} and overall degree k similarly):

$$k_{\text{in}}(i) = \sum_{j \neq i} a_{ji} \tag{1}$$

The *small-world* property [26] describes networks which are in between regular networks (where every node has the same degree, e.g., lattice graphs), and random graphs where edges are created uniformly at random (Erdős–Rényi (ER) model [9]). Small-world networks exhibit average path lengths that are short like in ER-graphs— in contrast to regular graphs, where average path length increases linearly with size— but comparatively strong clustering, which is low in ER-graphs but high in many regular networks. A measure for clustering, the *clustering coefficient CC*, is defined as the likelihood that a node's neighbors are also connected to each other in the undirected graph, i.e.:

$$CC(i) = \frac{\sum_{i \neq j, j \neq l, l \neq i} a_{ij} a_{jl} a_{li}}{2 \cdot k(i)(k(i) - 1)} \tag{2}$$

Secondly, the *scale-free* property [1] describes networks that have no characteristic scale, i.e., they have similarly structure no matter the size. In particular, the *in-degree* distribution of scale-free networks follows a power-law, for sufficiently large k_{in} (and up to a maximal degree for finite networks):

$$\Pr[k_{\text{in}} = x] \propto x^{-\alpha} \tag{3}$$

In empirical scale-free networks, α (which represents the slope on the linear log–log plot) is typically in the range of 2 to 3 [6].

The third important concept is the notion of self-similarity, which underlies many generative processes found in nature. In complex networks, self-similarity can be measured by the *fractal dimension* d_B [18], which is the exponent of the power-law relationship:

$$N_B(l_B) \propto l_B^{-d_B} \tag{4}$$

where N_B is the minimal number of connected components that the network can be split into such that each node is reachable from any other within l_B steps (in the undirected graph).

In addition to in- and out-degree, we use the following network centrality measures (cf. [12]) in this work: *Eigenvector centrality* C_λ is the (normalized) solution to $\boldsymbol{Ax} = \lambda \boldsymbol{x}$ with non-negative components where λ is the largest eigenvalue of the adjacency matrix (which uniquely exists since all entries are non-negative). Similar to PageRank [4], it yields the importance of a node in the network. *Closeness centrality* is the average distance from a node to every other node. However, since it is not defined for unconnected graphs, we use harmonic closeness centrality C_{dist}. With d_{ij} denoting the shortest distance between i and j (and n the total number of nodes):

$$C_{\text{dist}}(i) = \frac{1}{n-1} \sum_{j \neq i, d_{ij} < \infty} \frac{1}{d_{ij}} \qquad (5)$$

Betweenness centrality C_p measures how many shortest paths p_{ij} (from i to j) go through a node, i.e., it captures how much a node behaves as a hub in the network:

$$C_p(i) = \frac{|\{p_{kl} | i \in p_{kl}\}|}{|\{p_{kl}\}|} \qquad (6)$$

3   Complex Network Science in Software Engineering

For software systems, Valverde et al. [23] first discovered scale-free and small-world behavior and ascribed it to modularity requirements. Myers [11] found class graphs and procedure calls to be weakly scale-free, and attributed those properties to object-oriented (OO) design. However, rigorous statistical testing was not performed (and in fact, for most networks claimed to be scale-free, other distributions are more likely [5]).

Fractal dimension has been studied for software systems by Concas et al. [7], where d_B was found to usually be in the range of 3 to 5. Moreover, Turnu et al. [22] found the fractal dimension to be strongly correlated with OO software quality indicators and with the number of defects that later occurred, whereas correlations with OO quality metrics would not be significant at the common significance level of $p = 0.05$.

Network centrality measures have been found to be suitable for defect prediction as well: Zimmermann and Nagappan [27] found that out-degree and eigenvector centrality (among others) are more strongly correlated than OO metrics, but slightly worse than other software engineering metrics; overall, the correlation is of medium strength. However, for predicting central binaries, network centrality measures greatly outperformed other complexity measures. These results could later only be confirmed for larger-scale projects (on a source-code level) [21].

Šubelj and Bajec [19] have used complex network methodology for community detection, i.e., to find structurally grouped modules. Later, they successfully employed the detection on Java class graphs to infer the package structure with high accuracy as well as to separate the graph into structurally similar partitions, which could be used to generate refactoring recommendations [20,25].

4   Analysis of Formal Entity Networks

Generally, we denote as a *formal entity network* the graph of logically relevant entities in a formalization with their relationships. Such a network can be defined for any theorem prover system. For simply typed systems, nodes of the graph include types, constants and theorems. Depending on the logic and system used, other concepts may also be represented, for example type-classes. Edges model the relationships in between those entities, and between entities and the syntactical structure (e.g., theory files). Entity relationships exist between types, constants and types, theorems and constants, and theorems (proof dependencies). For Isabelle, we include locales and classes, locale dependencies, and model sort constraints as well as class relationships. Our syntactical structure consists of *defining positions* (i.e., the Isabelle source code from which entities are created), theories, and sessions. Depending on the research question, we consider sub-graphs of the full network, e.g., only entity nodes. Our analysis is mostly prover-agnostic, and could in principle be transferred to other systems.

The full dependency graph of the Isabelle distribution and the AFP consists of 2.3 million entity nodes with 206.5 million edges between them. When considering the 640 nonempty sub-graphs for individual AFP entries (excluding the `slow` group, which contains a few particularly computationally extensive entries), they make up for a total of 1.8 million nodes and 28.3 million edges. In the following, we analyze this body of data to answer the following research questions:

RQ1: Does the theorem dependency graph exhibit complex network properties?
RQ2: How does formalization quality reflect on the theorem dependency network?
RQ3: Can we detect important concepts within the dependency network automatically?

4.1   Dependency Graph as Complex Network

To assess complex network properties of Isabelle and the AFP, we analyze the full graph as well as the individual entry sub-graphs. However, for the average path length, paths to elements of the object logic would largely determine the result. Hence, we analyze the individual sub-graphs for small-world effects, and compare averages weighted by number of nodes. In the undirected graph, the average shortest path length is 3.6, which is quite close to the expected value of 3.0 for ER-graphs. On average, the clustering coefficient is 0.33 (i.e., about one third of a nodes' neighbors are also connected), which is orders of magnitudes larger than in ER-graphs (0.0052)— confirming the small-world hypothesis.

In-degree frequencies of the full dependency graph is shown in Fig. 1 as a scatter plot (left). The data roughly follows a linear line on the log–log scale, but there are some outliers. The indicator line marks the fitted power-law distribution with a slope of $\alpha = 1.79$. On the right of Fig. 1, the complementary cumulative distribution function (CCDF) of the empirical data and theoretical distributions (scaled to the empirical data at their minimal k_{in}) are shown. We compare the scale-free distribution with common alternatives, namely the log-normal, exponential, and Poisson distribution. The empirical data is close to the scale-free and log-normal distributions for smaller k_{in}, but towards the tail it does not follow the theoretical values of any distribution closely. This is

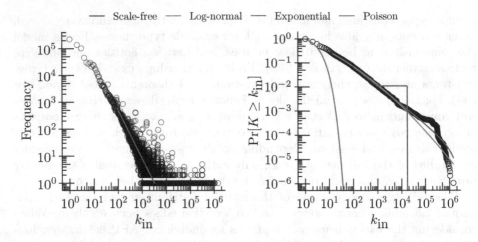

Fig. 1. In-degree distribution of full Isabelle/AFP dependency graph, on log–log scale. Scatter plot (left) with and CCDF (right), with fitted distributions.

not uncommon even for networks that are attributed scale-free structure: Datasets following a theoretical scale-free distribution very closely are found to be rare [5]. In a likelihood ratio test between the two plausible distributions (where *likelihood* is the probability of the data given the distribution, and $\mathcal{R}$ the log of the ratio between the likelihoods), the scale-free distribution is preferred with $\mathcal{R} = 5.8$.

For dependency graphs of the individual entries —which are quite heterogeneous— a scale-free distribution is preferred only in 44 out of 596 cases. Figure 2 shows the complementary cumulative distributions of a random selection of 20 entries (names truncated). The larger entries tend to follow the distribution of the overall network, whereas for the smaller entries, the tail of the distribution deviates. This indicates that finite-size effects might skew the distribution, which is often the case for scale-free networks [16]. Overall, the power-law parameter is similar, with a weighted average of $\alpha = 1.81$ for the individual entries.

While both small-world and scale-free effects don't provide easily digestible insights, they characterize the structure of the dependency graphs and indicate that complex network methodology is applicable. Hence, we expect that some results from complex network science carry over to our dependency graphs. Thus, to answer our next research questions, we utilize the centrality metrics introduced in Sect. 2 as they are helpful for complex networks in other settings. We first analyze those metrics on their own: Table 1 shows how the centrality metrics are correlated with each other in the networks of individual entries, as matrix of Spearman correlation coefficients s (i.e., strength of the monotonic relation between two variables)— all are significant with $p < .001$, $n = 1.8$ million. Some correlations are expected, since they would also occur in random graphs. In-degree is strongly correlated with closeness centrality (any directly connected

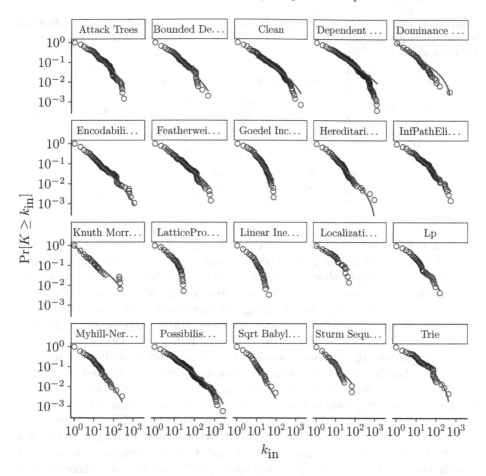

Fig. 2. CCDF of entity in-degrees for 20 randomly selected AFP entries (individual graphs), with fitted scale-free distribution (log–log scale).

node has a distance of one) as well as betweenness centrality (an incoming edge means that a node is likely in a shortest path), and moderately with eigenvector centrality (incoming edges contribute to the rank). Betweenness and closeness are also strongly related due to their similar nature. Notably, there is also a fairly strong correlation between eigenvector and closeness centrality.

4.2   Assessing Formalization Quality

Measuring formalization quality is a challenging task: Manual analysis is quite labor-intensive and difficult, as it is often not obvious on the surface why something was formalized in a certain way (many design decisions come from considerations that emerge during the proofs). Moreover, there are no defects (like in software systems) that would give an indication about quality. Metrics such

Table 1. Coefficients of spearman correlation between centrality metrics for individual formal entity graphs, all significant at $p < .001$, $n = 1.8$ million. Values above $s = 0.5$ are printed bold.

	k_{in}	k_{out}	CC	C_λ	C_{dist}	C_p
k_{in}		−.18	−.10	.50	**.97**	**.76**
k_{out}	−.18		−.13	−.31	−.19	.16
CC	−.10	−.13		−.13	−.08	−.09
C_λ	−.50	−.31	−.13		**.56**	.38
C_{dist}	**.97**	−.19	−.08	**.56**		**.77**
C_p	**.76**	.16	−.09	.38	**.77**	

as change frequencies would also not provide meaningful insights due to the journal character of the AFP: Functional changes are supposed to be monotone (i.e., existing interfaces shouldn't be changed even if they could be improved), and changes for maintenance reasons don't necessarily indicate poor quality, as they are often caused by proofs breaking due to re-naming or re-organization of upstream material. Therefore, we have to rely on static analysis, and utilize the recently developed *Isabelle linter* [10] for a rough indicator of formalization quality. The linter can identify problematic anti-patterns (*lints*), and implements 22 checks based on Isabelle style guides, focusing on maintainability and readability problems. For example, it would report a structured proof that starts with an invocation of `auto`, as that obfuscates the goal to be proven and makes the proof prone to break on minor changes such as improvements in the simplifier. However, almost all available lints are concerned with proofs, so they only uncover specific kinds of problems. Hence, lints do not indicate structural problems of specific entities, but rather serve as a general quality indicator for whole theories or entries. As quality metric, we use *lint frequency*, i.e., the number of lints per *source line of code* (SLOC). We compare the lint frequency to our centrality metrics in a range of scenarios, using the total, mean, and maximum as aggregation for the centrality scores on the formal entity network. For a thorough analysis, we also employ these metrics on the graph induced by theories and defining positions. Moreover, we utilize the fractal dimension (approximated by box covering [17]), which is found to be a particularly good quality indicator in software systems—Tosun et al. [21] use a linear regression in the log–log space to compute the power-law coefficient, which we include as d_B^{lin} in addition to the proper stochastic fit [6]. To evaluate how useful the centrality metrics are, we compare their performance to that of the SLOC metric. Since using the frequency already adjusts for size, there is no inherent correlation between theory size and lint frequency, but larger files could potentially be an indicator of lacking structure.

Table 2 shows the Spearman correlation for different aggregations of the network metrics. At our significance level of $p = 0.01$, most metrics are slightly correlated with lint frequency, however the strength peaks at $s = 0.30$ for between-

Table 2. Coefficients of Spearman correlation between centrality metrics and lint frequency in different scenarios, significant at the $p = 0.01$ level. Values above $s = 0.25$ are printed bold.

Lint Frequency	Subgraph	Aggregation	k_{in}	k_{out}	CC	C_λ	C_{dist}	C_p
per entry	entities	mean	.21	.21	−.25	−.20	.15	**.25**
		total	**.26**	**.26**	.21	.23	**.28**	**.28**
		max	.24	**.27**	−.17	−	.19	**.28**
	defining positions	mean	.20	.20	−.29	−.15	−	**.25**
		total	**.26**	**.26**	−	**.25**	.23	**.28**
		max	**.25**	**.25**	−	−	.12	**.28**
	entities	mean	.08	.15	.06	.07	.08	.23
		total	.15	.19	.07	.14	.18	.25
		max	.12	.23	.07	.09	.13	.24
	defining positions	mean	−	.07	−	.05	.05	.24
		total	.15	.19	.04	.19	**.27**	**.30**
		max	.11	.13	.04	.11	.12	**.27**
	theories		.15	.19	.06	.06	−	.16

ness centrality in defining positions, using the total centrality per theory; several other metrics are close contenders. Results for the global metrics are shown in Table 3. The fractal dimension performs similarly to the best centrality metrics, with a correlation coefficient of $s = 0.27$ in the optimal case. The difference between the computation methods of the fractal dimension is small, though the regression calculation is slightly better. In contrast, the simple SLOC metric performs significantly better: The coefficient of $s = 0.38$ (in the per-theory scenario) is stronger than any other correlation. Hence, the centrality metrics considered here are not good quality indicators.

Table 3. Coefficients for spearman correlation between global metrics and lint frequency in different scenarios. Values above 0.25 are printed bold.

Score	Aggregation	SLOC	d_B	d_B^{lin}
per entry	max	.17	.20	.23
	mean	−	−	−
	total	.25	**.26**	**.27**
			−	.15
per theory		**.38**	−	.05

4.3 Predicting Important Concepts

Knowing which concepts (types, constants, etc.) are important in a formalization is essential to understand the material quickly and effectively, but so far, such overviews are not available. For a small set of developments, this may be provided by hand. However, to be feasible on a large scale, and for interactive tooling, automatic extraction of important concepts is necessary. Research in software systems suggests that centrality metrics of the dependency network could be a good indicator for identifying the most important concepts, in particular closeness and in-degree centrality [27].

However, entities of a formal entity network do not directly relate to concepts. The higher-level commands one uses to define a concept can introduce many entities, e.g., the datatype command generates multiple types and constants to represent the potentially recursive datatype as a HOL type. Hence, we consider as a concept all defining positions that introduce a type, constant, class, or locale; from that, we select the entity that best matches the source code, e.g., the type entity with the user-supplied name for a recursive datatype definition.

To collect ground truth data, we asked AFP authors (who listed an e-mail address as affiliation) for their opinion on the five most important concepts and the five theorems that were most central to the reasoning in their formalization (as opposed to intermediate lemmas, main results, etc.). We included a list of unordered suggestions (generated by the top ten entities each of several graph metrics) to prompt responses, but prominently stated that elements not in this list should be considered as well. In total, we sent 219 e-mails (56 addresses were unreachable) and received replies from 41 authors, though not all answers were suitable for our analysis, i.e., listing specific entities. In total, the resulting ground truth dataset consists of data for 23 entries, containing 109 concepts and 96 theorems (multiple answers for same entries joined). While this is sufficient to evaluate simple models with few parameters, it would not allow training and evaluating more complicate models such as graph neural networks.

To evaluate our centrality metrics, we compute the match of the top-N elements with the ground truth in terms of precision (how many of the selected elements are relevant) and recall (how many of the relevant elements were selected). The harmonic mean of precision and recall (called F_1-score) is shown in Fig. 3 for a varying number of N, in four different scenarios. When predicting the entities representing important concepts, the F_1-score peaks for N in the range of 5 to 15 for all metrics. In-degree and closeness centrality consistently achieve the highest score, similar to what was found in software systems [27]. When only the defining positions of concepts are considered in the score calculations (to mitigate the error that comes with selecting an entity to define a concept), a slightly higher F_1-score of at most 0.34 is reached. Otherwise, the results are the same.

In contrast, the prediction for central theorems scores much lower: The F_1-score is well below 0.1 for most metrics. Only the betweenness centrality performs significantly better and peaks at about 0.19, both when predicting the full dependency graph, and when only the sub-graph of theorem entities is considered (the

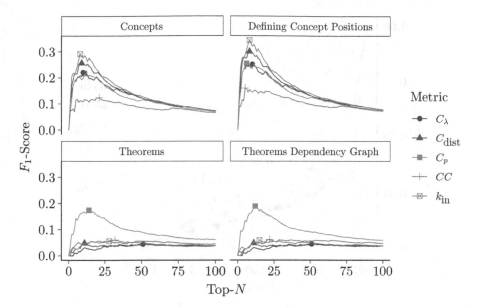

Fig. 3. F_1-scores for network centrality metrics in different scenarios by number of concepts selected, with marked maximum values.

only difference is that the peak is much sharper in the latter case). The absolute value may be somewhat skewed as the definition of "central theorem" was not clear to all participants who contributed to the ground truth data— some listed main results instead— but the large relative difference shows that betweenness centrality is the key metric for central theorems. An intuitive explanation is that central theorems are required in the transitive closure of proof dependencies of many theorems, and thus appear in many shortest paths.

Since predicting the correct defining positions is our most important use-case, we analyze the performance characteristic for that scenario in Fig. 4. Overall, the performance is strictly better where optimal F_1-score is higher in nearly all cases, i.e., in-degree is the best metric at every operation point. The in-degree curve starts out at a precision of 47.8% (recall 8.7%), then sharply decreases for increasing N. F_1-score is optimal for $N = 8$, at 27.7% precision and 51.7% recall. A second good operating point (optimal when recall is considered more important as precision, e.g., by a factor of two in the F_2-score) is at $N = 14$, where precision is only a little lower (21.1%) but recall much higher with 68.8%. However, the final choice of N may depend on other user interface design considerations. The relatively low precision is acceptable for a generated overview, and due to the nature of our ground truth data (restricting to the top few elements), many high-ranking concepts not in the ground truth set are likely to be relevant— several responses stated that it was difficult to limit themselves to such a small set.

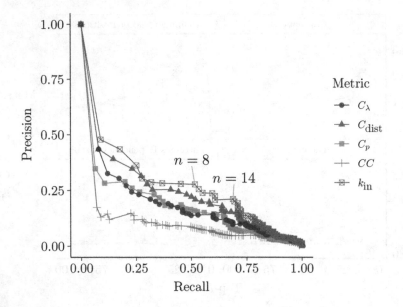

Fig. 4. Precision–recall curve for prediction of defining positions for important concepts, with varying number of top-N elements selected.

5   Discussion

We analyzed the formal entity networks induced by Isabelle/AFP entries and found that they exhibit small-world effects (i.e., have short average path lengths and high clustering coefficients) and are weakly scale-free. Hence, we attempted to connect those networks with research from complex networks of software systems, to obtain quality indicators and to extract important concepts. While network centrality metrics and fractal dimension (on a global level) work well as software defect indicators, we found that this is not the case for formalization networks, and the much simpler SLOC metric is superior to all the above. However, we note that in software systems, the absolute number of defects is used as ground-truth for quality, but frequently used components have a higher chance to incur a fault even if the underlying code quality is the same. Hence, defect count is dependent on centrality, whereas lint frequency is independent of such an effect. In contrast, extracting important concepts works quite well. We achieved best results (in terms of F_1-score) by selecting the top-8 entities with respect to in-degree, where 51.7% of defining positions identified as most important by article authors could be identified with a precision of 27.7%. A second sweet spot is attained for the top-14 entities, where 68.8% of the relevant defining positions can be retrieved with 21.1% precision.

5.1   Threats to Validity

The combinatorial explosion of parameter and experiment spaces poses the biggest threat to validity of this research. There are many ways of constructing formal entity graphs (for example, restricting the graph to just theorems and proof dependencies), multiple levels of abstractions (entities, defining positions, theories, sessions, entries), and many sub-graphs that can be considered (per-entry, full, full without base logic). Moreover, a large number of other graph metrics from social network analysis exist that one could utilize, though it is found that many of those are quite similar and strongly correlate with each other [8]. Furthermore, the large scale and variety of the involved formal entity networks make experiments quite time-consuming. We cover a reasonable portion of that parameter space and orientate ourselves at related research, but there might still be unexplored territory where better results are attainable.

5.2   Future Work

In this research, we only considered simple metrics and statistical correlations on Isabelle dependency graphs. With this basis, the rich body of data would also allow building and evaluating more intricate models, for instance deep learning models such as graph neural networks, towards which much of the research nowadays has gravitated. Recently, they were found to work quite well for defect prediction in software systems [24].

Additionally, the network analysis itself is not Isabelle-specific, and could be transferred to other systems as well. While we have not attempted that for the work presented here, one would expect the results to carry over at least to other classical prover systems. Verifying this assumption would be a worthwhile contribution—building the integration with other systems is the main challenge.

Next, while we could successfully extract important concepts from the dependency graph, more engineering work is required to put them in practice: Overviews for AFP entries need to be generated and added to the AFP website, which requires proper integration into the existing tooling and pipelines. Moreover, generating this overview dynamically in a running prover session would be a valuable add-on tool for Isabelle prover IDEs.

Furthermore, there are other applications of network structure that have not yet been examined. For instance, recommendations for lemma placement could be generated by utilizing community detection and finding entities that are not in the theory in which their community is located. Another application is in the automated theorem prover integration, where facts from the background theory that might be useful in solving a given goal need to be selected and ranked. This could be improved by utilizing graph knowledge.

Lastly, visualizations of the (aggregated) network graphs could be a valuable tool in making the vast formalization landscapes more navigable.

References

1. Albert, R., Jeong, H., Barabási, A.L.: Diameter of the world-wide web. Nature **401**(6749), 130–131 (1999). https://doi.org/10.1038/43601
2. Archive of formal proofs: statistics. https://www.isa-afp.org/statistics.html
3. Blanchette, J.C., Haslbeck, M., Matichuk, D., Nipkow, T.: Mining the archive of formal proofs. In: Kerber, M., Carette, J., Kaliszyk, C., Rabe, F., Sorge, V. (eds.) CICM 2015. LNCS (LNAI), vol. 9150, pp. 3–17. Springer, Cham (2015). https://doi.org/10.1007/978-3-319-20615-8_1
4. Brin, S., Page, L.: The anatomy of a large-scale hypertextual web search engine. Comput. Netw. ISDN Syst. **30**(1–7), 107–117 (1998)
5. Broido, A.D., Clauset, A.: Scale-free networks are rare. Nature Commun. **10**(1), 1–10 (2019). https://doi.org/10.1038/s41467-019-08746-5
6. Clauset, A., Shalizi, C.R., Newman, M.E.J.: Power-law distributions in empirical data, November 2009. https://doi.org/10.1137/070710111
7. Concas, G., Locci, M.F., Marchesi, M., Pinna, S., Turnu, I.: Fractal dimension in software networks. Europhysics Lett. **76**(6), 1221–1227 (2006). https://doi.org/10.1209/epl/i2006-10384-1
8. Concas, G., Marchesi, M., Murgia, A., Pinna, S., Tonelli, R.: Assessing traditional and new metrics for object-oriented systems. In: Proceedings - International Conference on Software Engineering, pp. 24–31. ACM Press, New York (2010). https://doi.org/10.1145/1809223.1809227
9. Erdös, P., Rényi, A.: On random graphs i. Publicationes Mathematicae Debrecen **6**, 290 (1959)
10. Megdiche, Y., Huch, F., Stevens, L.: A linter for isabelle: implementation and evaluation. In: Isabelle Workshop (2022)
11. Myers, C.R.: Software systems as complex networks: structure, function, and evolvability of software collaboration graphs. Phys. Rev. E **68**, 046116 (2003). https://doi.org/10.1103/PhysRevE.68.046116
12. Newman, M.E.J.: The structure and function of complex networks. SIAM Rev. **45**(2), 167–256 (2003). https://doi.org/10.1137/S003614450342480
13. Paulson, L.C.: Isabelle: the next seven hundred theorem provers. In: Lusk, E., Overbeek, R. (eds.) CADE 1988. LNCS, vol. 310, pp. 772–773. Springer, Heidelberg (1988). https://doi.org/10.1007/BFb0012891
14. Rudnicki, P.: An overview of the mizar project. In: Proceedings of the 1992 Workshop on Types for Proofs and Programs, pp. 311–330 (1992)
15. Scholze, P.: Half a year of the liquid tensor experiment: amazing developments, June 2021, https://xenaproject.wordpress.com/2021/06/05/half-a-year-of-the-liquid-tensor-experiment-amazing-developments/
16. Serafino, M., et al.: True scale-free networks hidden by finite size effects. In: Proceedings of the National Academy of Sciences of the United States of America vol. 118, no. 2, January 2021. https://doi.org/10.1073/pnas.2013825118
17. Song, C., Gallos, L.K., Havlin, S., Makse, H.A.: How to calculate the fractal dimension of a complex network: the box covering algorithm. J. Stat. Mech. Theor. Exp. **2007**(3), 3006 (2007). https://doi.org/10.1088/1742-5468/2007/03/P03006
18. Song, C., Havlin, S., Makse, H.A.: Self-similarity of complex networks. Nature **433**(7024), 392–395 (2005). https://doi.org/10.1038/nature03248
19. Šubelj, L., Bajec, M.: Generalized network community detection, October 2011
20. Šubelj, L., Bajec, M.: Clustering assortativity, communities and functional modules in real-world networks, vol. 87, Febuary 2012

21. Tosun, A., Turhan, B., Bener, A.: Validation of network measures as indicators of defective modules in software systems. In: ACM International Conference Proceeding Series, p. 1. ACM Press, New York (2009). https://doi.org/10.1145/1540438.1540446

22. Turnu, I., Concas, G., Marchesi, M., Tonelli, R.: The fractal dimension of software networks as a global quality metric. Inf. Sci. **245**, 290–303 (2013). https://doi.org/10.1016/j.ins.2013.05.014

23. Valverde, S., Ferrer Cancho, R., Solé, R.V.: Scale-free networks from optimal design. Europhysics Lett. **60**(4), 512–517 (2002). https://doi.org/10.1209/epl/i2002-00248-2

24. Šikić, L., Kurdija, A.S., Vladimir, K., Šilić, M.: Graph neural network for source code defect prediction. IEEE Access **10**, 10402–10415 (2022). https://doi.org/10.1109/ACCESS.2022.3144598

25. Šubelj, L., Bajec, M.: Software systems through complex networks science: review, analysis and applications. In: Proceedings of the First International Workshop on Software Mining, pp. 9–16. SoftwareMining 2012, Association for Computing Machinery, New York (2012). https://doi.org/10.1145/2384416.2384418

26. Watts, D.J., Strogatz, S.H.: Collective dynamics of 'small-world' networks. Nature **393**(6684), 440–442 (1998). https://doi.org/10.1038/30918

27. Zimmermann, T., Nagappan, N.: Predicting defects using network analysis on dependency graphs. In: Proceedings of the 30th International Conference on Software Engineering, ICSE 2008, pp. 531–540. Association for Computing Machinery, New York (2008). https://doi.org/10.1145/1368088.1368161

Re-imagining the Isabelle Archive of Formal Proofs

Carlin MacKenzie[1]([✉]) [iD], Fabian Huch[2] [iD], James Vaughan[1] [iD],
and Jacques Fleuriot[1] [iD]

[1] School of Informatics, University of Edinburgh, 10 Crichton Street,
Edinburgh EH8 9AB, UK
{s1724780,jvaughan,jdf}@ed.ac.uk
[2] Technische Universität München, Boltzmannstraße 3, 85748 Garching, Germany
huch@in.tum.de

Abstract. Since its inception in 2004 the Archive of Formal Proofs has
grown in size but its interface and functionality have only been minimally
improved. To transform the AFP into a more user-friendly and effective
resource, we redesigned the website to meet modern web standards and
practices. We ensure that our work is community-driven by basing the
redesign on results from a survey of the Isabelle community. The site
generation uses Hugo and is implemented as a proper Isabelle component, which also allows us to adapt the AFP metadata model to avoid
inconsistencies in the future. Notable improvements include a responsive
design, new theory browsing interface, integrated search, and enhanced
navigation.

Keywords: Isabelle · Archive of Formal Proofs · User interface · User
experience

1 Introduction

The Archive of Formal Proofs (AFP) is a large collection of heterogeneous
Isabelle developments that has been online and maintained since 2004. It is
organised in a journal-style fashion: submissions get reviewed by a team of AFP
editors, before they are added as articles (also called *entries*). With its super-
linear growth [2], the archive has amassed a total of 675 articles [1] at the time of
writing this paper. Hence, obtaining an overview and navigating its contents have
become increasingly difficult tasks. In fact, finding formalizations has become so
challenging that two standalone Isabelle/AFP search engines were introduced in
2020 [4,8]. However, the main AFP user interface, its website[1], has not kept up
with the times: The (mostly) static site is generated by a home-grown script-
ing solution rather than a modern static site generator (SSG) which would be
more maintainable. The resulting website does not have a responsive design, no

---

[1] https://isa-afp.org.

© The Author(s), under exclusive license to Springer Nature Switzerland AG 2022
K. Buzzard and T. Kutsia (Eds.): CICM 2022, LNAI 13467, pp. 162–167, 2022.
https://doi.org/10.1007/978-3-031-16681-5_11

search options or author pages, and its topic index has become cluttered due to the sheer size. Additionally, some of the archive metadata is out of shape: Author affiliations contain near-identical duplicates and many e-mail addresses that are no longer valid—56 out of 219 were unreachable in a recent analysis [3].

2   (Re)Design Considerations

At the time of evaluating usage, the Isabelle community primarily communicated through the Isabelle mailing list (where both beginners' problems and expert questions would be discussed alike) and we sought guidance from this community before we started this project. As we cared most about the general user experience, we decided that a survey would be most suitable. We created and tested it with a pre-study group from the *Artificial Intelligence Modelling Lab* at the University of Edinburgh before advertising it on the mailing list[2].

The survey [6] was open from 10–30 November 2020 and was completed by twenty-nine members of the mailing list that skewed towards long term and active users of the AFP. In general, respondents were satisfied with the AFP although they had specific criticisms about navigation, search, and theory browsing. Most respondents were neutral or negative towards a redesign of the user interface and user experience. Hence, we concluded that a completely new interface would not be welcome by the AFP users and so we chose to create a design that would maintain familiarity but rest on modern design conventions.

We used paper prototyping to first recreate the original design. Next we considered the placement and form of each component in turn. The navigation menu was kept on the side, and was made more cohesive by the addition of the theme colour as the background, linking the logo to the menu. The link to the search page was removed from this menu and was made into a prominent direct input on the home page (it is placed at the top-right of the page on other pages).

As for the archive metadata, we aimed for a scalable model that would make near-duplicates immediately obvious and allow authors to be treated as separate entities (e.g., to differentiate between authors with the same name, famously the two "Maximilan Haslbeck"), while being as close as possible to the original data model. Since the metadata is under version control and is maintained by AFP editors, it is kept in plain-text. For better tool support, we migrated the metadata from its original INI-like format to TOML[3].

For site generation, *Hugo* was chosen from the available template-based open-source components for its run-time performance (due to its statically compiled Go code) and rich feature set. For integration with Isabelle we created an Isabelle/Scala component that extracts the necessary data (e.g., dependencies between entries) and generates the website source code, ready for compilation by the SSG.

---

[2] https://lists.cam.ac.uk/sympa/arc/cl-isabelle-users/2020-11/msg00036.html.
[3] Tom's Obvious, Minimal Language https://toml.io/en/v1.0.0.

3   Implementation

The redesign started in late 2020 as a stand-alone project using metadata exported from the AFP. An implementation as a proper Isabelle component inside the AFP development repository followed in 2021. The website is fully integrated and was launched in the early summer of 2022. We provide the tools to extract required data (into JSON datasets, as often recommended [5]) and build and generate the site via the central Isabelle add-on component for the AFP. This also allows accessing the metadata by (statically typed) programmatic means. To transfer the metadata to the new model, we supplied a one-shot conversion tool, which also handles the initial author de-duplication.

To generate the site, our component reads metadata, Isabelle metadata files and theories. An Isabelle/AFP build is not needed, as the HTML presentation is dynamically loaded. When building the development snapshot of the site (which happens automatically when Isabelle or the AFP are changed), a status file containing information about the completed Isabelle/AFP build is read.

For the site generation, we extract the overall entry session structure and statistics with Isabelle/Scala functions, and we obtain keywords (for search and to relate entries) by utilising the Rapid Automatic Keyword Extraction algorithm [7] on the entry abstracts. From this data, the metadata, and templates, our component assembles the Hugo project, which is then compiled into the resulting static website.

The final design is shown in Fig. 2. It can now also be interactively explored at the original location (entry URLs have stayed the same as well). We discuss the individual aspects in the following.

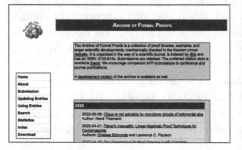

Fig. 1. Previous AFP home page

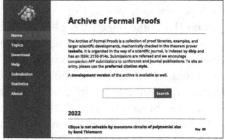

Fig. 2. Redesigned AFP website

3.1   Metadata Model

Previously, the data model of an entry was loosely defined by documentation and the implicit model of existing metadata and code. We improved on this by defining it explicitly inside Isabelle/Scala, over which it can be accessed programmatically (though an index is also available on the web page as JSON file).

As shown in Fig. 3, authors and affiliations are now handled as first-class entities, rather than just as names and URIs. To make duplications obvious and avoid inconsistencies, even when manually editing the TOML data, authors and affiliations are stored separately and only referenced by entries.

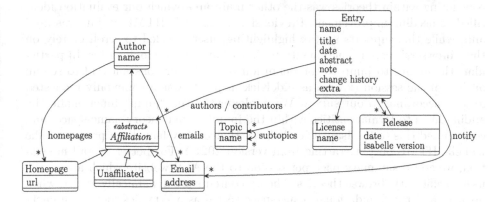

Fig. 3. Data model for the new AFP metadata.

3.2   Searching Capabilities

While the newly introduced Isabelle search engines (cf. Sect. 1) enable finding results in AFP theories, the previous AFP search facility redirected the user to Google search results restricted to the AFP domain. In our evaluation of the current AFP, users noted that search needed much improvement, as it was imprecise and would often return duplicates from different AFP versions.

Hence, the search capabilities had to be improved. A server-side search was not an option since the website had to be statically built to keep maintenance low. Therefore, we replaced this search with a client-side search using an optimised JavaScript search engine, namely the FlexSearch.js library[4]. We generate several search indices using keywords, topics, abstracts, titles, and author names. As the user types, the in-memory indices are searched in parallel, with auto-complete suggestions being provided by the keyword index. The search delay is hardly noticeable on a typical end-user machine and we deemed it not detrimental to user experience. On the results page, we primarily display the returned entries, but also include links to authors and topics.

The search was additionally integrated with FindFacts [4] (using the Find-Facts REST interface) which allows for searching within the theories of the AFP. Even though that search is server-side, the large amount of data searched in means that there is a small delay which made it necessary to de-bounce queries during user input. We display a short summary for those results (the number

---

[4] https://github.com/nextapps-de/flexsearch.

of matching constants, facts, and types) and link to the corresponding query on FindFacts.

3.3 Theory Browsing

Navigating within theories was the other main area which our evaluation identified as needing improvement. Previously, the Isabelle HTML output was used and, while this supports syntax highlighting, users needed to scroll or rely on their browsers' search functionality to find what they are looking for. In particular, the ability to jump to the definition of a type or constant and to see an outline of the session (known as SideKick in jEdit) was a commonly requested feature according to our survey. We implemented the "jump" functionality by adding links to formal entities within the foundational theory name-space, and contributed this to the Isabelle code-base. Hence, they have been present in the Isabelle HTML output starting from version 2021-1. For better overall navigation, we add a summary page per session to the AFP website which allows the user to efficiently browse theories: Theory contents are dynamically loaded when opened, and the foundational name-space serves as a side-kick view with clickable links that jump to definitions. The side-kick scrolls with the user, and the definition at the current scroll position is highlighted.

3.4 Site and Entry Navigation

Up until now, the AFP site only consisted of two content-focused pages, namely the "home" and "topic" pages. Using our improved author metadata, we added an author list and, with the help of Hugo taxonomies, we also created summary pages for each author, topic, and the dependencies of each entry. Using abstract keyword and topic similarity, we can now also display related entries—this makes it easier to browse developments that were split into multiple submissions, but also allows the discovery of new material. Additionally, we generate RSS feeds for those taxonomies, so AFP users can be notified of following updates.

4   Conclusion

For almost 20 years, the Archive of Formal Proofs has been a central resource for disseminating mathematical theories for the Isabelle theorem prover. However, during that time, its website has not been kept up-to-date with many features that users have come to expect. Our redesign maintains all the original functionality, whilst addressing user pain points that we have previously identified concerning search, navigation and code browsing. It remains familiar to existing users, but we believe the responsive and modern design will better reflect the continually evolving and improving Isabelle eco-system.

References

1. Archive of Formal Proofs: Statistics. https://www.isa-afp.org/statistics.html
2. Blanchette, J.C., Haslbeck, M., Matichuk, D., Nipkow, T.: Mining the archive of formal proofs. In: Kerber, M., Carette, J., Kaliszyk, C., Rabe, F., Sorge, V. (eds.) CICM 2015. LNCS (LNAI), vol. 9150, pp. 3–17. Springer, Cham (2015). https://doi.org/10.1007/978-3-319-20615-8_1
3. Huch, F.: Formal entity graphs as complex networks: assessing centrality metrics of the archive of formal proofs (2022). Unpublished
4. Huch, F., Krauss, A.: Findfacts: a scalable theorem search. In: Isabelle Workshop (2020)
5. Library of Congress: JSON (JavaScript object notation). Library of Congress (2014). https://www.loc.gov/preservation/digital/formats/fdd/fdd000381.shtml
6. MacKenzie, C., Fleuriot, J.D., Vaughan, J.: An evaluation of the archive of formal proofs. CoRR abs/2104.01052 (2021). https://arxiv.org/abs/2104.01052
7. Rose, S., Engel, D., Cramer, N., Cowley, W.: Automatic keyword extraction from individual documents. Text Min. Appl. Theory **1**, 1–20 (2010)
8. Stathopoulos, Y., Koutsoukou-Argyraki, A., Paulson, L.: Serapis: a concept-oriented search engine for the Isabelle libraries based on natural language. In: Isabelle Workshop (2020)

Injecting Formal Mathematics Into LaTeX

Dennis Müller$^{(\boxtimes)}$ (iD) and Michael Kohlhase (iD)

Computer Science, FAU Erlangen-Nürnberg, Erlangen, Germany
{dennis.mueller,michael.kohlhase}@fau.de
https://kwarc.info

Abstract. The paper presents the sTeX3 format for representing informal mathematics. sTeX3 acts as a surface language for two systems: the (presentation-oriented) LaTeX system to produce PDF and the semantics-aware MMT system for advanced knowledge management services. We discuss how the sTeX3 markup facilities allow in situ flexiformalization (and the necessary elaboration of complex structures), while staying presentationally neutral.

This paper uses sTeX3. The semantically annotated XHTML version of this paper is available at https://tinyurl.com/cicm22stexmmt.

1 Introduction

Most formal systems offer two kinds of representation formats for input/output: a standardized system language for efficient storage and loading of compiled content and various surface languages for interacting with users. While the former usually optimize fully explicit representations, the latter cater to the notational preferences (and cognitive abilities) of humans. As an effect, surface languages usually have to be elaborated into the system language by e.g. macro expansion, overloading resolution, and type inference. This setup works very well where there is only one formal system that consumes the surface language.

For sTeX [Koh08, sTeX] however, the LaTeX-based surface language for MMT/OMDoc presented in this paper, we have two such consumers: The (presentation-oriented) LaTeX system to produce PDF, and the semantics-aware MMT system [RK13, MMT] for advanced knowledge management services.

With our recent redesign and reimplementation of sTeX as sTeX3 (see our System Description at this very conference), we decided on two criteria regarding these two consumers:

LaTeX : sTeX3 should be compatible with arbitrary packages, document classes and general styles of presentation and typesetting, without impacting an authors preferred mode of expression or the final document layout, while still allowing for annotating text fragments with their (flexi-)formal semantics, and

Mmt : sTeX3 should act as a functionally complete surface language for the MMT system.

This implies two important challenges to solve:

© The Author(s), under exclusive license to Springer Nature Switzerland AG 2022
K. Buzzard and T. Kutsia (Eds.): CICM 2022, LNAI 13467, pp. 168–183, 2022.
https://doi.org/10.1007/978-3-031-16681-5_12

Informal→Formal: STEX3 needs to provide support for annotating mathematical content *as actually written* by authors in practice – i.e. it needs to support the linguistic phenomena ubiquitously employed by (informal) mathematicians, and appropriately translate them into (flexi-)formal presentations actionable by MMT.

Formal→Informal: STEX3 needs to allow for all the structural and object-level features provided by the MMT system, without being (too) invasive with respect to the final presentation of the content, and (importantly) without being semantics-aware and having access to the associated algorithms of a formal system (e.g. type inference), to remain consumable by LATEX.

In the following, we describe our approach to solving these two challenges in Sects. 3 and 4, respectively.

This paper itself uses STEX3 within the `llncs` document class. We therefore made the source files available on Overleaf at https://www.overleaf.com/read/bztqrgbjmkcd. The semantically annotated XHTML generated from it is linked above; the MMT/OMDoc declarations extracted and checked by MMT can be browsed at https://tinyurl.com/cicm22stexomdoc[1].

2   Preliminaries

Both STEX and MMT are based on the OMDoc [Koh06] format (Open Mathematical Documents), a conceptual ontology and XML-based representation format for the content and document structure of (informal) mathematical documents. STEX is the result of adding semantic macros to LATEX to annotate the OMDoc markup without changing the generated PDF, while MMT re-engineers and implements the formal core of OMDoc, using MMT/OMDoc as an ontology and XML-based system language and implements it in the MMT system (and corresponding surface language).

We will now give a brief overview over the MMT/OMDoc ontology and discuss how STEX3 now implements it in the form of a LATEX package. For system-related details, see the companion system description published at CICM 2022.

MMT/OMDoc: For the detailled ontology of MMT/OMDoc, we refer to [Mül19]. For now we restrict ourselves to the basics: content in MMT/OMDoc is organized in named theories, containing declarations (primarily constants). A constant has a name, optional aliases, an optional type and an optional definiens. Type and definiens are represented as OPENMATH expressions. Theories can include other theories, making their declarations available in the including theory. The semantics of MMT content is provided by rules written in Scala and activated in theories via dedicated rule constants, providing their fully qualified class path and (optionally) parameters. All theories and declarations are assigned globally unique MMT URIs for unambiguous referencing.

---

[1] Note that the latter requires a browser with MathML support to display formal expressions, such as Firefox, but notably not (vanilla) Chrome *yet* (intent-to-publish for MathML in Chromium has been announced, expected in 2023).

We will introduce and discuss the remaining concepts of the MMT/ OMDoc ontology in their respective dedicated sections below.

STEX : Omitting the details on how structured data can be represented in TEX, \begin{smodule}{name} opens a new module (the STEX equivalent of a MMT theory). It is represented internally as several macros storing data such as its name, its included modules, the names of contained symbols, and code to execute whenever the module is *activated* (e.g. when being included in some other module).

\symdecl{macroname}[name=symbolname,type=tp,def=df,args=arity] generates a new symbol with provided arity, name, type and definiens and generates a semantic macro \macroname that expands to \invoke_symbol{namespace?module? symname}. The command \notation{symname}[id]{...} introduces a new notation for a symbol with identifier id, that \invoke_symbol can defer to for typesetting, e.g. when using \macroname[id]. Importantly, "generating a new symbol" is distinct from generating a semantic macro, a symbol's name is distinct from its semantic macro, and notations are internally separate from the symbol they are attached to, allowing for them to be copied and inherited between symbols. The macro \symdef combines \symdecl and \notation for convenience.

STEX content can be formatted by pdflatex, and additionally be converted to semantically annotated XHTML, the annotations of which are entirely dictated by the STEX package itself. Most STEX macros and environments, when being converted to XHTML, introduce annotations in the form of XML attributes indicating their corresponding MMT/OMDoc concept – for example, a notation introduces annotations for the OPENMATH label (e.g. OMA, OMID, OMBIND), the full MMT URI of the symbol the notation belongs to, the argument positions, etc. As a result, the MMT system can import the generated XHTML by (to a large extent) directly mapping OPENMATH-related annotations to actual OPENMATH expressions, annotations generated by \symdecl to MMT constants, and those generated by \begin/\end{smodule} to MMT theories.

The macro \importmodule[some/archive]{path?Modulename} includes the module Modulename from the (optionally) provided archive into the current module, and exports an appropriate annotation with the full MMT URI in the XHTML, which MMT can directly translate to an include of the corresponding MMT theory.

The macro \MMTrule{classpath}{arguments} is largely ignored in STEX except for introducing a corresponding annotation that MMT can translate to a rule constant.

As in MMT, each module can be assigned a *meta theory*; if none is specified, a default meta theory is used. For the purposes of this paper, only two of its symbols are relevant: \bind $((\cdot) \to \cdot)$ behaves like (an informal variant of) a lambda or (dependent function type) pi operator, depending on context; \implicitbind $(\{\cdot\}_I \cdot)$ behaves largely the same way, but marks the variable(s) bound as representing *implicit* arguments.

The rest of this paper will now focus on the remaining features of STEX not directly covered by MMT/OMDoc, and the remaining features of MMT not covered by STEX by all of the above.

As a continuous example thoughout, we will work our way towards defining a *lattice* as a mathematical structure composing two *semilattices*, and declare the associated duality principle and the equivalence of the algebraic and order-theoretic definitions as theoey morphisms, such that the presentation remains "typical" for informal mathematics and the extracted formal content can be fully type checked by MMT, all in-situ in (the source code of) this very paper.

3   Translating Phenomena in Informal Mathematics to a Formal System

3.1   Formal Theories Vs. Document Fragments

The most fundamental difference between content representation formats for informal mathematics like OMDoc and for formal mathematical knowledge like MMT is that the former focuses on structured (fragments of) mathematical text whereas the latter focus on modular collections of assertions of properties of objects (called declarations in MMT). Both OMDoc and MMT use OPENMATH expressions to represent the (functional structure) of objects, i.e. mathematical formulae, and theories /inheritance to represent contexts.

While mathematical knowledge in MMT can be structured and sequentialized according to visibility criteria alone (i.e. which declarations can be referenced in some context), STEX needs to take the narrative structure of documents into account. In particular, there must be a facility to make module contexts available in document fragments to make the referenced concepts in them well-scoped. But unlike theories, they should not always "re-export" the used declarations, since this can lead to circular dependencies. For instance, in motivational or "preview" fragments, mathematical concepts are "used" before they are introduced. To account for this, STEX provides the \usemodule instruction which behaves like \importmodule but without "exporting" the contents of the used module when the current module is imported elsewhere. This distinction however does not exist in MMT.

Another important distinction of formal and informal systems is that informal mathematics relies on mathematical vernacular that comes in natural language variants, whereas formal systems are – a priori – (natural) language agnostic.

Signature and Language Theories. To account for this language diversity we generate two distinct MMT theories from an STEX-module: The *signature theory* is assigned the actual MMT URI ns?name of the module it is generated from, and contains exactly the formal, declarative parts of the module: The includes and constants generated from \importmodules and \symdecls, respectively.

An additional *language theory* is assigned the MMT URI ns/name?lang, where **lang** is the language the original document is in (by default english). This theory includes the signature theory, an explicit include for every usemodule, and for every (non-trivial) OPENMATH-expression occurring outside of symbol components a constant with that expression as definiens (which is checked by MMT by having its type inferred).

Besides allowing for representing all aspects of a document with a formal semantics, this distinction between language and signature theories is also very well suited for ſТEX3's approach to multilinguality in general (see [KM] for details).

3.2 Implicit Arguments and Variables

Another – often overlooked – difference between formal and informal systems is the use of variables. In Mmt, declarations are self-contained (i.e. closed) formal expressions, whereas OMDoc statements are text fragments containing (usually/often open) expressions. For instance, an english sentence like "For a natural number n, $n+n$ is even." contains the two open expressions n and $n+n$, whereas the corresponding formal declaration $\vdash \forall n : \mathbb{N}.\mathrm{even}(n + n)$ is closed.

Thus in Mmt, variables only ever occur on the object-level, i.e. in a type or definiens component of a term, and bound by some binding operator.

In OMDoc, the majority of symbols are arguably variables, by virtue of being unspecified, postulated or can be thought of as "universally quantified" over whole paragraphs, chapters or even full documents as scope. In the example above, the fragment "natural number n" acts as a variable declaration, whereas the "n" in the second expression only references the variable n.

Correspondingly, ſТEX3 allows for using variables in two ways: The command \svar[name=x]{some variable} marks up the text some variable as being a reference to some variable with name x. This is mostly useful for simple variable names, e.g. \svar{x}. Additionally, we can *declare* a variable including its notation, type, definiens etc. in the same manner as we do symbols, using \vardef{setA}[type=\collection]{\comp{A}}, which allows us to invoke the variable using the semantic macro \setA, producing A. This macro is local to the current TEX-group, determining its scope in the document. Such a variable declaration is exported to the XHTML and used by Mmt when resolving its subsequent occurrences.

For any expression we want to convert to Mmt/OMDoc, we look up any free variable in the current scope to determine their types and other attributes (if given), and abstract them away using a dedicated binding operator \implicitbind, effectively marking them as *implicit arguments*, which is handled thusly: Whenever a term with head o occurs in an XHTML document, Mmt looks up its type and/or definiens. If either is of the form \implicitbind{vars}{bd}, Mmt inserts fresh variables in their place and attempts to infer their values during type checking. This makes sure that all variables in an ſТEX document are bound and also allows us to conveniently specify implicit arguments. For example, we can declare typed equality like this:

```
1    \vardef{varx}[type=\setA]{\comp x}
2    \vardef{vary}[type=\setA]{\comp y}
3    \symdef{tpeq}[name=typed−equality,args=2,
4      type=\bind{\varx,\vary}\prop]{#1 \comp= #2}
```

The explicit type we wrote down here is simply $(x,y) \to \text{prop}$. However, by having declared x and y to be of type A, MMT considers A to be a free variable and abstracts it away, yielding the actual type $\{A \in \text{SET}\}_I(x \in A, y \in A) \to \text{prop}$.

This allows for writing the types of symbols in accordance with their (notational) arity, practically hiding implicit arguments from an author entirely.

While it might seem cumbersome to explicitly declare variables rather than just typing \svar{x} or even x directly (which both SLaTeX and the subsequent MMT-import allow for as well), in practice it has turned out that variables will be reused often enough (see e.g. the source file of this paper) to not cause too much overhead, while allowing for declaring their types and other attributes *once*.

3.3   Flexary Operators and Argument Sequences

While in formal settings, symbols are usually *required* to have a fixed arity, *flexary* operators (or, rather, notations) abound in informal settings. In most cases, this can be seen as a mere notational convention, e.g. if addition is known to be associative, then writing $a + b + c$ is unambiguous; similarly, by convention $A \to B \to C$ represents $A \to (B \to C)$ rather than $(A \to B) \to C$. Nevertheless, restricting ourselves to binary operators would mean forcing authors to type e.g. \plus{a}{\plus{b}{c}}, which would be prohibitively cumbersome. Even worse: If we wanted to write $a = b = c$, it is not even clear how to do that given a binary equality alone, given that e.g. the subexpression $a = b$ is a proposition, and the statement clearly does not mean to assert that *the proposition $a = b$* is equal to c.

[HKR14] attempts to solve this problem specifically in the context of MMT, however, here the notion that an operator is *flexary* (taking a sequence as an argument) is reflected both in the type and the definiens of a symbol – contrary to the idea that e.g. equality should "formally" be a *binary* operation with corresponding type, regardless of (flexary) "abuse" of notation.

Syntactically, SLaTeX handles this problem by allowing symbols to take a comma-separated list as (one or several) arguments, signified via an a in its key args= That allows us to specify equality more adequately as:

```
1   \symdef{eq}[name=equal,args=a,
2     type=\bind{\varx,\vary}\prop][#1]{##1 \comp= ##2}
```

We refer to [KM] for the details of the notation specification; suffice it to say that this allows us to write \eq{\varx,\vary,\varz} to obtain $x = y = z$.

A natural next step is to then also allow for *variable sequences* of unspecified length, which we can declare in SLaTeX via
\varseq{seqx}[type=\setA]{1}{\svar{n}}{{\comp x}_{#1}}, giving us the semantic macro \seqx. \seqx{i} then produces x_i, \seqx! produces $x_1, \ldots, x_n$, and \seqx can be used as argument for any flexary argument in a semantic macro – e.g. \eq\seqx yields $x_1 = \ldots = x_n$.

This poses the question how we translate all these representations into formal MMT/OMDoc, where equality is supposed to have type $(x \in A, y \in A) \to \text{prop}$.

In a first step, MMT wraps the argument sequence into an application of a dedicated constant seq(x,y,z), to clearly distinguish the boundaries of the argument sequence from (potential) neighbouring arguments. Additionally, we provide typical operations on sequences as constants, such as map, fold, head/tail etc. with corresponding simplification and typing rules. Lastly, we allow authors to specify the precise behaviour of a flexary operator via an additional key assoc= for \symdecl with the following possible values:

pre individually prefixed, e.g. $\forall x, y, z.P$ resolving to $\forall x.\forall y.\forall z.P$.
binr right-associative binary, e.g. $A \to B \to C$ resolving to $A \to (B \to C)$
binl analogously left-associative,
 bin associative binary,
conj conjunctive[2] binary, as in $A = B = C$ resolving to $A = C \wedge B = C$.

MMT attaches this information as meta data to the declaration, and subsequently uses it to insert (mainly) folds over the provided sequences in symbol applications to guarantee that typing checks succeed correctly. In the case where a sequence is *definite* (rather than variable), simplification rules on the sequence operations yield the actual desired outcome (e.g. $x = y = z$ resolving to $x = z \wedge y = z$).

3.4   Statements (Theorems, Definitions, Axioms, Proofs, ...)

Statements are among the most central notions in mathematics and should be adequately represented in STEX. But they are also the representational feature, where informal and formal practice differ most. In most formal systems, statements are represented as compact expressions, whereas in informal mathematics, they are usually type-labeled and (often) numbered text fragments (sometimes multiple paragraphs long) with (explicit or implicit) premises, may contain (sub-)definitions or lemmata, and in general contain a lot more structure than mere closed formal expressions. A single informal statement usually corresponds to multiple formal ones. Thus STEX needs to allow in-situ markup for the statement fragments in a way that this correspondence is made explicit without disrupting the (PDF) presentation.

We support doing so by allowing semantic macros outside of math mode to annotate arbitrary text, mark variable declarations inside statements as universally bound, and explicitly markup the *definiens* or *conclusion* of a statement. Similar support for *premises* and other statement components is planned. The MMT import is subsequently tasked with assembling all these individual components of a statement into the single (intended) formal expression.

For example, we can assert the reflexivity for our equality thusly:

```
1   \begin{sassertion}[type=axiom,name=equality−reflexive]
2      \conclusion{\\forall{The \symref{equal}{equation}
```

---

[2] This case requires a symbol for conjunction to be in scope, which can be marked as such using a dedicated parametric MMT rules. The expression $x = y = z$ above (and now here, too) type checks, because we include a module that provides one earlier.

```
3        $\arg[2]{\eq{\varx,\varx}}$ holds
4        \comp{for all} $\arg[1]{\inset\varx\setA}$}}.
5    \end{sassertion}
```

yielding:

Axiom 1. The equation $x = x$ holds for all $x \in A$.

Again, we refer to [KM] for syntactic details, and merely note that we use the already present semantic macros to annotate the statement, which is also marked-up as one single *conclusion* with no premises. For a more elaborate example, see Definition 2.

The *judgments-as-types* paradigm has been proven to be effective to model declaratively true propositions , axiomatic (via undefined declarations) or proven (via defined declarations), both in general as well as in MMT specifically. Within STEX, this is reflected by \begin{sassertion}[name=equality−reflexive] generating a symbol **equality-reflexive** with no semantic macro. MMT subsequently uses the contents of the \conclusion{}-macro and the semantic macros therein to construct the actual proposition \ *forall*{\inset\varx\setA}{\eq{\varx,\varx}} (i.e. $\forall(x \in A).(x = x)$) to assign **equality-reflexive** the type $\vdash(\forall x \in A.(x = x))$, where $\vdash$ maps propositions to types.

Definitions work similarly, except here, we do not generate a new constant. Instead, for each \definiendum[symbolname]{...}, MMT either attaches a definiens to **symbolname**, if **symbolname** was declared in the same module *and* was previously undefined, or generates an equality rule for the symbol and the new definiens otherwise.

Variables declared *within* a statement and marked with bind=forall are appropriately bound/abstracted away, as are previously declared variables marked with \varbindforall. In preparation for the next sections, this allows us to e.g. define associativity as a property of binary operations like this:

```
1    \vardef{varop}[args=a,op=\circ,assoc=bin,type=\funspace{\setA,\setA}\setA]
2        {#1}{##1 \comp\circ ##2}
3    \symdecl{associative}[args=1,type=\bind{\varop!}\prop]
4
5    \begin{sdefinition}[for=associative]
6        \varbindforall{varop}
7
8        An operation $\fun{\varop!}{\setA,\setA}\setA$ is called
9        \definame{associative}, if \definiens[associative]{\forall{$\arg[2]{
10           \eq{
11              \varop{(\varop{\varx,\vary}),\varz},
12              \varop{\varx,(\varop{\vary,\varz})}}
13        }
14    }$ \comp{for all} $\arg[1]{\inset{\varx,\vary,\varz}\setA}$}}.
15   \end{sdefinition}
```

Yielding:

Definition 1. *An operation* $\circ : A \times A \to A$ *is called* ***associative***, *if* $(x \circ y) \circ z = x \circ (y \circ z)$ *for all* $x, y, z \in A$.

MMT consequently generates the definiens

$$\{A \in \mathsf{SET}\}_I (\circ \in A \times A \to A) \to \forall (x, y, z \in A).((x \circ y) \circ z = x \circ (y \circ z)) \qquad \text{for} \qquad \text{the}$$

constant **associative**.

Notably, the sdefinition environment's for= key allows for a comma-separated list of multiple symbols, and may contain several \definiens-calls to allow for defining multiple symbols in the same statement. Additionally, every \definame, \definiendum and \definiens adds the referenced symbol to the for=-list automatically, and in the case where it contains a single element, the option [associative] for \definiens is optional, so the above example is redundant in several places. Also, note that sdefinition defers to the definition-environment provided by (in this paper) the llncs document classes for the actual typesetting (this behaviour can me customized).

4   Implementing Structural Features in sTEX

Having investigated how to translate informal phenomena to MMT/OMDoc, we will now discuss the reverse direction: Representing the language features available in MMT in sTEX.

MMT implements a mechanism for generic structural features [CICM20]. In MMT surface syntax,

featurename name ((parameters)∗) = (declarations)∗ ▌

behaves *internally* like an MMT heory, but generates a derived declaration with name name instead. After closing the declaration (with ▌), MMT will look for a structural feature (a special rule implemented in Scala) with name featurename, that processes the parameters and the body of the declaration and elaborates it into a set of primitive declarations computed from the derived declaration.

If we want to allow for structural features in sTEX, we consequently need to mimic the elaboration behaviour of the respective feature within LATEX, making a certain amount of code duplication unavoidable. Luckily, since LATEX does not need to be aware of the precise semantics of declarations in general, the elaboration in sTEX only needs to cover the names and notations of elaborated symbols, not the (computationally considerably more complicated) types and definientia.

More importantly however, structural features are an *extension principle* of MMT – users can in principle provide new feature anywhere. Consequently, sTEX needs to be similarly extensible. Where structural features in MMT need to be specified in Scala, for sTEX, their elaboration naturally needs to be specified using LATEX, and where MMT provides an *abstract class* for this purpose, sTEX provides a dedicated environment structural_feature_module that takes care of much of the boilerplate.

The rest of this section will now cover the most important and (in some cases) *primitive* structural features of MMT.

4.1   Mathematical Structures (Module Types)

Module types were introduced in [MRK18], already for the explicit purpose of mimicking the way mathematical structures – such as algebraic structures, topological spaces, ordered sets, and other models of a (usually first- or second-order) theory – are treated in informal mathematics.

Given a theory M, the type $\mathrm{MOD}(M)$ represents a dependent record type with manifest fields, with a field of name d for every declaration with name d in M. Given a record $m : \mathrm{MOD}(M)$, a field d is accessible via the record projection $m.d$. We refer to [MRK18] for the implementation details in MMT.

As an introductory example, the following code introduces the structured type of semilattices:

```
1   \begin{mathstructure}{semilattice}
2     \symdef{universe}[type=\collection]{\comp L}
3     \symdef{op}[type=\funspace{\universe,\universe}\universe,args=a,op=\circ,
4       assoc=bin]{#1}{##1 \comp{\circ} ##2}
5     \begin{sdefinition}[for=semilattice]
6       A structure $\mathstruct{\universe,\op!}$
7       with $\fun{\op!}{\universe,\universe}\universe$
8       is called a \definame{semilattice}, if
9       \inlineass[name=associative−axiom]{\conclusion{
10          \associative{$\arg{\op!}$ is \comp{associative}}}%
11      }},
12      \inlineass[name=commutative−axiom]{\conclusion{
13          \commutative{$\arg*{\op!}$}$\comp{commutative}}%
14      }} and
15      \inlineass[name=idempotent−axiom]{\conclusion{
16          \idempotent{$\arg*{\op!}$}$\comp{idempotent}}%
17      }}
18    \end{sdefinition}
19  \end{mathstructure}
```

This generates a new (nested) module semi-lattice-structure, with constants universe, op, and the three axioms (the command \inlineass behaves like the sassertion-environment), and generates the following output:

Definition 2. *A structure $\langle L, \circ \rangle$ with $\circ : L \times L \to L$ is called a **semilattice**, if $\circ$ is associative commutative and idempotent*

Since semilattice is defined as a structure with a sequence of axioms rather than a simple (object-level) definiens, we do not use the \definiens-macro here.

mathstructure also generates a new symbol semilattice. Subsequently, it allows us to *instantiate* semilattice via \instantiate or \varinstantiate, after which we can access the individual fields of the instance by their names. For example, the code

```
1   \varinstantiate{varL}{semilattice}{\mathcal{L}}
2   Let $\defeq{\varL!}{\mathstruct{\varL{universe},\varL{op}!}}$ a
3   \symname{semilattice}, then $\varL!$ is
4   \varL{commutative−axiom}{\comp{commutative}} by definition.
```

yields the following output:

Let $\mathcal{L}:=\langle L,\circ\rangle$ a semilattice, then $\mathcal{L}$ is commutative by definition.

Additionally, we can provide explicit definientia for some of the fields in an instantiation:

```
1   \varinstantiate{varLb}{semilattice}{\mathcal{L}_2}[universe=setA]
2   Let $\defeq{\varLb!}{\mathstruct{\varLb{universe},\varLb{op}!}}$
3   a \symname{semilattice} on $\setA$.
```

Let $\mathcal{L}_2:=\langle A,\circ\rangle$ a semilattice on A.

Internally, the mathstructure delegates to the generic environment structural_feature_module, which returns the list of the MMT URIs of all symbols declared in or imported into its body. mathstructure stores this information in a dedicated LATEX3 property list indexed by the name.

\instantiate and \varinstantiate then define a new property list the same way, but storing the respective code to invoke the assignment target, depending on whether it is a variable or a symbol, rather than the MMT URI. While this forbids assigning fields in an instantiation to arbitrary expressions, this is necessary to remain compatible with e.g. symbol references and notation usages. In practice, this limitation can be remedied by introducing a defined variable with the desired expressions as definiens. This also means, that \varLb{universe} does in fact expand to \setA directly.

For the unassigned fields, SΤEX instead generates new variables with names <instancename>.<fieldname>, copying the notations of the original declarations in the mathstructure-environment. For example, \varLb{op} ultimately expands to \stex_invoke_variable{varLb.op}, which inherits its notation (and all other required properties) from the \symdef{op} in semilattice−structure. The MMT importer is subsequently able to reconstruct the actual term varLb.op from the variable name for unassigned fields, and is provided with the simplification of the field projection already when using assigned fields.

4.2   Includes with Modification (MMT-structures)

In MMT, the name *structure* refers to a (usually named) include with modification – a theory morphism that copies the declarations of its domain into the codomain.

The contents of an MMT structure are internally represented with the same data structures as the declarations in a theory, and the surface syntax is identical in both cases; MMT however subsequently checks that every constant in a structure corresponds (by name) to one in the domain, and its components are interpreted as reassignments of those of the domain constant. The structure can thus assign new definientia, notations or aliases, provided judgments are preserved. Given a declaration d in a theory $ns_1?T$, its target in a structure s from some theory T_1 (including T) to $ns_2?T_2$ is given the

MMT URI $ns_2?T_2?s/[ns_1?T]/d$. Notably however, in MMT's surface syntax, the name of the generated MMT URI can not effectively be expressed (due to the full URI $ns_1?T$ occurring in its name), making it in practice impossible to refer to this declaration without using either its notation or an alias – it does, however, make it possible to disambiguate between multiple declarations with the same name but different full MMT URIs.

Since the name "structure" strongly conflicts with the usage of the word in informal mathematics, we instead refer to the corresponding sTEX feature as a copymodule. We also simplify the feature slightly by assuming that (as validated in practice) name clashes within one module (that is a meaningful domain of a copymodule) do not occur, and simply generate the name ?s/d instead.

Similarly, since reusing the \symdecl macro would be syntactically misleading with respect to its semantics, we instead introduce dedicated macros \renamedecl and \assign to provide aliases and new definientia in a copymodule.

For   \assign{symbolname}{term},   sTEX   merely resolves symbolname to its full MMT URI and exports the pair to the XHTML.
\renamedecl{symbolname}{newname} generates a new semantic macro \newname for s/symbolname; the variant
\renamedecl[name=anothername]{symbolname}{newname} additionally renames the symbol to anothername and is interpreted as an alias in the MMT import.

As a consequence, we can use copymodule environments in sTEX exactly like MMT structure, for example to assemble a module of lattices via combining two distinct copies of semilattice and providing appropriate new notations:

```
1    \begin{mathstructure}{lattice}
2      \begin{copymodule}{MathematicalStructures/semilattice−structure}{joinsl}
3        \renamedecl[name=universe]{universe}{universe}
4        % the target of "universe" is now called "lattice−structure?universe"
5        % rather than "lattice−structure?joinsl/universe"
6        \renamedecl[name=join]{op}{join}
7      \end{copymodule}
8      \begin{copymodule}{MathematicalStructures/semilattice−structure}{meetsl}
9        \assign{universe}{\symname{lattice−structure?universe}}
10       \renamedecl[name=meet]{op}{meet}
11     \end{copymodule}
12     \notation*{join}[op=\vee,vee]{#1}{##1 \comp\vee ##2}
13     \notation*{meet}[op=\wedge,wedge]{#1}{##1 \comp\wedge ##2}
```

We can then add the additional axioms using the techniques described above:

Definition 3. *A structure $\langle L,\vee,\wedge \rangle$ is called a **lattice**, if both $\langle L,\vee \rangle$ and $\langle L,\wedge \rangle$ are semilattice, and the equations $a\vee(a\wedge b) = a$ and $a\wedge(a\vee b) = a$ hold for all $a,b \in L$.*
*The operations $\vee$ and $\wedge$ are called **join** and **meet**, respectively.*

4.3   Parametric Theories, Views, Realizations and Realms

There are some special cases of MMT structure relevant for the following development:

1. *Unnamed* MMT structure are considered *implicit* – the targets of the individual declarations in the domain theory retain (and are referred to by) their original MMT URIs. In fact, MMT includes are internally represented simply as unnamed structures with no assignments. MMT requires however, that there is at most one implicit morphism between any two theories (which includes their compositions), and that the sub-diagram of theories with implicit morphism is acyclic.

2. *Total* MMT structures are those that explicitly assign all declarations in their domain to some valid term over the codomain. Structures marked as total are checked by MMT with respect to this requirement.

3. Structures that are both total and unnamed are referred to as *realizations* and are treated in a special manner: If s is a realization from S to T, S' has an implicit morphism m from S, and T' has an implicit morphism from T, then whenever a new implicit morphism from S to T occurs, it is replaced by the corresponding *pushout* along m and s.

While views, as the most general theoey morphisms, are conceptually fundamental to the MMT/OMDoc core design and ontology, in practice they are never actually *used* in the MMT surface language – instead, they have turned out to be always subsumed by a more appropriate kind of MMT structure; either a total structures or a realization.

Similarly, while MMT offers parametric theories and includes, the fact that an include of a parametric theory requires instantiating its parameters and is an implicit morphism implies that only one set of assignments for theory parameters can be supplied in the cone of outgoing implicit morphism of any theory that includes a parametric theory.

What this means is that both views and parametric theories are fully subsumed by total structures and realizations for views, and realizations for (includes of) parametric theories. Consequently, in STEX3 we decided to drop both entirely in favor of total structures (which we call interpretmodule in STEX) and realizations.

Realizations also allow us to implement *realms* [CFK14], at least in the special case where a canonical candidate for the *face* of the realm exists – a pillar, then, is any codomain of a realization with the face as its domain. In particular, realizations likely allow us to introduce *fully formal* foundations in STEX – or, rather, they allow for developing modules using the generic (informal) symbols in a natural way, while obtaining their counterparts in some formal setting via the induced pushout along realizations from informal modules into formal ones, whenever such a realization is possible.

The syntax of the interpretmodule and realization environments is perfectly analogous to that of copymodule, with the only difference being that interpretmodule explicitly checks that every declaration in the domain is being assigned, and realization explicitly modifies the symbols of the domain and adds the MMT URI of the domain to the list of imported modules. Since STEX already checks that no previously imported module is "activated" twice (for efficiency reasons), this is sufficient to mimic the behaviour of an MMT realizations in the informal setting of LATEX, where the precise type and definientia of symbol are semantically inaccessible and hence irrelevant anyway.

As a consequence, we can use interpretmodule to e.g. implement *endomorphisms* of modules, such as the *dual* of a lattice:

```
1   \begin{smodule}{DualLattice}
2     \importmodule{MMTStructures/lattice−structure}
3     \begin{interpretmodule}{MMTStructures/lattice−structure}{dual}
4       \assign{universe}{\symname{lattice−structure?universe}}
5       \assign{join}{\meet!} \assign{meet}{\join!}
6       \assign{joinsl/associative−axiom}{\symname{meetsl/associative−axiom}}
7       ...
8   \end{interpretmodule} \end{smodule}
```

And we can use a realization to represent the canonical (i.e. implicit in the MMT sense) isomorphism between the algebraic and order theoretic definitions of a lattice. For clarity, we will call the latter a *locally-bounded partial order*:

Definition 4. *A structure* $\langle S, \leq \rangle$ *is called a **locally-bounded partial order**, if* $\leq$ *is a partial order and for all* $a, b \in S$ *there is a **least upper bound** ($\mathtt{lup}(a, b)) \in S$ and a **greatest lower bound** ($\mathtt{glb}(a,b)) \in S$.*

We can then easily obtain the desired isomorphism like this:

```
1   \begin{sassertion}[type=theorem]
2     \begin{realization}{MMTStructures/lattice−structure}
3       \assign{universe}{\symname{lbporder−structure?universe}}
4       \assign{join}{\lup!} \assign{meet}{\glb!}
5     \end{realization}
6     Every \symref{lbporder}{locally−bounded partial order}
7     $\mathstruct{\universe,\porder!}$ is a \symname{lattice}
8     $\mathstruct{\universe,\join!,\meet!}$
9     with $\eq{\join{\vara,\varb},\lup{\vara,\varb}}$ and
10    $\eq{\meet{\vara,\varb},\glb{\vara,\varb}}$
11  \end{sassertion}
```

Theorem 1. *Every locally-bounded partial order* $\langle L, \leq \rangle$ *is a lattice* $\langle L, \vee, \wedge \rangle$ *with* $a \vee b = \mathtt{lup}(a,b)$ *and* $a \wedge b = \mathtt{glb}(a,b)$

5   Conclusion and Future Work

We have presented our approach to integrating the STEX package with the MMT system. As a result, we are now able to use the full suite of formal representation mechanisms of MMT/OMDoc from within LATEX, without sacrificing the latter's typesetting capabilities. Notably, while "properly" specifying fully formal content in STEX requires some expertise to a similar degree as other formal languages, subsequently *using* this content in a *seemingly* informal manner is no more difficult than most other LATEX macros. Additionally, STEX's module system allows for distributed and collaborative development of interconnected libraries.

A Formal Foundation for Informal Mathematics: Naturally, in order to apply useful services enabled by formal methods to informal mathematical documents (such as type checking) requires *some* formal foundation in terms of which the knowledge needs to be represented. As the examples in this paper demonstrate, we have already started experimenting with and implementing one such foundation by providing typing rules for some of the various symbols imported here. This poses the obvious question, what *"the"* adequate *formal foundation for informal mathematics* is, or rather, could be – or, more precisely, which typing and inference rules are necessary as primitives to support formal semantic services for mathematical concepts as defined and used in informal practice.

We are far from answering this question, if indeed a satisfying answer is even possible. Indeed, we conjecture that any plausible candidate for such a "foundation" is necessarily incomplete, context-dependent and likely even locally inconsistent , by virtue of informal practice often ignoring or glossing over technical aspects and details that distract from the matter at hand, but are technically needed for being formally rigorous .

However, we are now in a situation where we can empirically investigate this question by iteratively semantically annotating purely informal mathematical developments, and conveniently and easily experiment with plausible candidates by quickly providing ad-hoc rules where desirable.

References

[CFK14] Carette, J., Farmer, W.M., Kohlhase, M.: Realms: a structure for consolidating knowledge about mathematical theories. In: Watt, S.M., Davenport, J.H., Sexton, A.P., Sojka, P., Urban, J. (eds.) CICM 2014. LNCS (LNAI), vol. 8543, pp. 252–266. Springer, Cham (2014). https://doi.org/10.1007/978-3-319-08434-3_19

[CICM20] Müller, D., Rabe, F., Rothgang, C., Kohlhase, M.: Representing structural language features in formal meta-languages. In: Benzmüller, C., Miller, B. (eds.) CICM 2020. LNCS (LNAI), vol. 12236, pp. 206–221. Springer, Cham (2020). https://doi.org/10.1007/978-3-030-53518-6_13

[HKR14] Horozal, F., Rabe, F., Kohlhase, M.: Flexary operators for formalized mathematics. In: Watt, S.M., Davenport, J.H., Sexton, A.P., Sojka, P., Urban, J. (eds.) CICM 2014. LNCS (LNAI), vol. 8543, pp. 312–327. Springer, Cham (2014). https://doi.org/10.1007/978-3-319-08434-3_23

[KM] Kohlhase, M., Müller, D.: The sTeX3 package collection. Technical Report. https://github.com/slatex/sTeX/blob/main/doc/stex-doc.pdf. Accessed 24 Apr 2022

[Koh06] Kohlhase, M.: OMDoc – an open markup format for mathematical documents [version 1.2]. LNCS (LNAI), vol. 4180. Springer, Heidelberg (2006). https://doi.org/10.1007/11826095

[Koh08] Kohlhase, M.: Using LATEX as a semantic markup format. Math. Comput. Sci. **2**(2), 279–304 (2008). https://doi.org/10.1007/s11786-008-0055-5, https://kwarc.info/kohlhase/papers/mcs08-stex.pdf

[MMT] MMT – Language and System for the Uniform Representation of Knowledge. Project web site. https://uniformal.github.io/. Accessed 15 Jan 2019

[MRK18] Müller, D., Rabe, F., Kohlhase, M.: Theories as types. In: Galmiche, D., Schulz, S., Sebastiani, R. (eds.) IJCAR 2018. LNCS (LNAI), vol. 10900, pp. 575–590. Springer, Cham (2018). https://doi.org/10.1007/978-3-319-94205-6_38

[Mül19] Müller, D.: Mathematical knowledge management across formal libraries. PhD thesis. Informatics, FAU Erlangen-Nürnberg, December 2019. https://opus4.kobv.de/opus4-au/files/12359/thesis.pdf

[RK13] Rabe, F., Kohlhase, M.: A scalable module system. Inf. Comput. **0.230**, 1–54 (2013). https://kwarc.info/frabe/Research/mmt.pdf

[sTeX] sTeX: a semantic extension of TeX/LaTeX. https://github.com/sLaTeX/sTeX. Accessed 05 Nov 2020

[Wat+14] Watt, S.M., Davenport, J.H., Sexton, A.P., Sojka, P., Urban, J. (eds.): CICM 2014. LNCS (LNAI), vol. 8543. Springer, Cham (2014). https://doi.org/10.1007/978-3-319-08434-3

System Description sTeX3 – A LaTeX-Based Ecosystem for Semantic/Active Mathematical Documents

Michael Kohlhase(ID) and Dennis Müller(✉)(ID)

Computer Science, FAU Erlangen-Nürnberg, Erlangen, Germany
{michael.kohlhase,dennis.mueller}@fau.de
https://kwarc.info

Abstract. We report on sTeX3 – a complete redesign and reimplementation (using LaTeX3) from the ground up of the sTeX ecosystem for semantic markup of mathematical documents. Specifically, we present: i) The sTeX package that allows declaring semantic macros and provides a module system for organizing and importing semantic macros using logical identifiers. sTeX3 is a (now) standard LaTeX package with minimal dependencies and is compatible with arbitrary document class and package. ii) The RusTeX system, an implementation of the core TeX-engine in Rust. It allows for converting arbitrary LaTeX-documents to XHTML – for sTeX3-documents enriched with semantic annotations based on the OMDoc ontology. iii) An Mmt integration: The RusTeX-generated XHTML can be imported and served by the Mmt system for further semantic knowledge management services.

This paper uses sTeX3. The semantically annotated XHTML version of this paper is available at https://tinyurl.com/cicm22stex.

1 Introduction and History

In the sTeX project [sLX], we explore how established communication and publication workflows – this mainly means LaTeX in Mathematics and theoretical sciences – can be extended semantically for computer support. The central element of this endeavour is the sTeX package [Koh08,sTeX] which allows to *semantically preload* LaTeX documents via special (semantic) macros.

sTeX documents can be processed by `pdflatex` in the usual way. Additionally, in sTeX1 they could initially be processed by LaTeXML [LTX], a LaTeX-to-XML transformer, using a dedicated sTeX plugin producing OMDoc [Koh06]. Unfortunately, this plugin was elaborate, implementing the OMDoc-specific behaviour via dedicated Perl bindings for the majority of macros and was correspondingly difficult to maintain. It was also invasive with respect to the LaTeXML code base and quickly became incompatible with newer versions

K. Buzzard and T. Kutsia (Eds.): CICM 2022, LNAI 13467, pp. 184–188, 2022.
https://doi.org/10.1007/978-3-031-16681-5_13

of LaTeXML. Furthermore, conversion to OMDoc required the usage of dedicated document class, rendering sTEX incompatible with existing and established authoring workflows, and setting up sTEX to work in the first place was prohibitively difficult, involving manually changing core parameters of a user's TEX system.

Nevertheless, the sTEX package (and associated classes) have been used to produce extensive course materials (3000+ pages of slides and integrated narrative), ca. 2500 exercise/exam problems, and the SMGloM, a multilingual mathematical glossary [SMG], currently containing ≥2250 concepts in English (93%), German (71%) and Chinese (11%). This sTEX-**corpus**, together with the OMDoc format, have informed the development of the sTEX packages and document model. All sTEX content is available as *mathematical archives* [Hor+11] on https://MathHub.info and can be browsed on https://mmt.beta.vollki.kwarc.info/:sTeX.

While the original sTEX architecture and realization showed that semantic preloading of mathematical documents and the deployment of active documents based on this is possible given enough motivation, the technical/practical problems mentioned above quickly became a showstopper. Not surprisingly, the use of sTEX never quite gained much traction outside the authors' research group and collaborative projects. Additionally, the continuing development of the MMT system [RK13, MMT] over the last years similarly drove development of its own variant of the OMDoc format and ontology (MMT/OMDoc).

Consequently, we decided to rethink and reimplement sTEX from the ground up, using LaTeX3, with both the problems with sTEX1 and the developments of the MMT/OMDoc ontology in mind. The result is the sTEX3 package and system, which we present in this system description (extensive documentation available at [KM]).

Notably, this very document uses sTEX3 and its module system within the `llncs` document class, and is available (and compiles) on Overleaf at https://www.overleaf.com/read/tcnwysdzthwx. We occasionally refer to the source files available there for clarification. In this document, sTEX was configured such that every semantic macros generates a link to our document server, but it should be noted that this behavior can be fully customized.

2   The sTEX3 Package

The design of the sTEX3 system was based on the following guiding principles:

1. *Ease of set-up*: The sTEX3 package should work with a vanilla, unmodified TEX /LaTEX system – e.g. a sufficiently recent TEXLive installation – without the need of changing any TEX-parameters and without any external software.
2. *Universality*: The sTEX3 package should be compatible with arbitrary TEX document class, package, and authoring workflows. Semantically annotating existing environments (*theorems*, *definitions*, *proofs* etc.) should not impact document layout: their layout should be fully customizable.

3. *No code duplication*: The functionality of ꟃ℩EX3 macros and environments should be governed by the LATEX-code of the package alone (as opposed to dedicated macros bindings for OMDoc export that implement the same functionality with a different output format) to help maintainability.

4. MMT-*completeness*: ꟃ℩EX3 should be a full surface language for MMT/OMDoc.

Let us see how the current system is doing on these accounts.

1. Ease of set-up. Indeed, ꟃ℩EX3 now works with any unmodified TEX system with a LATEX3 kernel later than February 2022. For older, but not too outdated LATEX3 versions (up to TEX Live 2018 as running on Overleaf), the missing functionality can be easily added (in this document via the `stex-expl-compat` package).

2. Universality. The ꟃ℩EX3 package can be imported in the usual manner (via `\usepackage{stex}`) and only depends on three other packages, namely `ltxcmds`, `standalone` and `xspace`, all of which are ubiquitous, non-invasive and do not take package options that might lead to conflicts.

To allow for collaboration (e.g. via git) and compatibility with submission systems (e.g. arxiv.org) ꟃ℩EX3 can "persist" all semantic macros and other module content into a `.sms`-file during compilation (similar to the `.toc`-file), which can be used in subsequent compilations, obviating the need for the (potentially many) original modules to be physically present. This file can then be put under version control or distributed alongside the document.

To be adaptable to document styles, ꟃ℩EX3 determines the specific highlighting for symbols via four macros, which can be redefined , namely `\compemph{}`, `\symrefemph{}`, `\defemph{}` and `\varemph{}`. For this document, these are defined in `highlights.tex`.

While ꟃ℩EX1 declared its own environments `definition`, `example`, `theorem` etc., doing so necessarily made ꟃ℩EX1 incompatible with document class that predefine these environments (like `llncs`), or a user's preferences. However, this was necessary to allow for providing additional semantic information, e.g. as in `\begin{definition}[for=foo,type=inductive]`.

In ꟃ℩EX3, we instead use environments `sdefinition`, `sexample`, `sassertion` etc., that take care of the semantic information provided, but whose typesetting can be customized. For example, by setting `\stexpatchdefinition{\begin{definition}}{\end{definition}}`, every `\begin{sdefinition}[...]` will process the arguments provided, and then delegate to the `definition`-environments for layout and numbering. Analogously, `\stexpatchassertion[theorem]{\begin{theorem}}{\end{theorem}}` will delegate every `\begin{sassertion}[type=theorem,...]` to the `theorem`-environments.

3. No code duplication. This principle lead to the following design choice: Rather than converting ꟃ℩EX documents to OMDoc directly, we have the ꟃ℩EX package insert semantic annotations into a non-PDF output format; e.g. XHTML. The package itself determines the full MMT URLs for all symbols, governs the OPENMATH syntax tree and introduces annotations via merely three macros that a "backend" of choice should provide:

- `\stex_annotate:nnn{key}{value}{code}` annotates `code` with `key=value` (e.g. by wrapping `code` in a `<span property="key" resource="value">...`).
- `\stex_invisible:n{code}` exports `code`, but hides it in the presentation (e.g. by setting `style="display:none"`).
- Lastly, `\begin{stex_annotate_env}{key}{value}` acts like `\stex_annotate:nnn{key}{value}{code}`, but as an environments.

The file `stex-backend-pdflatex.cfg` contains the implementations for these macros for the standard `pdflatex` backend.

4. MMT *completeness* with respect to the ŞTEX3 package is a complex issue – at least when trying to avoid code duplication – since the MMT/OMDoc ontology supplies very powerful representational primitives, and will therefore be treated in a regular companion paper submitted to CICM.

3   OMDoc and Mmt

ŞTEX-XHTML: In ŞTEX1, translating ŞTEX content to OMDoc was achieved directly via LATEXML. In ŞTEX3, we instead translate ŞTEX content to XHTML, augmented with annotations via XML attributes corresponding to the OMDoc ontology. In principle, this workflow allows for a plurality of systems as translators, such as LATEXML or TEX4ht. In practice, unfortunately, it has turned out to be difficult to preserve the intended attribute annotations using a current version of LATEXML in math-mode, where they are most important. For now, we therefore implemented our own plain TEX-interpreter from the ground up using Rust, for converting LATEX documents to XHTML. The resulting RUSTEX software[1] uses a user's local LATEX system, keeping the number of required primitives to implement to a reasonable minimum , and can therefore handle (in principle) arbitrary TEX code to the virtually same degree as the user's TEX system (`pdflatex`, to be precise), at the cost of (a priori) no special treatment of higher-level LATEX macros (although RUSTEX allows for providing dedicated bindings for these, too). MMT bundles and interfaces with RUSTEX via the JNI to convert to XHTML using MMT's build system and cache ŞTEX modules across conversion tasks.

XHTML-MMT/OMDoc: Having obtained semantically annotated XHTML, we have implemented a new XHTML import in MMT to extract the semantic annotations and map them directly to the corresponding MMT/OMDoc concepts. In

---

[1] https://github.com/slatex/RusTeX.

addition to thus converting sTeX content to OMDoc, the MMT system can host the generated XHTML in a semantically informed manner and offer the full suite of available knowledge management services for sTeX, up to and including type checking and inference.

4   Ongoing and Future Work

Having solved many of the previous problems surrounding sTeX1 that discouraged users from using sTeX, the most pressing issues now are related to finding, managing, and reusing *existing* sTeX content. We are therefore working on a dedicated IDE for sTeX in the form of a Language Server Protocol server and a plugin for VS Code that bundles the MMT system, offers convenient interfaces to interact with it, allows for searching available sTeX content (online and locally) and generally helps with semantically annotating documents.[2]

References

[Hor+11] Horozal, F., Iacob, A., Jucovschi, C., Kohlhase, M., Rabe, F.: Combining source, content, presentation, narration, and relational representation. In: Davenport, J.H., Farmer, W.M., Urban, J., Rabe, F. (eds.) CICM 2011. LNCS (LNAI), vol. 6824, pp. 212–227. Springer, Heidelberg (2011). https://doi.org/10.1007/978-3-642-22673-1_15

[KM] Kohlhase, M., Müller, D.: The sTeX3 package collection. Tech. rep. https://github.com/slatex/sTeX/blob/main/doc/stex-doc.pdf. Accessed 24 Apr 2022

[Koh06] Kohlhase, M.: OMDoc–an open markup format for mathematical documents [Version 1.2] (2006). https://omdoc.org/pubs/omdoc1.2.pdf

[Koh08] Kohlhase, M.: Using LATEX as a semantic markup format. Math. Comput. Sci. **2**(2), 279–304 (2008). https://doi.org/10.1007/s11786-008-0055-5

[LTX] Miller, B.: LaTeXML: a LATEX to XML converter. https://dlmf.nist.gov/LaTeXML. Accessed 12 Mar 2021

[MMT] MMT - language and system for the uniform representation of knowledge. https://uniformal.github.io/. Accessed 15 Jan 2019

[RK13] Rabe, F., Kohlhase, M.: A scalable module system. Inf. Comput. **230**, 1–54 (2013). https://kwarc.info/frabe/Research/mmt.pdf

[sLX] sLaTeX: an ecosystem for semantically enhanced LATEX. https://github.com/sLaTeX. Accessed 11 Mar 2021

[SMG] SMGloM: a semantic, multilingual terminology for mathematics. https://smglom.mathhub.info. Accessed 21 Apr 2014

[sTeX] sTeX: a semantic extension of TeX/LaTeX. https://github.com/sLaTeX/sTeX. Accessed 11 May 2020

---

[2] Available at https://github.com/slatex/sTeX-IDE.

Theorem Proving and Expression Transformation

Lemmaless Induction in Trace Logic

Ahmed Bhayat[1]([✉]) [ID], Pamina Georgiou[2] [ID], Clemens Eisenhofer[2] [ID],
Laura Kovács[2] [ID], and Giles Reger[1] [ID]

[1] University of Manchester, Manchester, UK
{ahmed.bhayat,giles.reger}@manchester.ac.uk
[2] TU Wien, Vienna, Austria
{pamina.georgiou,clemens.eisenhofer,laura.kovacs}@tuwien.ac.at

Abstract. We present a novel approach to automate the verification
of first-order inductive program properties capturing the partial correctness of imperative program loops with branching, integers and arrays.
We rely on trace logic, an instance of first-order logic with theories, to
express first-order program semantics by quantifying over program execution timepoints. Program verification in trace logic is translated into
a first-order theorem proving problem where, to date, effective reasoning has required the introduction of so-called trace lemmas to establish
inductive properties. In this work, we extend trace logic with generic
induction schemata over timepoints and loop counters, reducing reliance
on trace lemmas. Inferring and proving loop invariants becomes an inductive inference step within superposition-based first-order theorem proving. We implemented our approach in the RAPID framework, using the
first-order theorem prover VAMPIRE. Our extensive experimental analysis shows that automating inductive verification in trace logic is an
improvement compared to existing approaches.

1 Introduction

Automating the verification of programs containing loops and recursive data
structures is an ongoing research effort of growing importance. While different
techniques for proving the correctness of such programs are in place [5,6,10,13],
most existing tools in this realm are heavily based on *satisfiability modulo theories* (SMT) backends [4,8] that come with strong theory reasoning but have limitations in quantified reasoning. In contrast, first-order theorem provers enable
quantified reasoning modulo theories [19,24,25], such as linear integer arithmetic
and arrays. First-order reasoning can thus complement the aforementioned verification efforts when it comes to proving program properties with complex quantification, as evidenced in our original work on the RAPID framework [11] which
utilised the VAMPIRE theorem prover [2,20].

At a high level, the RAPID framework [11] works by translating a program
into *trace logic*, adding a number of ad hoc trace lemmas, asserting a desired
property, and then running an automated theorem prover on the result. The
effectiveness of this approach depends on the underlying trace lemmas. This

K. Buzzard and T. Kutsia (Eds.): CICM 2022, LNAI 13467, pp. 191–208, 2022.
https://doi.org/10.1007/978-3-031-16681-5_14

paper focuses on building induction support into the VAMPIRE theorem prover to reduce reliance on these lemmas.

To understand the role of these trace lemmas (and therefore, what support must be added to the theorem prover) we briefly overview trace logic and the RAPID framework in a little more detail. Trace logic is an instance of first-order logic with theories, such that the program semantics of imperative programs with loops, branching, integers, and arrays can be directly encoded in trace logic. A key feature of this encoding is tracking program executions by quantifying over execution *timepoints* (rather than only over single states), which may themselves be parameterised by *loop iterations*. In principle, we can check whether a translated program entails the desired property in trace logic using an automated theorem prover for first-order logic. In our case, we make use of the saturation-based theorem prover VAMPIRE which implements the superposition calculus [3]. However, a straightforward use of theorem proving often fails in establishing validity of program properties in trace logic, as the proof requires some specific induction, in general not supported by superposition-based reasoning.

In our previous work [11], we overcame this challenge by introducing so-called *trace lemmas* capturing common patterns of inductive loop properties over arrays and integers. Inductive loop reasoning in trace logic is then achieved by generating and adding trace lemma instances to the translated program. However, there are two significant limitations to using trace lemmas:

1. Trace lemmas capture inductive patterns/templates that need to be manually identified, as induction is not expressible in first-order logic. As such, they cannot be inferred by a first-order reasoner, implying that the effectiveness of trace logic reasoning depends on the expressiveness of manually supplied trace lemmas.
2. When instantiating trace lemmas with appropriate inductive program variables, a large number of inductive properties are generated, causing saturation-based proof search to diverge and fail to find program correctness proofs in reasonable time.

In this paper we address these limitations by reducing the need for trace lemmas. We achieve this by introducing a couple of novel induction inferences. Firstly, *multi-clause goal induction* which applies induction in a goal oriented fashion as many safety program assertions are structurally close to useful loop invariants. Secondly, *array mapping induction* which covers certain cases where the required loop invariant does not stem from the goal. Specifically, we make the following contributions:

Contribution 1. We introduce two new inference rules, *multi-clause goal* and *array mapping* induction, for *lemmless induction* over loop iterations (Sects. 5–6). The inference rules are compatible with any saturation-based inference system used for first-order theorem proving and work by carrying out induction on terms corresponding to final loop iterations.

```
 1    func main() {
 2        const Int[] a;
 3        const Int[] b;
 4        Int[] c;
 5        const Int length;
 6        Int i = 0;
 7
 8        while (i < length) {
 9            c[2*i] = a[i]
10            c[(2*i) + 1] = b[i]
11            i = i + 1;
12        }
13    }
14    assert (∀pos_I.∃l_I.((0 ≤ pos < (2 × length))
15          → c(main_end, pos) = a(l) ∨ c(main_end, pos) = b(l)
          ))
```

Fig. 1. Copying elements from arrays a and b to even/odd positions in array c.

Contribution 2. We implemented our approach in the first-order theorem prover VAMPIRE [20]. Further, we extended the RAPID framework [11] to support inductive reasoning in the automated backend (Sect. 7). We carry out an extensive evaluation of the new method (Sect. 8) comparing against state-of-the-art approaches SEAHORN [12,13] and VAJRA/DIFFY [5,6].

2   Motivating Example

We motivate our work with the example program in Fig. 1. The program iterates over two arrays a and b of arbitrary, but fixed length length and copies array elements into a new array c. Each even position in c contains an element of a, while each odd position an element of b. Our task is to prove the safety assertion at line 14: at the end of the program, every element in c is an element from a or b. This property involves (i) alternation of quantifiers and (ii) is expressed in the first-order theories of linear integer arithmetic and arrays. Note that in the safety assertion, the program variable length is modeled as a logical constant of the same name of sort integer, whilst the constant arrays a and b are modeled as logical functions from integers to integers. The mutable array variable c is additionally equipped with a timepoint argument main_end, indicating that the assertion is referring to the value of the variable at the end of program execution.

Proving the correctness of this example program remains challenging for most state-of-the-art approaches, such as [5,6,10,12], mainly due to the complex quantified structure of our assertion. Moreover, it cannot be achieved in the current RAPID framework either, as existing trace lemmas do not relate the values of multiple program variables, notably equality over multiple array variables. In fact, to automatically prove the assertion, we need an inductive property/trace lemma formalizing that each element at an even position in c is an element of a or b at each valid loop iteration, thereby also restricting the bounds of the loop counter variable i. Naïvely adding such a trace lemma would be highly inefficient as automated generation of verification conditions would introduce many instances that are not required for the proof.

3   Related Work

Most of recent research in verifying inductive properties of array-manipulating programs focuses on quantified invariant generation and/or is mostly restricted to proving universally quantified program properties. The works [10,13] generate universally quantified inductive invariants by iteratively inferring and strengthening candidate invariants. These methods use SMT solving and as such are restricted to first-order theories with a finite model property. Similar logical restrictions also apply to [23], where linear recurrence solving is used in combination with array-specific proof tactics to prove quantified program properties. A related approach is described in [6], where relational invariants instead of recurrence equations are used to handle universal and quantifier-free inductive properties. Unlike these, our work is not limited to universal invariants but can infer and prove inductive program properties with alternations of quantifiers.

With the use of extended expressions and induction schemata, our work shares some similarity with template-based approaches [16,21,26]. These works infer and prove universal inductive properties based on Craig interpolation, formula slicing and/or SMT generalizations over quantifier-free formulas. Unlike these works, we do not require any assumptions on the syntactic shape of the first-order invariants. Moreover, our invariants are not restricted to the shape of our induction schemata. Rather, we treat inductive (invariant) inferences as additional rules of first-order theorem provers, maintaining thus the efficient handling of arbitrary first-order quantifiers. Our framework can be used in arbitrary first-order theories, even with theories that have no interpolation property and/or a finite axiomatization, as exemplified by our experimental results using inductive reasoning over arrays and integers.

Inductive theorem provers, such as ACL2 [17] and HipSpec [7], implement powerful induction schemata and heuristics. However these provers, to the best of our knowledge, automate inductive reasoning for only universally quantified inductive formulas using a goal/subgoal architecture, for which user-guidance is needed to split conjectures into subgoals. In contrast, our work can prove formulas of full first-order logic by integrating and fully automating induction in saturation-based proof search. By combining induction with saturation, we allow these techniques to interleave and complement each other, something that pure induction provers cannot do. Unlike tools such as Dafny [22], our approach is fully automated requiring no user annotations.

First-order theorem proving has been used to derive invariants with alternations of quantifiers in our previous work [11]. Our current work generalizes the inductive capabilities of [11] by reducing the expert knowledge of [11] in introducing inductive lemmas to guide the process of proving inductive properties.

4   Preliminaries

Many-Sorted First-Order Logic. We consider standard many-sorted first-order logic with built-in equality, denoted by $\simeq$. By $s = F[u]$ we indicate that the term u is a subterm of s surrounded by (a possibly empty) context F.

We use x, y to denote variables, l, r, s, t for terms and sk for Skolem symbols. A *literal* is an atom A or its negation $\neg A$. A *clause* is a disjunction of literals $L_1 \vee ... \vee L_n$, for $n \geq 0$. Given a formula F, we denote by $\text{CNF}(F)$ the clausal normal form of F.

For a logical variable x of sort S we write x_S. A *first-order theory* denotes the set of all valid formulas on a class of first-order structures. Any symbol in the signature of a theory is considered *interpreted*. All other symbols are *uninterpreted*. In particular, we use the theory of linear integer arithmetic denoted by $\mathbb{I}$ and the boolean sort $\mathbb{B}$. We consider natural numbers as the term algebra $\mathbb{N}$ with four symbols in the signature: the constructors 0 and successor suc, as well as pred and $<$ respectively interpreted as the predecessor function and less-than relation. Note that we do not define any arithmetic on naturals. We assume familiarity with the basics of saturation theorem proving.

4.1   Trace Logic $\mathcal{L}$

Trace logic, denoted as $\mathcal{L}$, is an instance of many-sorted first-order logic with theories. Its signature is $\Sigma(\mathcal{L}) := S_{\mathbb{N}} \cup S_{\mathbb{I}} \cup S_{\mathbb{L}} \cup S_V \cup S_n$, includes respectively the signatures of the theory of natural numbers $\mathbb{N}$ (as a term algebra), the in-built integer theory $\mathbb{I}$, a set $S_{\mathbb{L}}$ of timepoints (also referred to as *locations*), a set of symbols representing program variables S_V, as well as a set of symbols representing last iteration symbols S_n. For more details on trace logic, refer to [11].

4.2   Programming Model $\mathcal{W}$

We consider programs written in a WHILE-like programming language $\mathcal{W}$, as given in the (partial) language grammar of Fig. 2. Programs in $\mathcal{W}$ contain mutable and immutable integer as well as integer-array program variables and consist of a single top-level function main comprising arbitrary nestings of while-loops and if-then-else branching. We consider expressions over booleans and integers without side effects.

4.3   Translating Expressions to Trace Logic

Locations and Timepoints. We consider programs as sets of locations over time: given a program statement s, we denote its location by l_s of type $\mathbb{L}$, the location/timepoint sort, corresponding to the line of the program where the statement appears. When s is a while-loop the corresponding location is revisited at multiple timepoints of the execution. Thus, we model such locations

$$\begin{aligned}
\text{program} &::= \text{function} \\
\text{function} &::= \texttt{func main()\{ subprogram \}} \\
\text{subprogram} &::= \text{statement} \mid \text{context} \\
\text{context} &::= \text{statement; ... ; statement} \\
\text{statement} &::= \text{atomicStatement} \\
&\quad \mid \texttt{if(condition)\{ context \} else \{ context \}} \\
&\quad \mid \texttt{while(condition)\{ context \}}
\end{aligned}$$

Fig. 2. Grammar of $\mathcal{W}$.

as functions over *loop iterations* $l_s : \mathbb{N} \mapsto \mathbb{L}$, where the argument of sort $\mathbb{N}$ intuitively corresponds to the number of loop iterations. Further, for each loop statement s we model the last loop iteration by a symbol $nl_s \in S_n$ of target sort $\mathbb{N}$. Let p be a program statement or context. We use $start_\mathrm{p}$ to denote the location at which the execution of p has started and end_p to denote the location that occurs just after the execution of p. We use $main\_end$ to denote the location at the end of the main function.

Example 1. Consider line 6 of our running example in Fig. 1. Term l_6 corresponds to the timepoint of the first assignment of 0 to program variables i while $l_8(0)$ and $l_8(nl_8)$ denote the timepoints of the loop at the first and last loop iteration respectively. Further, we can quantify over all executions of the loops by quantifying over all iterations smaller than the last e.g. $\forall it_\mathbb{N}.it < nl_8 \to F[l_8(it)]$ where $F[l_8(it)]$ is some first-order formula.

Program Variables. Program variable are expressed as functions over timepoints. We express an integer variable v as a function $v : \mathbb{L} \mapsto \mathbb{I}$, where $v \in S_V$. Let tp be a term of sort $\mathbb{L}$. Then, $v(tp)$ denotes the value of v at timepoint tp. We model numeric array variables v with an additional argument of sort $\mathbb{I}$ to denote the position of an array access. We obtain $v : \mathbb{L} \times \mathbb{I} \mapsto \mathbb{I}$. Immutable variables are modelled as per their mutable counterparts, but without the timepoint argument.

Example 2. To denote program variable i at the location of the assignment in line 6, we use the equation $i(l_6) \simeq 0$. For the first assignment of c within the loop, we write $c(l_8(it), 2 \times i(l_8(it))) \simeq a(i(l_8(it)))$ for some iteration it. As a is a constant array, the timepoint argument is omitted.

Program Expressions. Let e be an arbitrary program expression. We write $[\![e]\!](tp)$ to denote the logical denotation of e at timepoint tp. We do not provide the full inductive definition of the denotation function $[\![\ ]\!](tp)$ here, just a few of its cases. If e is an integer variable v, then $[\![e]\!](tp) = v(tp)$. If e is an integer array access of the form $v[e_1]$, then $[\![e]\!](tp) = v(tp, [\![e_1]\!](tp))$. If e is an expression of the form $e_1 + e_2$, then $[\![e]\!](tp) = [\![e_1]\!](tp) + [\![e_2]\!](tp)$.

Common Abbreviations. Let e, e_1, e_2 be program expressions, tp_1, tp_2 be two timepoints and $v \in S_V$ denote the functional representation of a program variable. The trace logic formula $v(tp_1) \simeq v(tp_2)$ asserts that the variable v has the same value at timepoints tp_1 and tp_2. We introduce definitions for two formulas that are widely used in defining the axiomatic semantics of $\mathcal{W}$ in the next section. To ease the notational burden, we ignore array variables in the definitions provided. Firstly, we introduce a definition for the formula that expresses that the value of a variable v changes between timepoints tp_1 and tp_2 whilst the values of all other variables remain the same.

$$Update(v, e, tp_1, tp_2) \quad := \quad v(tp_2) \simeq [\![e]\!](tp_1) \wedge \bigwedge_{v' \in S_V \setminus \{v\}} v'(tp_1) \simeq v'(tp_2),$$

Secondly, we introduce a definition for the formula that expresses that the value of all variables stays the same between timepoints tp_1 and tp_2

$$EqAll(tp_1, tp_2) := \bigwedge_{v \in S_V} v(tp_1) \simeq v(tp_2)$$

4.4 Axiomatic Semantics of $\mathcal{W}$ in $\mathcal{L}$

The semantics of a program in $\mathcal{W}$ is given by the conjunction of the respective axiomatic semantics of each program statement of $\mathcal{W}$ occurring in the program. In general, we define reachability of program statements over timepoints rather than program states. We briefly recall the axiomatic semantics of assignments and while-loops respectively, again ignoring the array variable case.

Assignments. Let s be an assignment $v = e$, where v is an integer-valued program variable and e is an expression. The evaluation of s is performed in one step such that, after the evaluation, the variable v has the same value as e before the evaluation while all other variables remain unchanged. We obtain

$$[\![s]\!] := Update(v, e, start_s, end_s) \tag{1}$$

While-Loops. Let s be the while-statement `while(Cond){c}` where `Cond` is the *loop condition*. The semantics of s is given by the conjunction of the following properties: (2a) the iteration nl_s is the first iteration where `Cond` does not hold anymore, (2b) jumping into the loop body does not change the values of the variables, (2c) the values of the variables at the end of evaluating the loop s are equal to the values at the loop condition location in iteration nl_s. As such, we have

$$
\begin{aligned}
[\![s]\!] := \quad & \forall it_{\mathbb{N}}^s.\ (it^s < nl_s \to [\![\text{Cond}]\!](tp_s(it^s))) \\
\wedge \quad & \neg[\![\text{Cond}]\!](tp(nl_s)) & (2a) \\
\wedge \quad & \forall it_{\mathbb{N}}.\ (it < nl_s \to EqAll(start_c, tp_s(it))) & (2b) \\
\wedge \quad & EqAll(end_s, tp_s(nl_s)) & (2c)
\end{aligned}
$$

4.5   Trace Lemma Reasoning

Trace logic $\mathcal{L}$ allows one to naturally express common program behavior over timepoints. Specifically, it allows us to reason about (i) all iterations of a loop, and (ii) the existence of specific timepoints. In [11], we leveraged such reasoning with the use of so-called *trace lemmas*, capturing common inductive properties of program loops. Trace lemmas are instances of the schema of bounded induction for natural numbers

$$\Big(P(bl) \wedge \forall x_{\mathbb{N}}.\big((bl \leq x < br \wedge P(x)) \to P(\mathrm{suc}(x))\big)\Big) \to \\ \forall x_{\mathbb{N}}.(bl \leq x < br \wedge P(x)) \tag{3}$$

An example of a trace lemma would be the statement formalising that a certain program variable's value remains unchanged from a specific iteration to the end of loop execution. In this work, instead of adding instances of (3) statically to strengthen loop semantics, we move induction into the first-order prover. The advantage of adding instances of (3) dynamically is that during proof search we have more information available and can thus perform induction in a more controlled and goal oriented fashion.

Nonetheless, due to some limitations in our first-order prover, we are unable to completely do away with additional lemmas. Specifically, we need to nudge the prover to deduce that a loop counter expression will, at the end of loop execution, have the value of the expression it is compared against in the loop condition.

(A) Equal Lengths Trace Lemma We define a common property of loop counter expressions. We call a program expression e *dense* at loop w if:

$$Dense_{w,e} := \forall it_{\mathbb{N}}.\Big(it < nl_w \to \Big(\begin{array}{c} [\![e]\!](tp_w(\mathrm{suc}(it))) \simeq [\![e]\!](tp_w(it)) \vee \\ [\![e]\!](tp_w(\mathrm{suc}(it))) \simeq [\![e]\!](tp_w(it)) + 1 \end{array} \Big)\Big).$$

Let w be a while-statement, $C_w := $ e $<$ e' be the loop condition where e' is a program expression that remains constant during iterations of w. The *equal lengths trace lemma of* w, e *and* e' is defined as

$$\big(Dense_{w,e} \wedge [\![e]\!](tp_w(0)) \leq [\![e']\!](tp_w(0))\big) \to \tag{A}$$
$$[\![e]\!](tp_w(nl_w)) \simeq [\![e']\!](tp_w(nl_w)).$$

Trace lemma A states that a dense expression e smaller than or equal to some expression e' that does not change in the loop, will eventually, specifically in the last iteration, reach the same value as e'. This follows from the fact that we assume termination of a loop, hence we assume the existence of a timepoint nl_w where the loop condition does not hold anymore. As a consequence, given that the loop condition held at the beginning of the execution, we can derive that the loop counter value immediately after the loop execution $[\![e]\!](tp_w(nl_w))$ will necessarily equate to $[\![e']\!](tp_w(0)) = [\![e']\!](tp_w(nl_w))$. Note that a similar lemma can just as easily be added for dense but decreasing loop counters.

5   Multi-Clause Goal Induction for Lemmaless Induction

As mentioned above, the main focus of our work is moving induction into the saturation prover. We achieve this by adding inference rules that apply induction to loop counter terms. We leverage recent theorem proving effort on *bounded (integer) induction* in saturation [14, 15]. However, as illustrated in the following, these recent efforts cannot be directly used in trace logic reasoning since we need to (i) adjust bounded induction for the setting of natural numbers, and (ii) generalise to multi-clause induction. We discuss these steps using Fig. 1. Verifying the safety assertion of Fig. 1 requires proving the trace logic formula:

$$\forall pos_{\mathbb{I}}. \exists j_{\mathbb{I}}. (0 \leq pos < (2 \times length) \tag{4}$$
$$\rightarrow (c(main\_end, pos) \simeq a(j) \vee c(main\_end, pos) \simeq b(j))$$

For proving (4), it suffices to prove that the following, slightly modified statement is a loop invariant of Fig. 1:

$$\forall it_{\mathbb{N}}. it < nl_{\mathtt{w}} \rightarrow \forall pos_{\mathbb{I}}. \exists j_{\mathbb{I}}. (0 \leq pos < (2 \times i(tp_{\mathtt{w}}(it)))) \tag{5}$$
$$\rightarrow (c(tp_{\mathtt{w}}(it), pos) \simeq a(j) \vee c(tp_{\mathtt{w}}(it), pos) \simeq b(j))$$

where w refers to the loop statement in Fig. 1. As part of the program semantics in trace logic, we have formula (6) which links the value of c at the end of the loop to its value at the end of the program. Moreover, using the trace lemma A, we also derive formula (7) in trace logic:

$$\forall pos_{\mathbb{I}}.c(tp_{\mathtt{w}}(nl_{\mathtt{w}}), pos) \simeq c(main\_end, pos) \tag{6}$$
$$i(tp_{\mathtt{w}}(nl_{\mathtt{w}})) \simeq length \tag{7}$$

It is tempting to think that in the presence of these clauses (6)–(7), a saturation-based prover would rewrite the negated conjecture (4) to

$$\neg(\forall pos_{\mathbb{I}}. \exists j_{\mathbb{I}}. (0 \leq pos < (2 \times i(tp_{\mathtt{w}}(nl_{\mathtt{w}}))))$$
$$\rightarrow (c(tp_{\mathtt{w}}(nl_{\mathtt{w}}), pos) \simeq a(j) \vee c(tp_{\mathtt{w}}(nl_{\mathtt{w}}), pos) \simeq b(j)))$$

from which a bounded natural number induction inference (similar to the IntInd< rule of [15]) would quickly introduce an induction hypothesis with (5) as the conclusion, by induction over $nl_{\mathtt{w}}$. However, this is not the case, as most saturation provers work by first *clausifying* their input. The negated conjecture (4) would not remain a single formula, but be split into the following clauses where sk is a Skolem symbol:

$$a(x) \not\simeq c(main\_end, sk) \quad b(x) \not\simeq c(main\_end, sk)$$
$$\neg(sk \leq 0) \quad\quad\quad sk \leq 2 \times length$$

These clauses can be rewritten using (6)–(7). For example, the first clause can be rewritten to $a(x) \not\simeq c(tp_{\mathtt{w}}(nl_{\mathtt{w}}, sk))$. However, attempting to prove the negation of any of the rewritten clauses individually via induction would merely result in the addition of useless induction formulas to the search space. For example, attempting to prove $\forall it_{\mathbb{N}}. it < nl_{\mathtt{w}} \rightarrow (\exists x_{\mathbb{I}}. a(x) \simeq c(tp_{\mathtt{w}}(it), sk))$, is pointless as it is clearly false. *The solution we propose in this work is to use multi-clause induction*, whereby we attempt to prove the negation of the conjunction of multiple clauses via a single induction inference. For our running example Fig. 1, we can use the following rewritten versions of clauses from the negated conjecture $a(x) \not\simeq c(tp_{\mathtt{w}}(nl_{\mathtt{w}}, sk))$, $b(x) \not\simeq c(tp_{\mathtt{w}}(nl_{\mathtt{w}}, sk))$, and $sk \leq 2 \times i(tp_{\mathtt{w}}(nl_{\mathtt{w}}))$, with induction term $nl_{\mathtt{w}}$, to obtain the induction formula:

$$\neg \Big(\quad \forall x_{\mathbb{I}}. a(x) \not\simeq c(i(tp_{\mathtt{w}}(0)), sk) \qquad \forall it_{\mathbb{N}}. it < nl_{\mathtt{w}} \rightarrow$$
$$\wedge \ \forall x_{\mathbb{I}}. b(x) \not\simeq c(i(tp_{\mathtt{w}}(0)), sk)) \qquad \neg \Big(\quad \forall x_{\mathbb{I}}. a(x) \not\simeq c(i(tp_{\mathtt{w}}(it)), sk)$$
$$\wedge \ sk \leq 2 \times i(tp_{\mathtt{w}}(0)) \Big) \qquad \rightarrow \qquad \wedge \forall x_{\mathbb{I}}. b(x) \not\simeq c(i(tp_{\mathtt{w}}(it)), sk)$$
$$\wedge \ StepCase \qquad\qquad\qquad\qquad \wedge \ sk \leq 2 \times i(tp_{\mathtt{w}}(it)) \Big)$$

$$(8)$$

where *StepCase* is the formula:

$$\forall it_{\mathbb{N}}. it < nl_{\mathtt{w}} \wedge$$
$$\neg \Big(\quad \forall x_{\mathbb{I}}. a(x) \not\simeq c(i(tp_{\mathtt{w}}(it)), sk) \qquad \neg \Big(\quad \forall x_{\mathbb{I}}. a(x) \not\simeq c(i(tp_{\mathtt{w}}(\mathrm{suc}(it))), sk)$$
$$\wedge \ \forall x_{\mathbb{I}}. b(x) \not\simeq c(i(tp_{\mathtt{w}}(it)), sk) \qquad \rightarrow \qquad \wedge \forall x_{\mathbb{I}}. b(x) \not\simeq c(i(tp_{\mathtt{w}}(\mathrm{suc}(it))), sk)$$
$$\wedge \ sk \leq i(tp_{\mathtt{w}}(y)) \Big) \qquad\qquad\qquad \wedge \ sk \leq 2 \times i(tp_{\mathtt{w}}(\mathrm{suc}(it)))) \Big)$$

Using the induction formula (8), a contradiction can then easily be derived, establishing validity of (4). In what follows, we formalize the multi-clause induction principle we used above. To this end, we introduce a generic inference rule, called *multi-clause goal induction* and denoted as `MCGLoopInd`.

$$\frac{C_1[nl_{\mathtt{w}}] \qquad C_2[nl_{\mathtt{w}}] \quad \ldots \quad C_n[nl_{\mathtt{w}}]}{\mathrm{CNF}\left(\left(\left(\begin{array}{c} \neg(C_1[0] \wedge C_2[0] \wedge \ldots \wedge C_n[0]) \wedge \\ \forall it_{\mathbb{N}}. \left(\begin{array}{c} ((it < nl_{\mathtt{w}}) \wedge \neg(C_1[it] \wedge C_2[it] \wedge \ldots \wedge C_n[it])) \rightarrow \\ \neg(C_1[\mathrm{suc}(it)] \wedge C_2[\mathrm{suc}(it)] \wedge \ldots \wedge C_n[\mathrm{suc}(it)]) \end{array} \right) \end{array} \right) \\ \rightarrow (\forall it_{\mathbb{N}}. (it < nl_{\mathtt{w}}) \rightarrow \neg(C_1[it] \wedge C_2[it] \wedge \ldots \wedge C_n[it])) \right) \right)}$$

For performance reasons, we mandate that the premises $C_1 \ldots C_n$ be derived from trace logic formulas expressing safety assertions and not from formulas encoding the program semantics. The `MCGLoopInd` rule is formalised only as an induction inference over last loop iteration symbols. While restricting to $nl_{\mathtt{w}}$ terms is of purely heuristic nature, our experiments justify the necessity and usefulness of this condition (Sect. 8).

6 Array Mapping Induction for Lemmaless Induction

Multi-clause goal induction neatly captures goal-oriented application of induction. Nevertheless, there are verification challenges where `MCGLoopInd` fails to prove inductive loop properties. This is particularly the case for benchmarks

```
1        func main(){
2            const Int alength;
3            Int[] a;
4            Int i = 0;
5            const Int n;
6
7            while(i < alength){
8                a[i] = a[i] + n;
9                i = i + 1;
10           }
11
12           Int j = 0;
13           while(j < alength){
14               a[j] = a[j] - n;
15               j = j + 1;
16           }
17       }
18   assert (∀pos_I.((0 ≤ pos < alength)
19               → a(main_end, pos) = a(main_start, pos)))
```

Fig. 3. Adding and subtracting n to every element of array a.

containing multiple loops, such as in Fig. 3. We first discuss the limitations of
MCGLoopInd using Fig. 3, after which we present our solution, the *array mapping
induction* inference.

Let w_1 be the first loop statement of Fig. 3 and w_2 be the second loop. Using
MCGLoopInd, we would attempt to prove

$$\forall it_{\mathbb{N}}.\, it \leq nl_{w_2} \rightarrow \\ \forall pos_{\mathbb{I}}.\, (0 \leq pos < j(tp_{w_2}(it))) \rightarrow (a(tp_{w_2}(it), pos) \simeq a(main\_start, pos) \tag{9}$$

However, formula (9) is not a useful invariant for proving the assertion. Rather,
for w_2 we need a loop invariant similar to

$$\forall it_{\mathbb{N}}.\, it \leq nl_{w_2} \rightarrow \forall pos_{\mathbb{I}}.\, (0 \leq pos < j(tp_{w_2}(it))) \\ \rightarrow (a(tp_{w_2}(it), pos) \simeq a(tp_{w_2}(0), pos) - n \tag{10}$$

and a similar loop invariant for loop w_1. The loop invariant (10) is however not
linked to the safety assertion of Fig. 3, and thus multi-clause goal induction is
unable to infer and prove with it. To aid with the verification of benchmarks such
as Fig. 3, we introduce another induction inference which we call *array mapping
induction*. In this case, we trigger induction not on clauses and terms coming
from the goal, but on clauses and terms appearing in the program semantics.

The *array mapping induction* inference rule, denoted as AMLoopInd is given
below. Essentially, AMLoopInd involves analysing a clause set to heuristically
devise a suitable loop invariant. Guessing a candidate loop invariant is a difficult
problem. The AMLoopInd inference is triggered if clauses of the shapes of C_1 and

C_2 defined below are present in the clause set. Intuitively, C_2 can be read as saying that on each round of some loop w, some array a at position i is set to some function F of its previous value at that position. Clause C_1 states that i increases by m in each round of the loop. Together the two clauses suggest that the loop is mapping the function F to each mth location of the array starting from the array cell located at $i(tp_w(0))$. This is precisely what the induction formula attempts to prove. Note that for ease of notation, we present the inference for the case where the indexing variable is *increasing*. It is straightforward to generalise to the decreasing case. The `AMLoopInd` rule is[1]

$$C_1 = i(tp_w(\mathrm{suc}(x))) \simeq i(tp_w(x)) + m \ \lor \ \neg(x < nl_w)$$
$$\frac{C_2 = a\big(tp_w(\mathrm{suc}(x)), i(tp_w(x))\big) \simeq F[a\big(tp_w(x), i(tp_w(x))\big)] \ \lor \ \neg(x < nl_w)}{\mathrm{CNF}(StepCase \to Conclusion)}$$

where w is some loop and F an arbitrary non-empty context. Let i_0 be an abbreviation for $i(tp_w(0))$. Then:

$$
\begin{aligned}
StepCase: \quad & \forall it_\mathbb{N}. \big(\forall y_\mathbb{I}.\, it < nl_w \land \\
& y < i(tp_w(it)) - i_0 \land y \geq 0 \land y \bmod m = 0 \\
& \qquad \to a(tp_w(it), i_0 + y) \simeq F[a(tp_w(0), i_0 + y)] \big) \to \\
& (\forall y_\mathbb{I}.\, y < i(tp_w(\mathrm{suc}(it))) - i_0 \land y \geq 0 \land y \bmod m = 0 \\
& \qquad \to a(tp_w(\mathrm{suc}(it)), i_0 + y) \simeq F[a(tp_w(0), i_0 + y)]) \\
Conclusion: \quad & \forall x_\mathbb{I}.\, x < i(tp_w(nl_w)) - i_0 \land x \geq 0 \land x \bmod m = 0 \\
& \qquad \to a(tp_w(nl_w), i_0 + x) \simeq F[a(tp_w(0), i_0 + x)]
\end{aligned}
$$

To prove *StepCase*, it is necessary to be able to reason that positions in the array a remain unchanged until visited by the indexing variable. This can be achieved via the addition of another induction to the conclusion of the inference. We do not provide details of this induction formula here, but it is added to the conclusion by our implementation which we present in Sect. 7. The `AMLoopInd` inference is thus sufficient to prove the assertion of Fig. 3. While `AMLoopInd` is a limited approach for guessing inductive loop invariants, we believe it can be extended towards further, more generic methods to guess invariants, as discussed in Sect. 9. We conclude this section by noting that our induction rules are sound, based on trace logic semantics. Since both rules merely add instances of the bounded induction schema for natural numbers (3) to the search space, soundness is trivial and we do not provide a proof.

7   Implementation

Our approach is implemented as an extension of the RAPID framework, using the first-order theorem prover VAMPIRE.

---

[1] In the conclusion we ignore the base case of the induction formula as it is trivially true.

Extensions to Rapid. RAPID takes as an input a $\mathcal{W}$ program along with a property expressed in $\mathcal{L}$. It outputs the semantics of the program expressed in $\mathcal{L}$ using SMT-LIB syntax along with the property to be proven. For our "lemmaless induction" framework, we have extended RAPID as follows. Firstly, we prevent the output of all trace lemmas other than trace lemma A (Sect. 4.5). We added custom extensions to the SMT-LIB language to identify trace logic symbols, such as loop iteration symbols, program variables, within the RAPID encodings. This way, trace logic symbols to be used for induction inferences are easily identified and can also be used for various proving heuristics. We refer to this version (available online[2]) as RAPID^{l-}.

Extensions to Vampire. We implemented the `MCGLoopInd` inference rule and a slightly simplified version of the `AMLoopInd` rule in a new branch of VAMPIRE[3]. The main issue with the induction inferences `MCGLoopInd` and `AMLoopInd` is their explosiveness which can cause proof search to diverge. We have, therefore, introduced various heuristics in the implementation to try and control them. For `MCGLoopInd` we not only necessitate that the premises are derived from the conjecture, but that their derivation length from the conjecture is below a certain distance controlled by an option. The premises must be unit clauses unless another option `multi_literal_clauses` is toggled on. The option `induct_all_loop_counts` allows `MCGLoopInd` induction to take place on all loop counter terms, not just final loop iterators. In order for the `MCGLoopInd` and `AMLoopInd` inferences to be applicable, we need to rewrite terms not containing final loop counters to terms that do. However, rewriting in VAMPIRE is based on superposition, which is parameterised by a term order preventing smaller terms to be rewritten into larger ones. In this case, the term order may work against us and prevent such rewrites from happening. We implemented a number of heuristics to handle this problem. One such heuristic is to give terms representing constant program variables a large weight in the ordering. Then, equations such as $alength \simeq i(tp_{\mathtt{w}}(nl_{\mathtt{w}}))$ will be oriented left to right as desired. We combined these options with others to form a portfolio of strategies[4] that contains 13 strategies each of which runs in under 10s.

8 Experimental Results

Benchmarks. For our experiments, we use a total of 111 examples whose verification involved proving safety assertions of different logical complexity (quantifier-free, only universally/existentially quantified, and with quantifier alternations). Our benchmarks are divided into four groups, as indicated in Table 1: (i) the first 13 problems have quantifier-free proof obligations; (ii) the majority of benchmarks, in total 68 examples, contain universally quantified

---

[2] See commit `285e54b7e` of https://github.com/vprover/rapid/tree/ahmed-induction-support.

[3] See commit `4a0f319f` of https://github.com/vprover/vampire/tree/ahmed-rapid.

[4] `--mode portfolio --schedule rapid_induction.`.

Table 1. Experimental results.

Benchmark	(1)	(2)	(3)	(4)
atleast_one_iteration_0	✓	✓	✓	✓
atleast_one_iteration_1	✓	✓	✓	✓
count_down	✓	-	-	-
eq	✓	-	✓	-
find_sentinel	✓	✓	-	-
find1_0	✓	✓	✓	-
find1_1	✓	✓	✓	-
find2_0	✓	✓	✓	-
find2_1	✓	✓	✓	-
indexn_is_arraylength_0	✓	✓	✓	-
indexn_is_arraylength_1	✓	✓	✓	-
set_to_one	✓	✓	✓	✓
str_cpy_3	✓	✓	✓	-
add_and_subtract	✓	-	-	✓
both_or_none	✓	✓	-	✓
check_equal_set_flag_1	✓	✓	-	✓
collect_indices_eq_val_0	✓	✓	-	✓
collect_indices_eq_val_1	✓	✓	-	✓
copy	✓	✓	-	✓
copy_absolute_0	✓	✓	-	✓
copy_absolute_1	✓	✓	-	✓
copy_and_add	✓	-	-	✓
copy_nonzero_0	✓	✓	-	✓
copy_partial	✓	✓	-	✓
copy_positive_0	✓	✓	-	✓
copy_two_indices	✓	✓	-	-
find_max_0	✓	✓	-	✓
find_max_2	✓	✓	-	✓
find_max_from_second_0	✓	-	-	✓
find_max_local_2	-	-	-	-
find_max_up_to_0	-	-	-	-
find_max_up_to_2	-	-	-	-
find_min_0	✓	✓	-	✓
find_min_2	✓	✓	-	-
find_min_local_2	-	-	-	-
find_min_up_to_0	-	-	-	-
find_min_up_to_2	-	-	-	-
find1_4	-	✓	-	-
find2_4	✓	✓	-	-
in_place_max	✓	✓	-	✓
inc_by_one_0	✓	✓	-	✓
inc_by_one_1	✓	✓	-	✓
inc_by_one_harder_0	✓	✓	-	✓
inc_by_one_harder_1	✓	✓	-	✓
init	✓	✓	-	-
init_conditionally_0	✓	✓	-	-
init_conditionally_1	✓	✓	-	✓
init_non_constant_0	✓	✓	-	-
init_non_constant_1	✓	✓	-	✓
init_non_constant_2	✓	✓	-	✓
init_non_constant_3	✓	✓	-	✓
init_non_constant_easy_0	✓	✓	-	-
init_non_constant_easy_1	✓	✓	-	✓
init_non_constant_easy_2	✓	✓	-	✓
init_non_constant_easy_3	✓	✓	-	✓
init_partial	✓	✓	-	✓

Benchmark	(1)	(2)	(3)	(4)
init_prev_plus_one_0	✓	✓	-	-
init_prev_plus_one_1	✓	✓	-	-
init_prev_plus_one_alt_0	✓	✓	-	-
init_prev_plus_one_alt_1	✓	✓	-	-
insertion_sort	-	-	-	-
max_prop_0	✓	✓	-	✓
max_prop_1	✓	✓	-	✓
merge_interleave_0	✓	✓	-	✓
merge_interleave_1	✓	-	-	✓
min_prop_0	✓	✓	-	✓
min_prop_1	✓	✓	-	✓
partition_0	✓	✓	-	✓
partition_1	✓	✓	-	✓
push_back	✓	✓	-	✓
reverse	✓	✓	-	-
rewnifrev	✓	✓	-	✓
rewrev	✓	✓	-	-
skipped	✓	-	-	✓
str_cpy_0	✓	✓	-	-
str_cpy_1	✓	✓	-	-
str_cpy_2	✓	✓	-	-
swap_0	-	✓	✓	✓
swap_1	-	✓	✓	✓
vector_addition	✓	✓	-	✓
vector_subtraction	✓	✓	-	✓
check_equal_set_flag_0	✓	✓	-	-
find_max_1	-	-	-	-
find_max_from_second_1	✓	-	-	-
find1_2	✓	✓	-	-
find1_3	✓	✓	-	-
find2_2	✓	✓	-	-
find2_3	✓	✓	-	-
collect_indices_eq_val_2	-	✓	-	-
collect_indices_eq_val_3	✓	-	-	-
copy_nonzero_1	✓	✓	-	-
copy_positive_1	✓	✓	-	-
find_max_local_0	-	-	-	-
find_max_local_1	✓	-	-	-
find_max_up_to_1	-	-	-	-
find_min_1	-	-	-	-
find_min_local_0	-	-	-	-
find_min_local_1	✓	-	-	-
find_min_up_to_1	-	-	-	-
merge_interleave_2	✓	-	-	-
partition_2	✓	✓	-	-
partition_3	✓	✓	-	-
partition_4	-	-	-	-
partition_5	-	✓	-	-
partition_6	-	-	-	-
partition-harder_0	✓	✓	-	-
partition-harder_1	✓	✓	-	-
partition-harder_2	✓	-	-	-
partition-harder_3	✓	-	-	-
partition-harder_4	✓	-	-	-
str_len	✓	✓	-	-
Total solved	**93**	**78**	**13**	**47**

safety assertions; (iii) 7 problems come with the task of verifying existentially quantified assertions; (iv) and the last 23 programs contain assertions with alternation of quantifiers. The examples from (i)-(ii), a total of 81 programs, come from the array verification benchmarks of SV-COMP repository [1], with most of these examples originating from [9, 13].[5] These examples correspond to the set of those SV-COMP benchmarks which use the C fragment supported by RAPID; specifically, when selecting examples (i)-(ii) from SV-COMP, we omitted examples containing pointers or memory management. All SV-COMP examples from (i)-(ii) are adapted to our input format, as for example arrays in trace logic are treated as unbounded data structures. Further, the examples (iii)-(iv) are new examples crafted by us, in total 30 new examples. They contain existential and alternating quantification in safety assertions. We intend to submit these 30 examples from (iii)-(iv) to SV-COMP.

Experimental Setting. We used two versions of RAPID in our experiments. First, (1) RAPID^{l-} denotes our RAPID approach, using lemmaless induction `MCGLoopInd` and `AMLoopInd` in VAMPIRE. Further, (2) Rapid^{l+} uses trace lemmas for inductive reasoning, as described in [11]. We also compared RAPID^{l-} with other verification tools. In particular, we considered (3) SEAHORN and (4) VAJRA (and its extension DIFFY that produced for us exactly the same results as VAJRA). SEAHORN converts the program into a constrained horn clause (CHC) problem and uses the SMT solver Z3 for solving. VAJRA and DIFFY implement inductive reasoning and recurrence solving over loop counters; in the background, they also use Z3.

Rapid *Experiments*. Table 1 shows that RAPID^{l-} is superior to RAPID^{l+}, as it solves a total of 93 problems, while RAPID^{l+} only proved 78 assertions correct. Particularly, RAPID^{l-} can solve benchmark `merge_interleave_2` corresponding to our motivating example 1, and other challenging problems such as `find_max_local_1` also containing quantifier alternations.

While RAPID^{l-} can solve a total of ten problems more than RAPID^{l+}, it is interesting to look into which problems can now be solved. Many of the newly solved problems are structurally very close to the loop invariants needed to prove them. This is where multi-clause goal-oriented induction `MCGoalInd` makes the biggest impact. For instance, this allows RAPID^{l-} to prove the partial correctness of `find_max_ from_second_0` and `find_max_from_second_1`.

On the other hand, RAPID^{l-} also lost two challenging benchmarks that were previously solved by RAPID^{l+}, namely `swap_0` and `partition_5`. This could be for two reasons: (1) the strategies in the induction schedule of RAPID^{l-} are too restrictive for such benchmarks, or (2) the step case of the induction axiom introduced by our two rules are too difficult for VAMPIRE to prove. Strengthening lemmaless induction with additional trace lemmas from RAPID^{l+} is an interesting line of further work.

---

[5] Artifact evaluation: in order to reproduce the results reported in this section, please follow the instructions at https://github.com/vprover/vampire_publications/tree/master/experimental_data/CICM-2022-RAPID-INDUCTION.

Comparing with other tools. Both, SEAHORN and VAJRA/DIFFY require C code as input, whereas RAPID uses its own syntax. We translated our benchmarks to C code expressing the same problem. However, a direct comparison of RAPID, and in particular RAPID^{l-}, with most other verifiers requiring standard C code as an input is not possible as we consider slightly different semantics. In contrast to SEAHORN and VAJRA/DIFFY, we assume that integers and arrays are unbounded and that all array positions are initialized by arbitrary data. Further, we can read/write at any array position without allocating the accessed memory beforehand. Apart from semantic differences, RAPID can directly express assertions and assumptions containing quantifiers and put variable contents from different points in time into relation. In order to deal with the latter, we introduced history variables in the code provided to SEAHORN and VAJRA/DIFFY. Quantification was simulated by non-deterministically assigned variables and by loops. As a result, SEAHORN verified 13 examples, whereas VAJRA/DIFFY 47 of our benchmarks. As VAJRA/DIFFY restrict their input programs to contain only loops having very specific loop-conditions, several of our benchmarks failed. For example, $i < length$ is permitted, whereas $a[i] \neq 0$ is not. VAJRA/DIFFY could prove correctness for nearly all the programs satisfying these restrictions. SEAHORN, on the other hand, has problems with the complexity introduced by the arrays. It could solve especially those benchmarks whose correctness do not depend on the arrays' content.

9   Future Directions and Conclusion

We introduced lemmaless induction to fully automate the verification of inductive properties of program loops with unbounded arrays and integers. We introduced goal-oriented and array mapping induction inferences, triggered by loop counters, in superposition-based theorem proving. Our results show that lemmaless induction in trace logic outperforms other state-of-the-art approaches in the area. There are various ways to further develop lemmaless induction in trace logic. On larger benchmarks, particularly those containing multiple loops, our approach struggles. For loops where the required invariant is not connected to the conjecture, we introduced array mapping induction. However, the array mapping induction inference is limited in the form of invariants it can generate. We would like to investigate other methods, such as machine learning for synthesising loop invariants that are not too prolific. A completely different line of research that we are currently working on, is updating the trace logic syntax and semantics of $\mathcal{W}$ to deal with memory and memory allocation, aiming to efficiently reason about loop operations over the memory.

As shown in [18], the validity problem for first-order formulas of linear arithmetic extended with non-theory function symbols is Π_1^1-complete. Therefore, we do not expect any completeness result for inductive theorem proving. Proving relative completeness results for our verification framework is an interesting question.

Acknowledgements. This research was partially supported by the ERC consolidator grant ARTIST 101002685, the FWF research project LogiCS W1255-N23, the TU Wien SecInt doctoral program, and the EUProofNet Cost Action CA20111. Our research was partially funded by the Digital Security by Design (DSbD) Programme delivered by UKRI to support the DSbD ecosystem.

References

1. SV-comp repository. https://gitlab.com/sosy-lab/benchmarking/sv-benchmarks
2. Vampire website. https://vprover.github.io/
3. Bachmair, L., Ganzinger, H.: Resolution theorem proving. In: Robinson, A., Voronkov, A. (Eds.) Handbook of Automated Reasoning, vol. I, chap. 2, pp. 19–99. Elsevier Science (2001)
4. Barrett, C., et al.: CVC4. In: Gopalakrishnan, G., Qadeer, S. (eds.) CAV 2011. LNCS, vol. 6806, pp. 171–177. Springer, Heidelberg (2011). https://doi.org/10.1007/978-3-642-22110-1_14
5. Chakraborty, S., Gupta, A., Unadkat, D.: Verifying array manipulating programs with full-program induction. In: TACAS 2020. LNCS, vol. 12078, pp. 22–39. Springer, Cham (2020). https://doi.org/10.1007/978-3-030-45190-5_2
6. Chakraborty, S., Gupta, A., Unadkat, D.: Diffy: inductive reasoning of array programs using difference invariants. In: Silva, A., Leino, K.R.M. (eds.) CAV 2021. LNCS, vol. 12760, pp. 911–935. Springer, Cham (2021). https://doi.org/10.1007/978-3-030-81688-9_42
7. Claessen, K., Johansson, M., Rosén, D., Smallbone, N.: Automating inductive proofs using theory exploration. In: Bonacina, M.P. (ed.) CADE 2013. LNCS (LNAI), vol. 7898, pp. 392–406. Springer, Heidelberg (2013). https://doi.org/10.1007/978-3-642-38574-2_27
8. de Moura, L., Bjørner, N.: Z3: an efficient SMT solver. In: Ramakrishnan, C.R., Rehof, J. (eds.) TACAS 2008. LNCS, vol. 4963, pp. 337–340. Springer, Heidelberg (2008). https://doi.org/10.1007/978-3-540-78800-3_24
9. Dillig, I., Dillig, T., Aiken, A.: Fluid updates: beyond strong vs. weak updates. In: Gordon, A.D. (ed.) ESOP 2010. LNCS, vol. 6012, pp. 246–266. Springer, Heidelberg (2010). https://doi.org/10.1007/978-3-642-11957-6_14
10. Fedyukovich, G., Prabhu, S., Madhukar, K., Gupta, A.: Quantified invariants via syntax-guided synthesis. In: Dillig, I., Tasiran, S. (eds.) CAV 2019. LNCS, vol. 11561, pp. 259–277. Springer, Cham (2019). https://doi.org/10.1007/978-3-030-25540-4_14
11. Georgiou, P., Gleiss, B., Kovács, L.: Trace logic for inductive loop reasoning. In: 2020 Formal Methods in Computer Aided Design (FMCAD), pp. 255–263. IEEE (2020)
12. Gurfinkel, A., Kahsai, T., Komuravelli, A., Navas, J.A.: The SeaHorn verification framework. In: Kroening, D., Păsăreanu, C.S. (eds.) CAV 2015. LNCS, vol. 9206, pp. 343–361. Springer, Cham (2015). https://doi.org/10.1007/978-3-319-21690-4_20
13. Gurfinkel, A., Shoham, S., Vizel, Y.: Quantifiers on demand. In: Lahiri, S.K., Wang, C. (eds.) ATVA 2018. LNCS, vol. 11138, pp. 248–266. Springer, Cham (2018). https://doi.org/10.1007/978-3-030-01090-4_15
14. Hajdú, M., Hozzová, P., Kovács, L., Schoisswohl, J., Voronkov, A.: Induction with generalization in superposition reasoning. In: Benzmüller, C., Miller, B. (eds.)

CICM 2020. LNCS (LNAI), vol. 12236, pp. 123–137. Springer, Cham (2020). https://doi.org/10.1007/978-3-030-53518-6_8

15. Hozzová, P., Kovács, L., Voronkov, A.: Integer induction in saturation. In: Platzer, A., Sutcliffe, G. (eds.) CADE 2021. LNCS (LNAI), vol. 12699, pp. 361–377. Springer, Cham (2021). https://doi.org/10.1007/978-3-030-79876-5_21

16. Karpenkov, E.G., Monniaux, D.: Formula slicing: inductive invariants from preconditions. In: Bloem, R., Arbel, E. (eds.) HVC 2016. LNCS, vol. 10028, pp. 169–185. Springer, Cham (2016). https://doi.org/10.1007/978-3-319-49052-6_11

17. Kaufmann, M., Moore, J.S.: An industrial strength theorem prover for a logic based on common Lisp. In: IEEE Transactions on Software Engineering, pp. 203–213 (1997)

18. Korovin, K., Voronkov, A.: Integrating linear arithmetic into superposition calculus. In: Duparc, J., Henzinger, T.A. (eds.) CSL 2007. LNCS, vol. 4646, pp. 223–237. Springer, Heidelberg (2007). https://doi.org/10.1007/978-3-540-74915-8_19

19. Kovács, L., Robillard, S., Voronkov, A.: Coming to terms with quantified reasoning. In: POPL, pp. 260–270 (2017)

20. Kovács, L., Voronkov, A.: First-order theorem proving and VAMPIRE. In: Sharygina, N., Veith, H. (eds.) CAV 2013. LNCS, vol. 8044, pp. 1–35. Springer, Heidelberg (2013). https://doi.org/10.1007/978-3-642-39799-8_1

21. Larraz, D., Rodríguez-Carbonell, E., Rubio, A.: SMT-based array invariant generation. In: Giacobazzi, R., Berdine, J., Mastroeni, I. (eds.) VMCAI 2013. LNCS, vol. 7737, pp. 169–188. Springer, Heidelberg (2013). https://doi.org/10.1007/978-3-642-35873-9_12

22. Leino, K.R.M.: Dafny: an automatic program verifier for functional correctness. In: Clarke, E.M., Voronkov, A. (eds.) LPAR 2010. LNCS (LNAI), vol. 6355, pp. 348–370. Springer, Heidelberg (2010). https://doi.org/10.1007/978-3-642-17511-4_20

23. Rajkhowa, P., Lin, F.: Extending VIAP to handle array programs. In: Piskac, R., Rümmer, P. (eds.) VSTTE 2018. LNCS, vol. 11294, pp. 38–49. Springer, Cham (2018). https://doi.org/10.1007/978-3-030-03592-1_3

24. Bjoner, N., Reger, G., Suda, M., Voronkov, A.: AVATAR modulo theories. In: GCAI, pp. 39–52 (2016)

25. Reger, G., Schoisswohl, J., Voronkov, A.: Making theory reasoning simpler. In: TACAS 2021. LNCS, vol. 12652, pp. 164–180. Springer, Cham (2021). https://doi.org/10.1007/978-3-030-72013-1_9

26. Srivastava, S., Gulwani, S.: Program verification using templates over predicate abstraction. In: PLDI, pp. 223–234 (2009)

Unified Decomposition-Aggregation (UDA) Rules: Dynamic, Schematic, Novel Axioms

Alan Bundy[✉] and Kwabena Nuamah

School of Informatics, University of Edinburgh, Edinburgh, UK
{A.Bundy,K.Nuamah}@ed.ac.uk

Abstract. We introduce *Unified Decomposition-Aggregation (UDA) Rules*. They are a family of axiom schemata that are instantiated at run-time to add new axioms to a logical theory. These new axioms are implications, whose preconditions will be constructed from an analysis of the goal to be proved and the theory in which it is to be proved. We illustrate their application to query answering using the FRANK system.

Keywords: UDA rules · query answering · hybrid reasoning

1 Introduction

Unified Decomposition-Aggregation (UDA) rules[1] were invented as part of the FRANK (Functional Reasoner for Acquiring New Knowledge) query answering system [5,10]. They have the potential for wider application. We define UDA rules in Definition 3.

UDA rules decompose a goal into a set of subgoals. The answers returned by the subgoals are then aggregated into an answer to the original goal using an *aggregation function*. This decomposition process can recur. What distinguishes a UDA rule from other implications is that it dynamically computes the number and form of the subgoals. For instance, to answer the query *"What will be the population of the UK in 2030?"* regression over past census data is used via a *temporal UDA rule*. The subgoals might represent the sub-queries *"What will be the population of the UK in t_i?"*, where each t_i is a year for which the UK population can be retrieved from prior census data sources. Regression is then applied to these previous population numbers and the resulting function is extrapolated to the year 2030. Another distinction is that the answers to queries are incrementally both constructed and evaluated by aggregation functions during their propagation from leaves to the root of the inference graph. A comparison between FRANK's inference process and that of traditional automated reasoning systems is given in [5].

---

[1] Previously called decomposition rules.

K. Buzzard and T. Kutsia (Eds.): CICM 2022, LNAI 13467, pp. 209–221, 2022.
https://doi.org/10.1007/978-3-031-16681-5_15

The main novel contribution of this paper is to demonstrate that axiomatic theories can be dynamically constructed by adding correct new axioms to a theory on an as-needed basis.

There are many kinds of UDA rules: parent goals can be decomposed into child sub-goals in many different ways and the corresponding child answers can be aggregated in many different ways to return a value to their parent goal. There is not a 1–1 correspondence between decompositions and aggregations, there are many legitimate combinations depending on the original query, retrieved data and user preferences, e.g., speed *vs* accuracy.

Instantiation of a schematic UDA rule may be incremental. For instance, the choice of aggregation function may be delayed, e.g., the kind of regression to be used may depend on the available data. This means that the precise form of a UDA rule may not be known until inference is complete.

To illustrate the use of UDA rules, suppose the query is:

What will be the population in 2030 of the African country that will have the largest GDP in 2025?.

First, a *partition UDA rule* can be used to formulate a query to predict the 2025 GDP of each African country and return the country with the largest one. Temporal decomposition can be used to make this prediction for each country. Then, temporal decomposition will be used again to predict the 2030 population of the selected country.

Partition rules can also be used to estimate the total population of a continent by summing the populations of its individual countries. Or, the cost of a compound object might be estimated by summing the costs of each of its components.

A precursor to UDA rules can be found in the first author's Mecho project [3], which solved mechanics problems stated in English. Part of the solution process consisted of instantiating the laws of Physics, such as $F = m.a$, where m is the mass of an object, a its acceleration and F the sum of the forces acting on it. Consider the task of instantiating F. This required Mecho to partition this sum of forces into each of the individual forces, e.g., gravitational attraction, tension in a string, friction from an inclined plane, etc. The partition UDA rule would have been ideal for this task, but since we had not then invented such rules, Mecho dealt with it in a more ad hoc way. The potential for additional applications of UDA rules can be found in Sect. 7.

These examples illustrate the potentially wide applicability of UDA rules, especially as automated reasoning broadens its remit beyond the traditional axiomatic theories of pure mathematics and formal methods, to reason about the wider environment.

2   Association Lists

FRANK uses UDA rules to construct an inference graph in which each decomposition of a goal into subgoals represents an AND branch in the graph. The leaves of the graph are subgoals that can be instantiated by matching facts from web-based knowledge sources. It uses a diverse and dynamically chosen set of knowledge sources whose facts are represented in a diverse number of formalisms. In order to combine these facts, they are translated into a common formalism: association lists, abbreviated as *alists*[2] [10]. Alists are defined in Definition 1.

Definition 1 (Alist). *An* alist *is a set of pairs* $\{\langle A_i, a_i \rangle | 1 \leq i \leq n\}$, *where each* A_i *is an attribute and* a_i *is its value. This will sometimes be written as* $\{\langle A_1, a_1 \rangle, \ldots, \langle A_n, a_n \rangle\}$ *or abbreviated as* $\mathcal{A}$.

- *One attribute must be Predicate. This allows an alist:*

$$\{\langle Predicate, P \rangle, \langle A_1, a_1 \rangle, \ldots, \langle A_n, a_n \rangle\}$$

  to be represented as the typed relation $P(a_1, \ldots, a_n)$ *where its type is* $P :$ $A_1 \times \ldots \times A_n \mapsto Bool$, *i.e., the* A_i *attributes are interpreted as the types of their values.*
- *Typical attributes are Subject, Object, Predicate, Time, etc., abbreviated as* s, o, p *and* t. *Values can be names of entities, numbers, functions, etc.*
- *We will use the notation* $\mathcal{A}(t)$ *to indicate that* $\mathcal{A}$ *contains a distinguished term* t *at some unspecified argument position, e.g., an attribute or its value.*
- *We also need to interpret Alists as functions from some of its values to others. For this purpose we use Hilbert's Epsilon Operator*[3], *written:* $\epsilon x. \mathcal{A}(x)$ *where the alist* $\mathcal{A}$ *contains the variable* x. *This is read as the value of* x *such that* $\mathcal{A}(x)$ *is true. If there is more than one such value, an arbitrary choice is made. If there is no such value then* $\mathcal{A}(x)$ *is false, so it cannot be the child alist of a successful decomposition, and can, therefore, be ignored.*

Further details about the semantics of alists are discussed in the second author's thesis [10].

3 Variables in Alists

Alists can represent goals by specifying a variable as the value for at least one of its attributes. A goal will be satisfied during inference by instantiating its variable to a concrete value. Instantiation occurs when a leaf alist is matched to a fact in a knowledge base. The values of variable are then propagated through the inference graph from the leaves to the root alist. At each stage the values of the child alists are aggregated to give a value for the parent alist. Aggregation is performed by the *aggregation functions* of the UDA rules used. Aggregation functions range from the identity function, through various arithmetic functions

---

[2] See https://en.wikipedia.org/wiki/Association_list (last accessed: 02-02-2022). Alists are not lists but sets, but the 'alist' terminology has, unfortunately, become standard.
[3] https://en.wikipedia.org/wiki/Epsilon_calculus accessed on 02.02.2022.

(e.g., Σ, max, min) to statistical operations, such as regression. FRANK's range of statistical methods has recently been significantly extended [8].

Definition 2 describes the different kinds of variables used in an alist as defined in [10].

Definition 2 (Kinds of Variables).

Projection Variables: Those variables in a child alist whose values are to be projected to its parent. *They become operation arguments of the aggregation function in the parent alist. They are prefixed with a* ?, *e.g.,* ?x *denotes a projection variable. In general, a child alist may have several projection variables, so we use vector notation to denote them all, e.g.,* ?$\vec{x}$.

Auxiliary Variables: Those variables in a child alist whose values are used locally, but which are not projected to its parents. *They are prefixed with a* \$, *e.g.,* \$$x$ *denotes an auxiliary variable. In general, an alist may have several auxiliary variables, so we use vector notation to denote them all, e.g.,* \$$\vec{x}$.

Operation Variables: Those variables that are used as arguments for an $\mathcal{A}$'s aggregation function h. *An operation variable must be either a projection or an auxiliary variable, so we omit any prefix. They must exist as an attribute value in* $\mathcal{A}$.

Projection variables are instantiated to values that are aggregated from child alists and projected to their parents or by matching against facts in a knowledge base. The value projected to the root of the inference graph becomes the answer to the original query.

For instance, consider the question from Sect. 1:

What will be the population in 2030 of the African country that will have the largest GDP in 2025?.

The inference graph that FRANK generates to answer this query will consist of several nodes labelled by alists. Two such alists are (1) and (2). Alist (1) represents the sub-query where FRANK predicts the GDP ?g of Ghana in 2025. This alist is one of many generated for the different countries in Africa in order to find the one with the largest predicted GDP in 2025. Alist (2) represents the sub-query where FRANK predicts the population ?p of country \$$c$ in 2030.

$$\{\langle p, gdp \rangle, \langle s, Ghana \rangle, \langle o, ?g \rangle, \langle t, 2025 \rangle\} \tag{1}$$
$$\{\langle p, population \rangle, \langle s, \$c \rangle, \langle o, ?p \rangle, \langle t, 2030 \rangle\} \tag{2}$$

where p is the predicate, s the subject, o the object and t the time. \$$c$ is an auxiliary variable that will be instantiated to the country with the largest GDP. ?g and ?p are projection variables whose values will be propagated to their parents. In the case of ?g it will then become an auxiliary variable to the max function, to determine the country with the largest GDP. In the case of ?p it will be the population of country \$$c$. Note that, in one of the children of (2), \$$c$ will have been a projection variable ?c, to be propagated to alist (2).

Each alist has an aggregation function attribute h with a value h_τ, say. This function h_τ is applied to the projection variables of the child alists to instantiate the projection variable of the parent. This aggregation function is associated with the UDA rule on the AND branch connecting the parent alist to its children. The aggregation operation requires each alist to be regarded as a function from the projection variables of the children to the projection variable of the parent. As described in Sect. 2, we use the Hilbert Epsilon operation ϵ to convert alists from relations to functions for this purpose.

4 UDA Rules

UDA rules are implications applied by backwards reasoning to a parent alist to create a set of child alists, i.e., the logical implication is from children to parent. The size of this set and the anatomy of its members is described by: the type τ of the rule, the parent alist that it is applied to and the environment in which it is applied. τ is incrementally defined during the inference process. For instance, when a choice is made to use a partition rule, τ will be instantiated to $part(v)$, where $part$ refers to the partition rule and v is a variable that will eventually be instantiated to specify the type of aggregation. When this function is chosen to be the summation of the values of the child values, the rule type, $part(v)$, will be further instantiated to $part(\Sigma)$. In general, τ will be different for each UDA rule application. For instance, the children of a partition UDA rule will depend on the available partitions of the parent, e.g., by $partOf$ relations. The children of a temporal UDA rule will depend on the available values of the parent's predicate for earlier values of its time attribute, e.g., by $before$ relations. The general form of a UDA rule is given in Definition 3. Their correctness is defined in Subsect. 4.3 and their application is illustrated in Sect. 5.

Definition 3 (UDA rule). *A UDA rule is an implication of the form:*

$$Decompose(\mathcal{A}(\vec{x}), \tau) = [\mathcal{A}_j(?\vec{x}_j)|1 \le j \le m]$$
$$\implies \mathcal{A}(\vec{h_\tau}(\epsilon?\vec{x}_1.\ \mathcal{A}_1(?\vec{x}_1), \dots, \epsilon?\vec{x}_m.\ \mathcal{A}_m(?\vec{x}_m)))$$

where:

- *Decompose is a function that takes the parent alist $\mathcal{A}(\vec{x})$ and the type of decomposition τ and returns a list of m child alists $[\mathcal{A}_j(?\vec{x}_j)|1 \le j \le m]$.*
- *The various alists differ only in the attribute values singled out as their arguments, e.g., $\vec{x}$ in $\mathcal{A}(\vec{x})$. To suppress clutter, their attributes and other values are not indicated, but are identical in each alist.*
- *$\vec{h_\tau}$ is the aggregation function that takes the values $\epsilon?\vec{x}_j.\ \mathcal{A}_j(?\vec{x}_j)$ assigned to the projection variables $?\vec{x}_j$ of the child alists $\mathcal{A}_j(?\vec{x}_j)$ and calculates an aggregated value:*

$$\vec{h_\tau}(\epsilon?\vec{x}_1.\ \mathcal{A}_1(?\vec{x}_1), \dots, \epsilon?\vec{x}_m.\ \mathcal{A}_m(?\vec{x}_m))$$

to be projected back to the parent alist $\mathcal{A}(\vec{x})$ as the value(s) of the variable(s) $\vec{x}$. Note that the $?\vec{x}_j$ are projection variables as their values will be projected back to their parent alist $\mathcal{A}(\vec{x})$. The $\vec{x}$ may or may not be projection variables.

- *h_τ is selected during inference as an operation to aggregate the values returned by the new child alists. Note that the choice of aggregation function might depend on the values to be aggregated, the query intent, context and the decomposition. For instance, the choice of an aggregation function for prediction during temporal decomposition will depend on, say, the number of data points available and their underlying statistical distribution. So, the choice may be left as a variable to be instantiated only after the UDA rule has been applied.*
- *$[\mathcal{A}_j | 1 \leq j \leq m]$ is a form of list composition, that we have invented, which is analogous to set comprehension. Lists are used, rather than sets, because argument order may matter[4] for the aggregation function h_τ.*
- *Vector notation is used for the variables $\vec{x}_j$ and $\vec{x}$, and for the aggregation function h_τ because more than one variable and/or function may be involved in a rule.*
- *Note that the logical implication of UDA rules is from right to left: the values of the projection variables of the child alists determine the values of the operands of the parent alist. But FRANK builds the inference graph by applying the UDA rules left to right, i.e., from the goal alist to the leaf node alists. The projection variables of the leaf alist are then instantiated by matching them against facts in the knowledge sources.*

The above UDA rule can be represented graphically as shown in Fig. 1.

Another novel feature of UDA rules is that they define two kinds of inference: firstly, the replacement of parent goals with child sub-goals via decomposition and secondly, the propagation of answers from leaves to root via aggregation. Unlike most automated reasoning, where variables are just instantiated to compound terms, aggregation evaluates these compound terms to return values, e.g., numbers, countries.

4.1   The New Axiom

We can describe the new axiom being added by a UDA rule using the alternative first-order logic (FOL) notation introduced in Sect. 2. Let P be the value of the compulsory *Predicate* attribute of $\mathcal{A}(\vec{x})$ and P_j its value in $\mathcal{A}_j(?\vec{x}_j)$ for $1 \leq j \leq m$. Then the new axiom is:

$$P_1(?\vec{x}_1) \wedge \ldots \wedge P_m(?\vec{x}_m) \implies P(\vec{h}_\tau(?\vec{x}_1, \ldots, ?\vec{x}_m)) \tag{3}$$

For the sake of readability, we have omitted any additional arguments of P and the P_j.

If the UDA rule was used to add this axiom directly into the logical theory, then any FOL theorem prover, e.g., resolution [1], could be used to derive the required answer to the query.

---

[4] Although not for any of the examples in this paper.

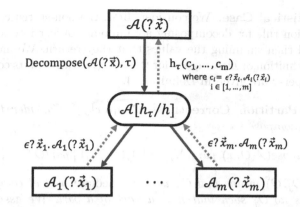

Fig. 1. A graphical representation of a UDA rule. It shows a single step of inference in FRANK and is applied recursively to construct inference graphs. Solid red arrows show the direction of decomposition while the dashed green ones show the direction of aggregation during upward propagation of projection variable instantiations from children to their parent. (Color figure online)

4.2 Representing Uncertainty

FRANK applies both arithmetic and statistical aggregation functions, e.g., summation and regression. Statistical aggregation introduces uncertainty into the reasoning, so we cannot, in general, demand logical correctness of the axioms that UDA rules introduce into a theory. There is also noise in the knowledge sources. So, the validity of FRANK's inference is statistical as well as logical. It, therefore, returns an error interval with its conclusions [10,12].

Noise in the answer is assumed to be a Gaussian[5]. For quantitative queries, FRANK uses the mean as the answer and the coefficient of variation (cov), which is the standard deviation normalised by the mean, as the measure of uncertainty. The cov can be interpreted as an error interval around the answer. To calculate the answer's cov, each $\mathcal{A}$ and each $\mathcal{A}_j$ is given an additional cov attribute/value pair and an uncertainty propagation function, similar to $\vec{h_\tau}$, is used to propagate the uncertainties of the child alists $\mathcal{A}_j$s to the uncertainty of their parent $\mathcal{A}$.

4.3 The Correctness of the UDA Rules

We can associate a *correctness property* with each UDA rule. In the case of rules that use a statistical aggregation function, the correctness property is also statistical. The proofs of these correctness properties would be conducted in a theory of sets, and the definition of the corresponding aggregation functions. Their automation is currently future work.

---

[5] Or similar, depending on the statistical methods used.

The Non-Statistical Case. We consider, first, the non-statistical case. Consider the partition rule for decomposing a compound object into a partition of sub-objects and then summing the values that they return. We name this rule $part(\Sigma)$. The definition of *Decompose* for the $part(\Sigma)$ rule and its corresponding correctness property are given in Definition 4.

Definition 4 (Partition Correctness Property). *Consider the following definition of Decompose.*

$$Decompose(\mathcal{A}(O, x), part(\Sigma)) = [\mathcal{A}_j(O_j, x_j) \mid partOf(O_j, O)]$$

where $partOf(O_j, O)$ means that O_j is an immediate part of compound object O, i.e., there is no O_j such that it is a part of a part. We assume that the call to partOf returns all such parts, so the O_js form a partition. We have distinguished two attribute values in $\mathcal{A}(O, x)$ and the $\mathcal{A}_j(O_j, x_j)$: the objects O and their properties x. Otherwise, the alists $\mathcal{A}$ and the $\mathcal{A}_j$ are identical.

Note that we have dropped the vector notation as, in this rule, the x and x_j are singleton variables.

The $part(\Sigma)$ rule correctness property is now:

$$x = \Sigma([x_j \mid \mathcal{A}_j(O_j, x_j) \wedge partOf(O_j, O)]) \implies \mathcal{A}(O, x) \qquad (4)$$

where the x_j are formed into a list to which Σ is applied and the result x gives a value that makes $\mathcal{A}(O, x)$ true.

Sometimes the partition rule is used not to sum the child scores, but to find a maximum, minimum or apply another arithmetic operation. Different correctness properties are needed for these cases, such as:

$$x = max([x_j \mid \mathcal{A}_j(O_j, x_j) \wedge partOf(O_j, O)]) \implies \mathcal{A}(O, x)$$
$$x = min([x_j \mid \mathcal{A}_j(O_j, x_j) \wedge partOf(O_j, O)]) \implies \mathcal{A}(O, x)$$

The Statistical Case. We now consider the statistical case, illustrated by the regression aggregation function, *regress*, used by the temporal rule, which we name the $temp(reg)$ rule[6]. Definition 5 defines *Decompose* for the $temp(reg)$ rule and its corresponding correctness property.

Definition 5 (Temporal Correctness Property).

$$Decompose(\mathcal{A}(T, y), temp(reg)) = [\mathcal{A}_j(T_j, y_j) \mid before(T_j, T)]$$

where $before(T_j, T)$ means that time T_j occurs before time T. We have distinguished two attribute values in the alists: the times T and the alists' property's

---

[6] Note that the temporal rule can also use non-statistical aggregation functions, e.g., $temp(max)$ could be used to find the maximum value of a property among a set of times.

values y at the corresponding times. We use y instead of x, as, traditionally, the x-axis would be time and the y-axis is the value of the property over time. Otherwise, the alists $\mathcal{A}$ and the $\mathcal{A}_j$ are identical.

$$\rho : \langle f, st \rangle = regress([y_j \mid \mathcal{A}_j(T_j, y_j) \wedge before(T_j, T)])$$
$$\implies \exists y. \, y - sd(T) \leq F(T) \leq y + sd(T) \wedge \mathcal{A}(T, y)$$

regress returns the pair of the function f that it creates together with the first standard deviation function sd. $f(T)$ and $sd(T)$ extrapolate the functions f and sd to the time T. y is the value that makes $\mathcal{A}(T, y)$ true. The correctness property asserts that $f(T)$ lies within the error interval, i.e., within one standard deviation of y. ρ is the probability that $f(T)$ lies within this interval, which can be calculated using the **68-95-99.7** *rule[7], which gives the value 0.6827. $\langle T_j, y_j \rangle$ are the points used in the regression.*

FRANK's inference graph is constructed by applying correct UDA rules, but there are many branching points, so search is controlled by heuristics. For instance, FRANK prefers to consult knowledge sources containing facts with low uncertainty and to use decompositions that result in a representative list of child alists and that respect user preferences [13].

5   Worked Example

To illustrate the use of these UDA rules in FRANK, consider the following query.

What will be the population of Africa in 2030?

This can be represented as the following alist:

$$\{\langle p, population \rangle, \langle s, Africa \rangle, \langle o, ?x \rangle, \langle t, 2030 \rangle\} \tag{5}$$

where the 4 attributes p, s, o, t mean *predicate, subject, object* and *time*, respectively and $?x$ is a projection variable.

To answer this query FRANK can use two UDA rules: $part(\Sigma)$ and $temp(reg)$. $part(\Sigma)$ is used to partition *Africa* into its constituent countries and $temp(reg)$ is used to estimate the 2030 populations for each country, which are then summed using the Σ aggregation function of $part(\Sigma)$.

FRANK's GUI interface for this query is given in Fig. 2. FRANK's search strategy is to apply correct UDA rules respecting the heuristics outlined in Subsect. 4.3. One possible inference path proceeds as follows.

1. FRANK tries to match alist (5) to various knowledge sources, but finds no match because Africa's population for 2030 is not yet known.
2. It then makes an unsuccessful attempt to apply $temp(reg)$ to alist (5), which fails due to the lack of census records for the whole of Africa.

---

[7] https://en.wikipedia.org/wiki/68-95-99.7_rule accessed on 12.5.22.

3. Instead, it makes a successful attempt to apply $part(\Sigma)$ to alist (5) which creates the following 54 child alists:

$$\{\langle p, population \rangle, \langle s, Algeria \rangle, \langle o, ?x_1 \rangle, \langle t, 2030 \rangle\}$$
$$\ldots \{\langle p, population \rangle, \langle s, Zimbabwe \rangle, \langle o, ?x_{54} \rangle, \langle t, 2030 \rangle\}$$

4. FRANK tries to match each of these alists to various knowledge sources, but finds no matches because population data for 2030 is not yet known.
5. It then makes successful attempts to apply $temp(reg)$ to each of these 54 child alists. The $regress$ aggregation function returns a population growth function and a standard deviation. Both are extrapolated to 2030 and the resulting 2030 population estimates and their standard deviations are propagated back to their shared parent alist.
6. $part(\Sigma)$'s aggregation function, Σ, is used to sum the 54 population estimates. Their standard deviations are turned into $covs$ so that they can be combined both together and with a cov for the regression. The resulting cov is also propagated to the parent alist to provide the uncertainty estimate for the population estimate.
7. Both the population and uncertainty estimates are returned to the user.

6   Related Work

We have not found anything in the literature directly related to UDA rules. For instance, although axioms of the form (3) are commonly used in automated reasoning [1], they are usually provided manually by human experts, rather than automatically generated to meet the needs of the current (sub)goal.

There are also logics in which uncertainty measures are associated with axioms [7,9], but these measures are usually probabilities or similar. We found probabilities to be unsuitable as an uncertainty measure for quantitative answers. The probability that a particular number is the true answer to a query is usually 0. For instance, the precise annual population of a country is inherently uncertain, given the long period and whether tourists, illegal immigrants, babies being born, people in a vegetative state, people in hiding, etc. should be or can be counted. The range given by an error interval is a much better uncertainty measure. Probabilities are, however, an appropriate measure for *qualitative* answers, e.g., multi-hop information retrieval [6]. We are exploring their use in FRANK, including situations where a query requires mixing probabilities with error intervals, e.g., consider the example in Sect. 1, where the $covs$ on each African country's GDP must be propagated into the probability that a particular country has the largest GDP.

The way in which UDA rules dynamically construct the inference search space, e.g., by determining the branching rate of a rule, is reminiscent of *proof planning* [2], whereby meta-level rules are used to pre-plan the shape of a proof[8]. The main difference is that proof planning uses a two-level inference

---

[8] We are grateful to an anonymous reviewer for pointing out this analogy and suggesting that we discuss it here.

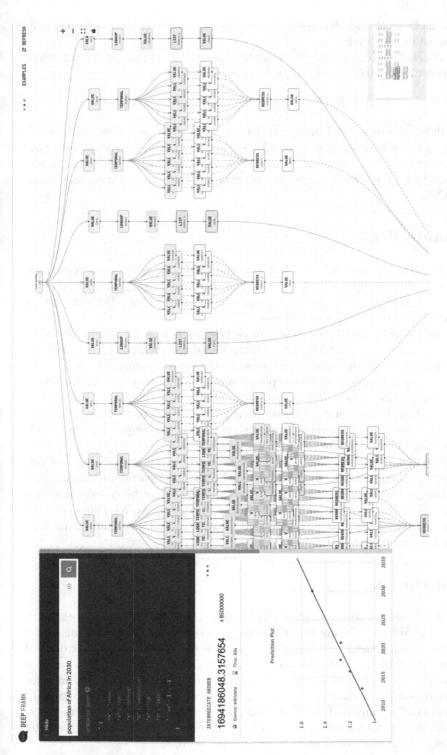

Fig. 2. The web browser-based user interface for FRANK. On the left side is shown the user's question, the generated alist query, the intermediate answer, error intervals on the answer and the knowledge sources used. On the right is shown the corresponding inference graph.

process: meta-level and object-level rules, whereas UDA rules are dynamically constructed object-level rules.

An alternative to an alist representation might be one based on functions. For instance, the $part(\Sigma)$ rule might be represented as:

$$F(y) = \Sigma(\{F_j(y_j)|partOf(y, y_j) \wedge 1 \leq j \leq m\})$$

where, using the functional interpretation of alists, F is associated with $\mathcal{A}$, F_j with $\mathcal{A}_j$, y is the compound object and the y_j are its parts. The projection variables are now the outputs of the F and F_j functions. This functional representation might be used, for instance, in a functional programming implementation of FRANK.

7   Future Work

We plan to develop some theories in which the UDA rule correctness properties can be proved, initially manually, but ultimately automatically by FRANK. We also plan to develop our correctness properties for particular UDA rules into a general theory of UDA rule correctness.

We are considering the use of a choice-style UDA rule for meta-level inference, i.e., to automate FRANK's engineering decisions, such as which knowledge sources and reasoning methods to use to solve the current query. This will assist us in our vision of whole-system explanations of AI systems [4,11]. By representing these choices in a declarative form, FRANK's explanation system will be able to describe not only the object-level inference it uses to answer a query, but also the meta-level inference it used to construct the object-level inference process from knowledge sources and reasoning methods.

We are experimenting with a probabilistic uncertainty measure for qualitative queries. These uncertainty measures need to be interleaved during inference as a quantitative value propagated from a child alist may be converted into a quantitative value propagated by its parent and vice versa, e.g., $covs$ on each African country's GDP may be propagated into the probability that a particular country has the largest GDP.

8   Conclusion

We have introduced the concept of UDA rules. They are implication schemata that are instantiated during the inference process to create new axioms to be added to a logical theory and then used in a proof. These new axioms depend on: the type τ of the rule, the parent alist that it is applied to, the original goal and the theory in which it is applied. They are useful whenever the number of preconditions of the new axiom depends on these factors, e.g., the number of components of a compound object, the number of available values of a time varying attribute, the forces acting on an object, the number of applicable reasoning methods and the number of relevant knowledge sources.

UDA rules have been implemented in the FRANK query answering system. We have proposed correctness properties for the different kinds of UDA rules. Some of FRANK's UDA rules have statistical aggregation functions, such as regression. The correctness properties for these rules give only a probability that FRANK's answers lie within the first standard deviation of the true answer. The knowledge bases from which FRANK draws its facts are also potentially noisy. Given these two sources of uncertainty, FRANK returns both an estimated answer and an error interval.

Acknowledgements. Thanks to Nicholas Ferguson, Thomas Fletcher, Xue Li, Ruqui Zhu and five anonymous reviewers for feedback on an earlier draft. This research has been supported by Huawei grants CIENG4721/LSC and HO2017050001B8s. For the purpose of open access, the author has applied a Creative Commons Attribution (CC BY) licence to any Author Accepted Manuscript version arising from this submission.

References

1. Bachmair, L., Ganzinger, H.: Resolution theorem proving. In: Robinson, J.A., Voronkov, A. (eds.) Handbook of Automated Reasoning, vol. 1, vol. I, chap. 2, pp. 19–199. Elsevier (2001)
2. Bundy, A.: A Science of Reasoning, pp. 178–198. MIT Press, Cambridge (1991)
3. Bundy, A., Byrd, L., Luger, G., Mellish, C., Milne, R., Palmer, M.: Solving mechanics problems using meta-level inference. In: Buchanan, B.G. (ed.) Proceedings of IJCAI-79, pp. 1017–1027. International Joint Conference on Artificial Intelligence (1979)
4. Bundy, A., Nuamah, K.: Combining deductive and statistical explanations in the FRANK query answering system. In: Gong, Z., Li, X., Oguducu, S.G. (eds.) Proceedings of the 12th IEEE International Conference on Big Knowledge (ICBK), IEEE, Auckland, New Zealand, December 2021
5. Bundy, A., Nuamah, K., Lucas, C.: Automated reasoning in the age of the internet. In: Fleuriot, J., Wang, D., Calmet, J. (eds.) AISC 2018. LNCS (LNAI), vol. 11110, pp. 3–18. Springer, Cham (2018). https://doi.org/10.1007/978-3-319-99957-9_1
6. Das, R., et al.: Multi-step entity-centric information retrieval for multi-hop question answering. In: Proceedings of the 2nd Workshop on Machine Reading for Question Answering, pp. 113–118 (2019)
7. Fierens, D., et al.: Inference and learning in probabilistic logic programs using weighted boolean formulas. Theor. Pract. Logic Program. **15**(3), 358–401 (2015)
8. Fletcher, T., Bundy, A., Nuamah, K.: Statistics automation in a query-answering system. Technical report, The University of Edinburgh (2022)
9. Nilsson, N.J.: Probabilistic logic. Artif. Intell. **28**(1), 71–87 (1986)
10. Nuamah, K.: Functional inferences over heterogeneous data, unpublished Ph.D. Dissertation, University of Edinburgh (2018)
11. Nuamah, K.: Deep algorithmic question answering: towards a compositionally hybrid AI for algorithmic reasoning. In: Workshop on Knowledge Representation for Hybrid and Compositional AI (2021)
12. Nuamah, K., Bundy, A.: Calculating error bars on inferences from web data. In: Arai, K., Kapoor, S., Bhatia, R. (eds.) IntelliSys 2018. AISC, vol. 869, pp. 618–640. Springer, Cham (2019). https://doi.org/10.1007/978-3-030-01057-7_48
13. Nuamah, K., Bundy, A., Jia, Y.: A context mechanism for an inference-based question answering system. In: AAAI2021 Workshop on CSKGs, vol. 8 (2021)

Working with Families of Inverse Functions

David J. Jeffrey[1,2] and Stephen M. Watt[1,2(✉)]

[1] Department of Applied Mathematics, University of Western Ontario,
London, Canada
`djeffrey@uwo.ca`
[2] David R. Cheriton School of Computer Science, University of Waterloo,
Waterloo, Canada
`smwatt@uwaterloo.ca`

Abstract. When evaluating or simplifying mathematical expressions, the question arises of how to handle inverse functions. The problem is that for a non-injective function $f\colon D \to R$, the inverse is generally not a function $R \to D$ since there may be multiple pre-images for a given point. The majority of work in this area has fallen into two camps: either the inverse functions, and expressions involving them, are treated as multi-valued objects, or inverse functions are taken to have one principal value. Both these approaches lead to difficulties in evaluation and simplification. It is possible to define the inverse as a function from R to sets of elements of D, but then the algebra of expressions involving the inverse becomes overly complicated. This article extends previous work based on a different approach: instead, the inverse of a function is taken to be a labelled family of functions, with the label specifying the pre-image in the original function's domain. This convention is already used by some authors for logarithms, but it can be applied more generally. In some cases, the branch indices can appear in identities that give more broadly applicable simplification rules. In this paper we survey how this approach can be applied to elementary functions, including the Lambert W.

Keywords: Inverse functions · Simplification Rules · Branch Cuts

1 Introduction

Consider the following integral evaluated in Maple.

$$\int \sqrt{2+x}\sqrt{1-x}\,dx = -\frac{\sqrt{2+x}\,(1-x)^{\frac{3}{2}}}{2} + \frac{3\sqrt{2+x}\,\sqrt{1-x}}{4}$$
$$+ \frac{9\sqrt{(1-x)\,(2+x)}\,\arcsin\!\left(\frac{2x}{3}+\frac{1}{3}\right)}{8\sqrt{1-x}\,\sqrt{2+x}}\,. \tag{1}$$

The fraction $\sqrt{(1-x)\,(2+x)}/\sqrt{1-x}\,\sqrt{2+x}$ in the last term appears unnecessary. Abramowitz & Stegun [1] give the solution of the cubic equation

© The Author(s), under exclusive license to Springer Nature Switzerland AG 2022
K. Buzzard and T. Kutsia (Eds.): CICM 2022, LNAI 13467, pp. 222–237, 2022.
https://doi.org/10.1007/978-3-031-16681-5_16

$x^3 + 3px - 2q = 0$ as

$$x = (q + \sqrt{q^2 + p^3})^{\frac{1}{3}} + (q - \sqrt{q^2 + p^3})^{\frac{1}{3}},$$

as did early Maple, but now Maple gives

$$x = \left(q + \sqrt{p^3 + q^2}\right)^{\frac{1}{3}} - \frac{p}{\left(q + \sqrt{p^3 + q^2}\right)^{\frac{1}{3}}}.$$

For $p = -2, q = 1$ the solutions agree, but for $p = 2, q = 1$ they do not. Can a computer system compare the expressions symbolically? The Lambert W function has the unlikely simplification $W_1(\frac{1-i\pi}{e}) = -1 + i\pi$. We ask how such simplifications can be programmed.

This paper is organized as follows: Sect. 2 introduces the issues relating to expressions involving inverse functions and some of the literature on the topic. Section 3 develops the ideas of inverse function families for the elementary functions and Sect. 4 presents some applications. Sections 3 and 4 recapitulate previous work. Section 5 explores these concepts when the function is not periodic and showcases the importance of considering the desired range of the inverse. Section 6 presents two new theorems where the proofs need the handling of inverse functions as described here. Section 7 shows how generalizing branch indices to non-integer values can provide useful pseudoinverses. Finally, Sect. 8 gives some conclusions.

2   Basic Concepts

In mathematical computation, symbolic or numeric, one encounters expressions consisting of nested application of functions to constants and variables drawn from some domains. These expressions may be viewed as functions producing values or as symbolic expressions in a free algebra. These two views are not incompatible—it is often convenient to view expressions as functions that operate on values that are symbolic expressions in a free algebra. In both views we talk about *evaluation* and *simplification*.

When expressions are viewed as functions, evaluation means replacing the variables with values from their domains and applying the composed functions to produce a result. Simplification means producing an equivalent function that is simpler according to some criterion, such as having smaller expression size, lower evaluation cost or better approximation properties. Here the notion of function equivalence is the usual one, meaning that both functions have the same domains and both produce equivalent results for all values in their domain. What is meant by "equivalent results" will depend on the setting.

When viewed as symbolic expressions from a free algebra, evaluation means replacing variables with free algebra terms to produce well-formed expressions. Simplification again means to produce equivalent alternatives that are simpler according to some criterion, usually expression size. Here the notion of equivalence is more subtle, however. When all the free algebra axioms are universally quantified, then expressions are equivalent if each can be transformed to the other by application of the algebraic axioms. When function arguments may represent values outside their domain, such as fields not allowing division by zero, then equivalence must require that transformations are applied only when the subexpressions for function arguments could never evaluate to excluded values. Alternatively, evaluation may produce exceptions, or simplification may introduce restrictions on the variables. All this can be stated more precisely, but the present explanation suffices for our purposes.

In some settings we may have "multivalued functions". These have been formulated in the literature in various ways. A well-known example is Carathéodory's statement [2] that $\ln(AB) = \ln A + \ln B$, which he interprets as each ln evaluating to a set and that the addition requires selecting values from each set to make the equation true. Selecting other values does not give equality. A second example is $\ln(x^2) = 2\ln x$. Letting $\ln_0$ denote the principal branch[1], the left-hand side must be the set

$$\{\ln_0(x^2) + 2\pi i k, \ k \in \mathbb{Z}\}$$

whereas the right-hand side is

$$2 \times \{\ln_0 x + 2\pi i k, \ k \in \mathbb{Z}\} = \{2\ln_0 x + 4\pi i k, \ k \in \mathbb{Z}\}.$$

In working with expressions, the domain of the function arguments and variables is essential. We focus on the situation in which there is one domain. That is, for some domain D, all functions have signature $D^n \to D$ for some n. Multivalued functions are then functions mapping to $D^m, m > 1$. This situation arises naturally when expressions include non-injective functions and it is desired to construct expressions involving their inverses. Many authors allow such functions to occur in expressions, implying that the expressions now represent values in $D^* = \bigcup_{i=0}^{\infty} D^i$. This approach has several problems, including that the resulting sets may contain extraneous or infeasible points and that the usual axioms used for simplification may no longer be valid.

The problem of multivalued functions cannot be ignored, however, as it remains necessary to treat expressions with function inverses.

In general, a function $f\colon D \to D$ will have an inverse $\text{inv}(f)\colon f(D) \subseteq D \to D$ only if f is injective (1–1). For non-injective functions, we may partition D as $D = \bigcup_{i=1}^{n} D_i$, where $D_i \cap D_j = \emptyset$ if $i \neq j$, and such that $f|_{D_i} : D_i \to D$ is injective for each i, introducing a *family of inverse functions* $\text{inv}(f)_i = \text{inv}(f|_{D_i})$. The choice of partitioning is not generally unique, and if D is a metric space, then it is usual to take D_i as connected components and to call $f(\partial D_i)$, the images of the

---

[1] When different authors use the same symbol to mean different things, the discussion of notation becomes problematic. For this example, ln is a set; $\ln_0$ is a unique value. Below, we shall change to notation in which ln is also a unique value.

partition boundaries, *branch cuts*. Even if f is continuous on D then $\mathrm{inv}(f)_i$ will in general not be continuous on the branch cuts. Viewing the graphs functions $D \to D$ as subsets of $D \times D$, we may interpret $\bigcup_i \mathrm{inv}(f)_i$ as a Riemann surface.

The approach of using a family of inverse functions, as opposed to a single multivalued function, is that it becomes possible to write identities that hold over larger domains. For example, with $\mathrm{inv}(\exp)_k z = \ln z + 2k\pi i$,

$$\mathrm{inv}(\exp)_m A + \mathrm{inv}(\exp)_n B = \mathrm{inv}(\exp)_s (AB) \ ,$$

$$s = m + n + \mathcal{K}(\mathrm{inv}(\exp)_0 A + \mathrm{inv}(\exp)_0 B) \ ,$$

where $\mathcal{K}$ is the unwinding number [3] and ln is the usual principal branch.

The implementation of multivalued functions in computational environments, whether numerical (e.g. Matlab) or symbolic (Maple, Mathematica, Sage, etc.), has been a topic of ongoing discussions between mathematicians and system implementers [2,3,5,8]. One can say that mathematicians prefer an *ad hoc* interactive approach in which a function evaluation would return a set of possible values, which the mathematician can employ as the problem demands. In contrast, computer systems require deterministic rules which predetermine what will be returned by a function evaluation. This dichotomy has been adequately discussed in the above references, and is not repeated here.

Some readers may point out that there are examples of computer systems being ambiguous in the meaning of a multivalued function, similar to mathematicians. In Maple, the help page for `RootOf` describes it as a placeholder for all the roots of an equation in one variable. In contrast `evalf(RootOf`$(x^2 - 4)$ evaluates to 2 only. This special case, however, distracts from the main ideas here, and will not be pursued further.

One theme of this review is the roles played by the function domain and range in the understanding of multivalued functions. Given a function $f(z)$, earlier discussions have focused on the domain of f, and detailed the branch cuts present in the domain. For example, the DLMF [11] defines complex logarithm, and complex inverse trigonometric functions showing diagrams of their domains containing branch cuts, but no diagrams of the ranges. The view here is that the ranges of these functions show the nature of the multi-valuedness more clearly than the domain alone.

3   The Elementary Functions

The basis for the discussion is an extension of the notation for the natural logarithm that was introduced in [4]. We first consider each of the elementary functions in pairs, a function and its inverse, using this approach.

3.1   Exponential and Logarithm

We begin my declaring that $\ln z$ denotes the uniquely defined principal branch function seen in all computational systems, numerical or symbolic. Figure 1 shows the domain and range of $\ln z$. The solid red line denotes the branch cut in the domain, and its image in the range. We note that if $z = x + iy$ then the branch cut is $\{z = x| - \infty < x < 0\}$ and its image in the range is $\{\ln(-x) + i\pi| - \infty < x < 0\}$. The dashed green line in the range is the line $\{\ln(-x) - i\pi| - \infty < x < 0\}$ and marks the lower boundary of the range, but does not belong to the range. It is important to note that the values taken by this logarithm are confined to the region between the red and green lines. Points outside this region cannot be reached.

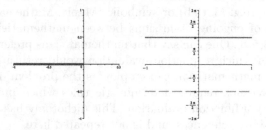

Fig. 1. Domain (left) and range (right) of the principal branch logarithm. The range is confined between the horizontal lines.

This is the motivation for defining a family of logarithm functions which together allow all points in the range to be reached. Using a subscript $k \in \mathbb{Z}$, we have:

$$\ln_k z = \ln z + 2\pi i k .$$

Although k is here an integer, there are possibilities for non-integer values being useful. In order to cover the range completely and without overlap, however, integers are necessary. We can describe $\ln_k z$ as the kth branch of logarithm. An immediate benefit of this notation is a precise statement regarding a well-known property of logarithm, which otherwise is justified ambiguously by Carathéodory [2]. We state

$$\forall k \in \mathbb{Z}, \exists m, n \in \mathbb{Z}, \text{ such that } \ln_k z_1 z_2 = \ln_m z_1 + \ln_n z_2 .$$

An application of the definition allows us to describe compactly the asymptotic behaviour of the Lambert W function.

$$W_k(z) \approx \ln_k(z) - \ln \ln_k(z) ,$$

where it should be noted that two different branches of logarithm are used. A Maple implementation of the extended function is

```
Ln := proc (z::algebraic) local branch;
            if nargs <> 1 then
                error "Expecting 1 argument, got", nargs
            elif type(procname, 'indexed') then
                branch := op(procname); ln(z) + 2*Pi*I*branch;
            else ln(z)
            end if
        end proc;
```

Maple's names for the function, `log` and `ln` are left unchanged and the new name `Ln` created. Note that the name `Log` is not used, because there could be confusion with the subscript indicating the base of the logarithm. Thus `Ln[2](5.)` = `1.60944+12.566 i` .

3.2   Sine and Arcsine

As with logarithm, the range of the arcsine function is confined to a strip in the complex plane, in this case parallel to the imaginary axis, and the remainder of the range cannot be reached by the function. Branch cuts in the domain of arcsine are $\{z = x| -\infty < x < -1 \bigcup 1 < x < \infty\}$. In the range, the vertical lines are $\{\sin(\pm\pi/2 + iy)| - \infty < y < \infty\}$ with the solid red boundaries belonging to the range of the function and the dashed green parts showing the boundary without being part of the range (Fig. 2).

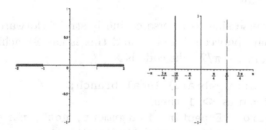

Fig. 2. Domain and range of arcsine. The solid red lines in the range correspond to the solid red lines in the domain. The dashed green lines mark the border of the range, but do not belong to it. (Color figure online)

As with logarithm, the only way to reach the parts of the range outside the principal strip is by defining additional branches of arcsine. As with logarithm, the only way to reach the parts of the range outside the principal strip is by defining additional branches of arcsine. In order to avoid clashes between Maple's standard notation of arcsin, the extensions here, are denoted `invsin` This name is modeled on Maple's `invfunc` notation, but the use differs in that `invfunc` is a table, whereas we use invsin as a function. Using this, we extend the definition to a branched inverse sine by

$$\text{invsin}_0 z = \arcsin z \ , \tag{2}$$

$$\text{invsin}_k z = (-1)^k \text{invsin}_0 z + k\pi \ . \tag{3}$$

The principal branch now has the equivalent representation $\text{invsin}_0 z = \text{invsin} z = \arcsin z$. It has real part between $-\pi/2$ and $\pi/2$. Notice that the branches are spaced a distance π apart in accordance with the antiperiod[2] of sine, but the repeating unit is of length 2π in accord with the period of sine.

The Maple code for the function is

```
invsin := proc (z::algebraic) local branch;
            if nargs <> 1 then
               error "Expecting 1 argument, got", nargs
            elif type(procname, 'indexed') then
                branch := op(procname);
                branch*Pi+(-1)^branch*arcsin(z)
            else arcsin(z)
            end if
         end proc;
```

Examples of its use appear below.

3.3   Inverse Cosine

The shift from inverse sine to inverse cosine is straightforward. The principal branch has real part between 0 and π, and this is easiest achieved by setting $\text{invcos}_k z = \text{invsin}_{k+1} z - \pi/2$. The code is

```
invcos := proc (z::algebraic) local branch;
              if nargs <> 1 then
                 error "Expecting 1 argument, got", nargs
              elif type(procname, 'indexed') then
                  branch := op(procname);
                  invsin[branch+1](z)-Pi/2
              else arccos(z)
              end if
           end proc;
```

---

[2] An antiperiodic function is one for which $\exists \alpha$ such that $f(z + \alpha) = -f(z)$, and α is then the antiperiod. This is a special case of a quasi-periodic function [10], namely one for which $\exists \alpha, \beta$ such that $f(z + \alpha) = \beta f(z)$.

3.4   Inverse Tangent

The principal branch has real part from $-\pi/2$ to $\pi/2$, and the kth branch is $\mathrm{invtan}_k\, z = \mathrm{invtan}\, z + k\pi$. As code:

```
invtan := proc (z::algebraic) local branch;
            if nargs <> 1 then
             error "Expecting 1 argument, got", nargs
            elif type(procname, 'indexed') then
               branch := op(procname); branch*Pi+arctan(z)
            else arctan(z)
            end if
          end proc;
```

The two-argument inverse tangent function has been implemented in many computer languages. It is a synonym for arg, meaning the argument or phase of a complex number, in that $\arg(x + iy) = \arctan(y, x)$ for $x, y \in \mathbb{R}$. It can be described using the branches of invtan as

$$\arctan(y, x) = \mathrm{invtan}_k(y/x) \,,$$

where $k = H(-x)\,\mathrm{sgn}\, y$, and H is the Heaviside step function.

3.5   Fractional Powers

The principal branch of $z^{1/n}$ is defined by $\exp(\frac{1}{n}\ln z)$, and replacing $\ln z$ by $\ln_k z$ gives the branched function. The standard notation for roots and fractional powers does not leave an obvious place for the branch label, and most obvious names are already used by Maple or Mathematica. We use the name invpw, meaning inverse (integer) power. The Maple code defines invpw[k](z,n), where the subscript is the branch, as usual, while the fractional power is $1/n$. Thus it is modelled on the Maple surd function. Unlike the other inverse functions, there are only n distinct values, but we allow k to be any integer.

Since square root is so common, it is coded separately as invsq[k](z), and it can be displayed in traditional notation as $(-1)^k \sqrt{z}$.

3.6   Definitions in Terms of Logarithms

Kahan [9] prefers to give definitions of the inverse trigonometric functions in terms of logarithms. We may extend these definitions to the families of inverse functions as follows:

$$\mathrm{invsh}_k\, z = (-1)^k \ln_{(-1)^k \frac{k}{2}} \left(z + \sqrt{1+z^2}\right) \qquad \mathrm{invsin}_k\, z = \frac{1}{i}\,\mathrm{invsh}_k\, iz$$

$$\mathrm{invch}_k\, z = \frac{1}{i}\,\mathrm{invcos}_k\, z \qquad\qquad\qquad \mathrm{invcos}_k\, z = \frac{\pi}{2} - \mathrm{invsin}_{-k}\, z$$

$$\mathrm{invth}_k\, z = \tfrac{1}{2}\left(\ln_{\frac{k}{2}}(1+z) - \ln_{-\frac{k}{2}}(1-z)\right) \qquad \mathrm{invtan}_k\, z = \frac{1}{i}\,\mathrm{invth}_k\, iz$$

These cover the domains of the functions inverted. The details of how to ensure continuity on a path between branches is left for a future article.

4   Applications

We now demonstrate some uses of the new notation.

4.1   Plotting

With the new functions, we can easily plot branches. Figure 3 shows plots produced by the Maple commands

```
> plot([invsin[-1](x),invsin(x),invsin[1](x)],x=-1 .. 1,
    linestyle=[dot,solid,dash]);
> plot([invtan[-1](x),invtan(x),invtan[1](x),invtan[2](x)],
    x=-5..5,linestyle=[dot,solid,dash,dashdot]);
```

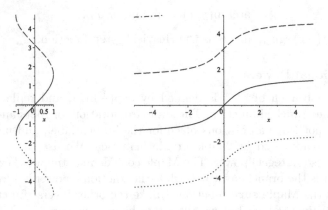

Fig. 3. The branches of inverse sine and inverse tangent plotted taking advantage of branch notation.

4.2   Identities

In order to express identities containing inverse functions correctly, we need the unwinding number,

$$\mathcal{K}(z) = \left\lceil \frac{z - \pi}{2\pi} \right\rceil ,$$

defined in [3] (rather than in [4] where the sign is different). Note that the unwinding number is a built-in function in Maple, called **unwindK**. This immediately gives us

$$\ln_k e^z = z - 2\pi i \mathcal{K}(z) + 2\pi i k . \tag{4}$$

Note the special case $\ln_{\mathcal{K}(z)} e^z = z$.

Consider an identity one might see in a traditional treatment:

$$\cos x = \sqrt{1 - \sin^2 x} , \qquad (5)$$

where the author would add "and the branch of the root is chosen appropriately". Using the branched root, we write the more precise

$$\cos x = \mathrm{invsq}[\mathcal{K}(2ix)](1 - \sin^2 x) = (-1)^{\mathcal{K}(2ix)} \sqrt{1 - \sin^2 x} . \qquad (6)$$

We can contrast the two approaches in Maple with the command

```
> plot([ sqrt(1-sin(x)^2), invsq[unwindK(2*x*I)](1-sin(x)^2)],
      x = -7 .. 7, linestyle = [dot,solid]);
```

The resulting plot is given in Fig. 4.

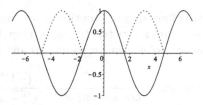

Fig. 4. The graph of $\sqrt{1 - \sin^2 x}$ using branch notation for square root. The dotted line uses an unbranched sqrt, meaning it is plotted using the built-in sqrt, which leaves sign choices to the user; the solid line uses a sqrt function that allows the sign choice to be part of the definition, rather than a separately added sign.

The Abramowitz and Stegun [1] 'identity' for adding arctangents is

$$\mathrm{Arctan}\, x + \mathrm{Arctan}\, y = \mathrm{Arctan}\, \frac{x + y}{1 - xy}$$

The more precise identity is

$$\mathrm{invtan}(x) + \mathrm{invtan}(y) = \mathrm{invtan}_k \frac{x + y}{1 - xy}, \text{ where } k = H(xy - 1)\,\mathrm{sgn}(x) , \qquad (7)$$

and H is the Heaviside step. A more complicated example from [1] is their identity for $\mathrm{Arcsin}\, x + \mathrm{Arcsin}\, y$, which becomes

$$\mathrm{invsin}\, x + \mathrm{invsin}\, y = \mathrm{invsin}[k] \left(x\sqrt{1 - y^2} + y\sqrt{1 - x^2} \right) , \qquad (8)$$

$$k = H(x^2 + y^2 - 1)(\mathrm{sgn}\, x + \mathrm{sgn}\, y)/2 .$$

Here the branch of invsin is allowed to vary, but there might be another formula which includes variable branches of square root.

As a final identity, we consider formula (4.4.39) in [1].

$$\text{Arctan}(x + iy) = k\pi + \frac{1}{2}\arctan\frac{2x}{1 - x^2 - y^2} + \frac{i}{4}\ln\frac{x^2 + (y+1)^2}{x^2 + (y-1)^2} .$$

To turn this identity into something that computer-algebra systems can use, one should decide what to do with k. This can be replaced by

$$\text{invtan}_k(x + iy) = \frac{1}{2}\text{invtan}_n\frac{2x}{1 - x^2 - y^2} + \frac{i}{4}\ln\frac{x^2 + (y+1)^2}{x^2 + (y-1)^2} ,$$

where $n = 2k + \text{sgn}(x)H(x^2 + y^2 - 1)$.

4.3 Calculus

Calculating the derivative of an inverse function is a standard topic in calculus. The results in the textbooks are restricted to the principal branches of the functions. It is possible, however, to generalize results to any branch. For example

$$\frac{d}{dx}\text{invsin}_k x = \frac{1}{\cos(\text{invsin}_k x)} = \frac{(-1)^k}{\sqrt{1 - x^2}} .$$

Integration by substitution is a well-known application of inverse functions. A specific difficulty has been the application of the substitution $u = \tan\frac{1}{2}x$ in integrals such as

$$\int\frac{3\,dx}{5 - 4\cos x} = \int\frac{6\,du}{1 + 9u^2} = 2\arctan(3\tan\frac{1}{2}x) . \tag{9}$$

The right-hand side is discontinuous, as has been pointed out in [6,7]. The correction to the usual integration formula [7] can be rewritten in the new notation as

$$\int\frac{3\,dx}{5 - 4\cos x} = 2\,\text{invtan}_{\mathcal{K}(ix)}(3\tan\frac{1}{2}x) . \tag{10}$$

The contrast is illustrated in Fig. 5 by the plot

```
> plot([ 2*invtan[unwindK(I*x)](3*tan((1/2)*x)),
    2*arctan(3*tan((1/2)*x))], x=-3..9,linestyle=[dot,solid],
        discont=true);
```

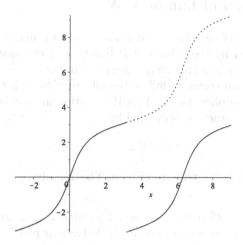

Fig. 5. A graph of the discontinuous and continuous integral expressions. The solid curves are the discontinuous expression (9). The dotted curve follows (10), which coincides with (9) for $x < \pi$, but then extends the integral continuously. The benefits of the dotted curve are discussed in [7].

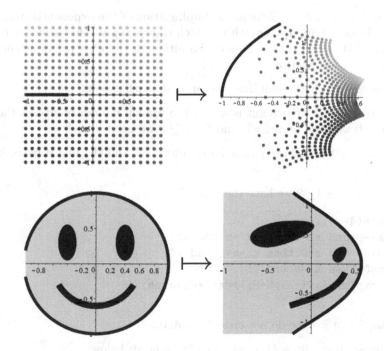

Fig. 6. The W map of the complex plane. The line segment in the top example has domain $x \in [-1, -1/e]$. The range of the second example is clipped.

5   Simplification of Lambert W

An important point to remember when working with branches is that the relevant branch is determined by the value of the function or expression, more than by the value of the argument. This means working and thinking in the range of the expression, rather than trying to follow the effects of branch cuts in the domain. Here we use this to simplify the Lambert W function for special values. We recall that the Lambert W function is defined by

$$W(z)e^{W(z)} = z, \quad z \in \mathbb{C}. \tag{11}$$

The principal branch is illustrated in Fig. 6. The domain contains the branch cut $z = \{(x,0) \mid x \le -1/e\}$.

The simplification question is as follows: Consider $x \in \mathbb{R}$, and ask when $W = x - i\pi$, with $\in \mathbb{C}$. That is, we ask about the behaviour of W on horizontal lines, as shown in Fig. 7 We have

$$We^W = (x - i\pi)e^{x-i\pi} = e^x(-x + i\pi) \ .$$

More generally,

$$(x + \alpha i\pi)e^{x+\alpha i\pi} = e^x(x + \alpha i\pi)e^{\alpha \pi i} \ ,$$

and for $\alpha = n/2$ with $n \in \mathbb{Z}$ we get a simplification of the exponential term. The difficulty is identifying the branch to which things apply. This is determined by the value of W, not by its argument. So although the e^x attracts attention, it does not affect the branch.

Since $x + \alpha i\pi$ is a horizontal line, we can start with $x = 0$.

For $x = 0, \alpha = 1/2$, that point is where the branch boundary crosses the axis. So for $x < 0$ we need branch=1, and for $x \ge 0$ we need branch 0.

For $x = 0, \alpha = 1, 3/2, 2$ the points lie in branch=1 and the line stays in branch 1 for all x.

So for $\alpha = (4m + 1)/2$ we have the rule

```
alph:=(4*m+1)/2;
if m>=0 and x>=0 then k:=m end if;
if m>=0 and x<0 then k:=m+1 end if;
if m<0 then k:=m end if;
W(k,exp(x)*(x*I-alph*Pi))=x+I*Pi*alph;
```

Note that for $m < 0$ we do not cross boundary.

Now if we set $\alpha = (4m - 1)/2$, we cross boundaries below.

```
alph:=(4*m-1)/2;
if m<=0 then if x>=0 then k:=m; else k:=m-1 end if;
else          k:=m
end if;
W(k,exp(x)*(-x*I+alph*Pi))=x+I*Pi*alph;
```

There will be similar rules for $\alpha = (2m \pm 1)/2$, but without the x check.

Also, if we substitute $\ln x$ for x, we get

$$W(k, x(\ln x + \alpha i\pi)e^{i\alpha\pi} = \ln x + \alpha i\pi$$

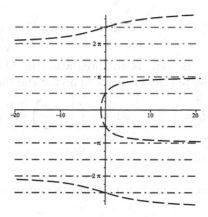

Fig. 7. The range of W showing the branches and the simplification contours. The black curves are the branch boundaries; the red and blue lines $x + \alpha i\pi$ (Color figure online)

6   Proof with Multivalued Functions

Multivalued inverse functions require new ways of thinking of relations. For example, the following theorem is new.

Theorem 1. *For all $a, b \in \mathbb{R}$ and for all $z \in \mathbb{C} \setminus \mathbb{R}$, we have*

$$\sqrt{a+z}\sqrt{b-z} = \sqrt{(a+z)(b-z)} \ .$$

Proof. Without loss of generality, we can assume that z lies in the upper half plane, i.e., we assume that $\Im z > 0$, as shown in Fig. 8. For all a, $0 < \arg(a + z) < \pi$. Therefore $0 < \arg(\sqrt{a+z}) < \pi/2$. For all b, $-\pi < \arg(b - z) < 0$.

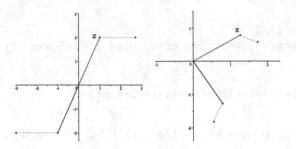

Fig. 8. The imaginary part of z is assumed positive, and $a > 0$ and $b < 0$. Thus $a + z$ is a point to the right of z and $b - z$ is to the left. After mapping under square root, the points both lie in the principal range, and their product also lies in the principal range.

Therefore $-\pi/2 < \arg(\sqrt{b-z}) < 0$. Therefore $-\pi/2 < \sqrt{a+z}\sqrt{b-z} < \pi/2$. Also $-\pi < \arg((a+z)(b-z)) < \pi$, and therefore $-\pi/2 < \sqrt{(a+z)(b-z)} < \pi/2$.

An application of this theorem is to integration in Maple. The integral of $\sqrt{2+x}\sqrt{1-x}$ was presented in (1). After applying the theorem, we obtain

$$\int \sqrt{2+x}\sqrt{1-x}\,dx = \sqrt{2+x}\,\sqrt{1-x}\,\left(\frac{1}{4}+\frac{x}{2}\right) + \frac{9}{8}\arcsin\left(\frac{2x}{3}+\frac{1}{3}\right)$$

7   Pseudoinverses

Altering branch structures using rational indices can result in useful pseudoinverses. For example

$$\exp\left(\ln_{-1/2}(z)\right) = \exp\left(\ln z - i\pi\right) = -z\;.$$

Here the subscript labels the $-1/2$ branch of the natural logarithm and should not be confused with a logarithm base.

This is what is needed to express the asymptotic behaviour of the Lambert W function in the neighbourhood of the origin. Since the defining equation is

$$W(z)\exp(W(z)) = z\;,$$

it is clear that for large z, we need $W(z)$ is large. In fact $W_k(z) \approx \ln_k z$. In addition, however, it is possible that W can become large and negative when z is small. This can be obtained if $\exp(W(z)) \to -\infty$. Thus we consider the possibility of

$$W(z) \approx \ln_{-1/2}(z) + v\;.$$

Substituting into the defining equation, and abbreviating $L(z) = \ln_{-1/2}(z) = \ln z - i\pi$, we obtain

$$W\exp(W) = (L(z)+v)\exp(L(z)+v) = z$$

This gives $(L(z) + v)(-z)e^v = z$. This can be simplified to

$$-L(z) - v = e^{-v} .$$

8   Conclusions

We have reviewed the usual approaches to handling multivalued functions, and inverse functions in particular, and have detailed an alternative approach that allows clear treatment of branches and the relations among them. A particular feature of our approach is to emphasize the need to place more emphasis on the range of the inverse function, because it is in the range that the branches are defined. Rather than treating the multivalued inverses as awkward artifacts that make expressions ambiguous, we are able to make precise statements, allow identities to have maximal extent, and do not need *ad hoc* interpretation.

References

1. Abramowitz, M., Stegun, I.J.: Handbook of Mathematical Functions. Dover (1965)
2. Carathéodory, C.: Theory of functions of a complex variable, 2nd edn. Chelsea, New York (1958)
3. Corless, R.M., Davenport, J.H., Jeffrey, D.J., Watt, S.M.: According to Abramowitz and Stegun. SIGSAM Bull. **34**, 58–65 (2000)
4. Corless, R.M., Jeffrey, D.J.: The unwinding number. Sigsam Bull. **30**(2), 28–35 (1996)
5. Davenport, J.H.: The challenges of multivalued "Functions". In: Autexier, S., et al. (eds.) CICM 2010. LNCS (LNAI), vol. 6167, pp. 1–12. Springer, Heidelberg (2010). https://doi.org/10.1007/978-3-642-14128-7_1
6. Jeffrey, D.J.: The importance of being continuous. Math. Mag. **67**, 294–300 (1994)
7. Jeffrey, D.J., Rich, A.D.: The evaluation of trigonometric integrals avoiding spurious discontinuities. ACM Trans. Math. Softw. **20**, 124–135 (1994)
8. Jeffrey, D.J., Norman, A.C.: Not seeing the roots for the branches. SIGSAM Bull. **38**(3), 57–66 (2004)
9. Kahan, W.M.: Branch cuts for complex elementary functions, or, much ado about nothing's sign bit. In: Powell, M.J.D., Iserles, A. (eds.), The state of the art in numerical analysis: Proceedings of the Joint IMA/SIAM Conference. Oxford University Press, April 1986
10. Lawden, D.F.: Elliptic Functions and Applications, Springer, New York (1989). https://doi.org/10.1007/978-1-4757-3980-0
11. Lozier, W.D., Olver, F.W.J., Boisvert, R.F.: NIST Handbook of Mathematical Functions. Cambridge University Press, Cambridge (2010)

Satisfiability, QBF, and SMT Solving

Formal Methods for NFA Equivalence: QBFs, Witness Extraction, and Encoding Verification

Edith Hemaspaandra[1] and David E. Narváez[2]([✉])

[1] Department of Computer Science, Rochester Institute of Technology, Rochester, NY 14623, USA
eh@cs.rit.edu
[2] Department of Computer Science, University of Rochester, Rochester, NY 14627, USA
david.narvaez@rochester.edu

Abstract. Nondeterministic finite automata (NFAs) are an important model of computation. The equivalence problem for NFAs is known to be PSPACE-complete, and several specialized algorithms have been developed to solve this problem. In this paper, we approach the equivalence problem for NFAs by means of another PSPACE-complete problem: quantified satisfiability (QSAT). QSAT asks whether a quantified Boolean formula (QBF) is true. We encode the NFA equivalence problem as a QBF which is true if and only if the input NFAs are not equivalent. In addition to determining the equivalence of NFAs, we are also able to decode strings that witness the inequivalence of two NFAs by looking into the solving certificate. This is a novel application of certified QSAT solving. Lastly, we formally verify key aspects of the encoding in the Isabelle/HOL theorem prover.

Keywords: QSAT · QBF · Finite automata · Interactive theorem proving · Isabelle/HOL · Formal methods

1 Introduction

Finite automata are fundamental concepts in computer science. In theoretical computer science education, for example, the contrast between deterministic and nondeterministic finite automata is commonly used to introduce students to the idea of nondeterminism. In practical settings, examples are lexical analysis in compilers [1] and bounded model checking [19].

Equivalence of two finite automata is a classic computational problem. For *deterministic finite automata* (DFAs), Hopcroft's algorithm [12] runs in near-linear time. For *nondeterministic finite automata* (NFAs), the equivalence problem is PSPACE-complete. Recent work by Bonchi and Pous [3,4] uses bisimulation and coinduction to determine the equivalence of two NFAs. From the

Research supported in part by NSF grant DUE-1819546 and NSF grant CCF-2030859 to the Computing Research Association for the CIFellows Project.

theoretical point of view, NFA equivalence being in PSPACE means it can be encoded as an instance of any other problem that is PSPACE-complete. One such problem is QSAT: the problem of determining whether a quantified Boolean formula (QBF) is true. The ideas behind the proof of PSPACE-completeness of QSAT were used by Jussila and Biere [19] to generate short QBFs encoding the bounded model checking problem where given one automaton one wants to check that no bad state is reachable. Ultimately, their findings suggested that QSAT solvers were not a feasible tool at the time. Nevertheless, in the last decade there has been increasing interest in developing new ideas for QSAT solving and much has been advanced in terms of algorithms and tools for this problem. In particular, new circuit-based formula formats have been introduced which aim at overcoming the limitations of clause-based QSAT solving [18]. Part of our research studies whether the state of the art in non-clausal QSAT solving supports an alternative method to solve NFA equivalence via a workflow that includes encoding the problem as a QBF, solving it through QSAT solvers, and obtaining domain-specific information from the solving process. Our results indicate that, despite the great advances in the area, most solvers struggle with even small instances of NFA equivalence problems. On the other hand, we show that in the case when two NFAs are found to be not equivalent using this method, it is possible to use current technology for QSAT certification to extract a witness string.

Another part of our research seeks to provide a verified encoder of the NFA equivalence problem which can output formulas that can then be passed to solvers. Such a development belongs to the very active area of research that seeks to prove not only that the conceptual ideas of encoding a problem using (in our case, quantified Boolean) constraints is correct but also providing an implementation that matches the conceptual ideas [6,14].

The rest of this paper is structured as follows. Section 2 goes through the definitions of DFAs and NFAs, plus the basic definitions of quantified Boolean formulas. Section 3 explains the details of the QBF encoding of the problem of inequivalence of NFAs. Section 4 explains the process of extracting a *witness string* out of the solving certificate generated by a QSAT solver. Section 5 presents some experimental results regarding solving these formulas in practice. Section 6 discusses the details of the proof of correctness of a key part of the encoding. Finally, we conclude in Sect. 7 and give some directions for future work.

2   Background

A *finite automaton* is a computational model defined by a 5-tuple $M = (Q, \Sigma, \delta, q_0, F)$ where Q is a set of states, Σ is the alphabet, δ is a transition function to move between the states as the automaton reads an input string, $q_0 \in Q$ is the initial state, and $F \subseteq Q$ is the set of final states. M accepts string w if M ends in a final state after reading w. The *language* $L(M)$ of a finite automaton M is the set of all strings that M accepts, and two automata M and M' are *equivalent* if $L(M) = L(M')$. If two automata are not equivalent, there is a *witness string* that one automaton accepts, but the other automaton rejects. The key difference between a *deterministic finite automaton* (DFA) and a *nondeterministic finite automaton* (NFA) is the nature of the transition function δ to move between the states. For DFAs, $\delta : Q \times \Sigma \rightarrow Q$ indicates the next state after reading a symbol σ at a state q. For NFAs, $\delta : Q \times (\Sigma \cup \{\epsilon\}) \rightarrow \mathcal{P}(Q)$ indicates the set of possible next states after either reading a symbol $\sigma \in \Sigma$ at a state q for $\delta(q, \sigma)$, or not reading a symbol from the input string for $\delta(q, \epsilon)$ (these are called ϵ-transitions). Despite their difference, DFAs and NFAs accept the same class of languages since an NFA can be transformed into a DFA that accepts the same language, though this transformation may incur in an exponential blow-up in the size of the state set.

Boolean *satisfiability* (SAT) is the problem of determining whether a propositional Boolean formula is satisfiable, i.e., whether there exists an assignment of the variables of a formula φ that makes φ evaluate to *true*. This is arguably the most famous NP-complete problem. Propositional Boolean formulas are typically described, as we do in Sect. 3, using conventional operations like $\wedge$ (logical and), $\vee$ (logical or), and $\rightarrow$ (implication). An alternative way to think about propositional Boolean formulas is as combinational circuits. The inputs of a circuit representing a propositional Boolean formula φ are the variables of φ, and the satisfiability problem translates to determining whether there is an assignment of values to the inputs of the circuit such that the output of the circuit is *true*.

By including the universal quantifier $\forall$ (meaning *for all*) and the existential quantifier $\exists$ (meaning *there exists*), we get a *quantified boolean formula* (QBF). The problem of determining whether a QBF is satisfiable (QSAT) is complete for PSPACE, a problem class that is assumed to strictly contain NP [30].

As we mentioned before, determining if two NFAs are equivalent is a PSPACE problem, so QBFs can be used to encode this problem. We are interested in comparing the bisimulation approach to the QSAT approach for different classes of NFAs, starting with randomly generated NFAs.

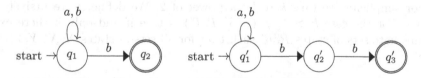

Fig. 1. NFAs N (left) and N' (right). The alphabet Σ is $\{a, b\}$, and final states are drawn with a double circle. These NFAs are inequivalent since N accepts the witness string ab, but N' does not.

3   QBF Encoding

As mentioned in the introduction, NFA equivalence is PSPACE-complete, and so this problem can be encoded in polynomial time as a QBF. In this section, we will show that it is remarkably straightforward to encode NFA inequivalence into a QBF formula. This is particularly remarkable since NFA inequivalence is a nondeterministic problem at heart. Our formula is closely related to the one Stockmeyer and Meyer used to prove PSPACE-hardness of QSAT in their seminal 1973 paper [30]. That paper points out the similarity between their construction and the one from Savitch's proof showing that NPSPACE = PSPACE [29], and the approach we use reveals that the construction from [30] also shows that QBF is NPSPACE-hard, and so provides an alternative proof of PSPACE = NPSPACE. We point out that the formula we use is closely related to the QBF encoding of Savitch's Theorem from [27] (the naïve encoding of Savitch's Theorem produces an existential formula of exponential length). Finally, for readers with background in bounded model checking, this encoding will resemble a double application of the *non-copying iterative squaring method* [19] as we are interested in simultaneously solving reachability problems in two automata.

Consider two NFAs $N = (Q, \Sigma, \delta, q_0, F)$ and $N' = (Q', \Sigma, \delta', q_0', F')$. For simplicity, in this explanation we assume that these NFAs do not have ϵ-transitions (those can easily be handled by a quick preprocessing step that does not increase the number of states). For $S \subseteq Q$ and $\sigma \in \Sigma$ we let $\delta(S, \sigma) = \bigcup_{s \in S} \delta(s, \sigma)$ (and similarly for N'), and we extend the definition of δ and δ' to strings in the obvious and standard way.

For $S, T \subseteq Q$ and $S', T' \subseteq Q'$, we can construct a QBF $\varphi_k(S, S', T, T')$ that is true if and only if there exists a string w of length at most k such that $\delta(S, w) = T$ and $\delta(S', w) = T'$.

N and N' are inequivalent if and only if there is a witness string w of length at most ℓ, the choice of which is discussed below, that is accepted by one NFA and not by the other. This is to say that N and N' are inequivalent if and only if

$$\exists T. \exists T'. \ [\varphi_\ell(\{q_0\}, \{q_0'\}, T, T') \wedge (T \cap F = \emptyset \text{ iff } T' \cap F' \neq \emptyset)].$$

For simplicity, we take k to be a power of 2. We define φ_k recursively as follows. For the case $k > 1$, $\varphi_k(S, S', T, T')$ is true if and only if there exist intermediate sets of states R, R' such that for all sets of states X, X', Y, Y', we have[1]

$$\varphi_k(S, S', T, T') := \exists R, R'. \ \forall X, X', Y, Y'. \ [((S, S', R, R') = (X, X', Y, Y')$$
$$\vee \ (X, X', Y, Y') = (R, R', T, T'))$$
$$\rightarrow \varphi_{k/2}(X, X', Y, Y')]. \quad\quad (1)$$

---

[1] When we write $A = B$ for sets A and B, we mean that the standard representation of A and B as arrays of Boolean values is equal, i.e., $A = B \equiv \bigwedge(a_i \leftrightarrow b_i)$. Similarly, when we quantify over a set we mean that we quantify over the Boolean variables representing the set.

For the base case of $k = 1$, we have

$$\varphi_1(S, S', T, T') := \left[S = T \wedge S' = T' \right]$$

$$\vee \bigvee_{\sigma \in \Sigma} \left[\delta(S, \sigma) = T \wedge \delta'(S', \sigma) = T' \right]. \tag{2}$$

Equation 2 encodes that either the sets S, T (resp. S', T') are equal (and thus $\delta(S, \epsilon) = S = T$ and $\delta(S', \epsilon) = S' = T'$) or that for some $\sigma \in \Sigma$, $\delta(S, \sigma) = T$ and $\delta'(S', \sigma) = T'$.

It is easy to see that $2^{n+n'} - 1$ is an upper bound for ℓ, where n and n' are the number of states in N and N', respectively. Note that this implies that φ_ℓ can be computed in polynomial time, since the depth of recursion is at most $n + n'$ and the size of φ_k is at most $c(n + n')$ plus the size of $\varphi_{k/2}$, where c is a fixed constant.

We obtain this $2^{n+n'} - 1$ upper bound by converting N and N' to equivalent DFAs (of size at most 2^n and $2^{n'}$, respectively) using the subset construction [26] and then using the Cartesian product construction (see [13]) to obtain a DFA for the symmetric difference of size at most $2^n 2^{n'} = 2^{n+n'}$. If this DFA accepts anything, it will accept a string of length at most $2^{n+n'} - 1$. It is important to note that the upper bounds on the number of states are tight (for the subset construction [21,23] and for the Cartesian product construction [34]).

We can do a bit better by not converting the NFAs to DFAs. Though the Cartesian product construction does not give an NFA for the symmetric difference of two NFAs, it does work just fine for the intersection of two NFAs. Note that x is in the symmetric difference of $L(N)$ and $L(N')$ if and only if $x \in L(N) \cap \overline{L(N')}$ or $x \in \overline{L(N)} \cap L(N')$. An NFA that accepts $\overline{L(N)}$ has size at most 2^n. This gives an NFA of size at most $n2^{n'}$ for $L(N) \cap \overline{L(N')}$ and an NFA of size at most $2^n n'$ for $\overline{L(N)} \cap L(N')$, and so a witness string of length at most $\max(n2^{n'}, 2^n n') - 1$. Here also the upper bounds on the number of states are tight (for computing the complement of an NFA [17] and for the intersection of two NFAs [11]).

Our arguments above use the number of states - 1 as an upper bound on the length of a shortest witness. State complexity, whether deterministic or nondeterministic, is well-studied. But it is not inconceivable that the length of a shortest witness is less than the number of states - 1. Compared to state complexity, little is known about this very natural problem. The only paper that looks at this problem for basic operations on finite automata is [2], where it is shown that for all $m, m' \geq 1$, there exist two deterministic finite automata M and M' with m and m' states respectively such that the length of a shortest string in $L(M) \cap L(M') = mm' - 1$. It is an interesting open question to look at the length of a shortest witness string for the other operations we are looking at, such as disjoint union or complementation.

4   Reading a Witness String from the Certificate

The QBF described in Sect. 3 is satisfiable if and only if there is a witness string that is accepted by one of the two input NFAs and rejected by the other. In the event that the solver finds the formula to be satisfiable, it would be desirable to obtain such a witness string from the solving process. Nevertheless, unlike SAT solvers, which will output a satisfying assignment if an input formula is found to be satisfiable, the yes/no output of a QSAT solver does not really convey the reason why a formula is satisfiable. (Some QSAT solvers do output the settings for the top-level variables in the case that the top-level quantifier is existential.) This, in the context of our application, means that decoding a witness string solely from the solver output is impossible.

Fortunately, the QSAT community, in an effort to improve the reliability of solvers and tools, has invested in developing ways to certify the execution of QSAT solvers [9, 20]. This is akin to, and draws from, the similar effort that happened in the SAT solver community in order to certify unsatisfiability [7, 10, 33].[2] The additional information obtained from the certificate of satisfiability output by a QSAT solver execution on our QBFs contains the information necessary to construct a witness string. In this section we detail the process of extracting such a string.

4.1   What is a QSAT Certificate?

In order to certify the satisfiability of a QBF it is enough to provide so-called Skolem functions to replace variables quantified by an existential quantifier. The certification process consists of carrying out the replacement and verifying that the resulting formula (which would only contain variables quantified by universal quantifiers, but essentially a propositional Boolean formula) is a tautology.

In practice, the Skolem functions of the certificate are output as a (combined) circuit. The inputs of the circuit are the universal variables, and the outputs of the circuit are the existential variables. The output corresponding to an existential variable x depends only on the values of the inputs corresponding to universal variables that precede x in the quantifier. In particular, if the top-level quantifier of a QBF is existential (as is the case for the formulas we deal with in this paper), then the outputs corresponding to top-level variables depend on no input and are thus constant.

4.2   Processing the Certificate

Recall that φ_k was defined as follows

$$\varphi_k := \exists R, R'. \ \forall X, X', Y, Y'. \ [((S, S', R, R') = (X, X', Y, Y')$$
$$\lor \ (X, X', Y, Y') = (R, R', T, T'))$$
$$\to \varphi_{k/2}(X, X', Y, Y')].$$

---

[2] It is relevant to point out that satisfiable (propositional) Boolean formulas do not require certificates as the satisfying assignment is in itself the certificate.

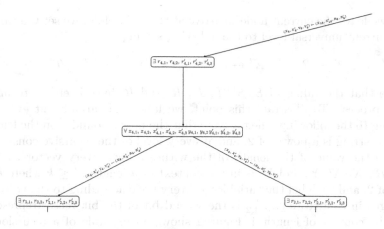

Fig. 2. Detail of the recursion tree for the example in Fig. 1

where at every level of the recursive call we are essentially trying to find intermediate state sets R and R' and, once we find them, we will only recurse into $\varphi_{k/2}$ for the specific values of X, X', Y and Y' that coincide with either S, S', R, and R', or with R, R', T, and T'. The values of X, X', Y, and Y' will take the roles of S, S', T, and T' in the next level, and so on until reach φ_1.

Since at every universal level we are only interested in two settings of its variables, we can think of the processing of the certificate of satisfiability as a tree where every node is associated with a quantification level, every existential node has one child and every universal node has two children. This tree has the following invariant: every universal node has $2(n + n')$ variables, its parent node has $n + n'$, and its grandparent node has $2(n + n')$ variables (recall n and n' are the number of states in N and N'). One exception is the top-level: the total number of variables under the top level existential quantifier is $3(n + n')$ but for practical purpose we break this into two levels, one with $2(n + n')$ variables representing the bit vectors $\{q_0\}$, T, $\{q'_0\}$, and T', and another level of $n + n'$ variables representing the sets R and R' from φ_k where k is the upper bound on the length of the witness string. The variables in levels with $n + n'$ variables encode two bit vectors, one for the state set R of the first machine and one for the state set R' of the second machine. Levels with $2(n + n')$ variables encode four bit vectors for either S, S', T, and T', or X, X', Y, and Y', depending on how they are being interpreted (recall the roles of universal variables change at every recursion level).

We process the tree depth-first. At every universal level, we distinguish the left and right children. As we leave the node to arrive at the left child, we set the variables in the current universal level to the following settings:

$$X \leftarrow S \qquad X' \leftarrow S' \qquad Y \leftarrow R \qquad Y' \leftarrow R', \qquad (3)$$

and as we leave the current node to arrive at the right child, we set the variables in the current universal level to the following settings:

$$X \leftarrow R \qquad X' \leftarrow R' \qquad Y \leftarrow T \qquad Y' \leftarrow T'$$

Note that the values of S, S', T, T', R, and R' have been set recursively by this process. To illustrate this point, we label the variables at every level according to the following scheme: Note that the upper bound k on the length of a witness string is a power of 2, and at every level of the recursive construction we halve the value of the length of the witness. Then every vector S, S', T, T', R, R', X, X', Y, and Y' is in the context of a length $k' \leq k$ where k' is a power of 2, and we label the variables at every node according to the vector and the length in context, e.g., $x'_{4,2}$ is the second bit of the bit vector representing X' in the context of length 4. Figure 2 shows an example of a recursion tree corresponding to the processing of the certificate of the example in Fig. 1. Then the assignments in Eq. 3 at level 4 traversing into the left child turn into the following assignments:

$X_4 \leftarrow S_8$	$X'_4 \leftarrow S'_8$	$Y_4 \leftarrow R_4$	$Y'_4 \leftarrow R'_4$
$x_{4,1} \leftarrow x_{8,1}$	$x'_{4,1} \leftarrow x'_{8,1}$	$y_{4,1} \leftarrow r_{4,1}$	$y'_{4,1} \leftarrow r'_{4,1}$
$x_{4,2} \leftarrow x_{8,2}$	$x'_{4,2} \leftarrow x'_{8,2}$	$y_{4,2} \leftarrow r_{4,2}$	$y'_{4,2} \leftarrow r'_{4,2}$
	$x'_{4,3} \leftarrow x'_{8,3}$		$y'_{4,3} \leftarrow r'_{4,3}$

The process continues until a leaf node is reached. At the leaf node, the formula is roughly of the following form. (Note that this formula includes additional constraints we did not include in Formula 2 for simplicity, but that are needed to precisely extract the witness string from the certificate.)

$$\varphi_1(S, S', T, T') = [(S, S') = (T, T')]$$
$$\vee \exists \sigma_1, \sigma_2, \ldots, \sigma_{|\Sigma|}.$$
$$\left(\text{ExactlyOne}(\{\sigma_1, \ldots, \sigma_{|\Sigma|}\}) \wedge \delta(S, \sigma_i) = T \wedge \delta(S', \sigma_i) = T' \right),$$

where ExactlyOne is the Boolean constraint that sets exactly one of the variables in the parameter set to *true*. (Here we overload the transition function δ to take Boolean variables σ_i as parameters in the place of a symbol in the alphabet Σ, i.e., we assume a one-to-one mapping $\mu : \Sigma \rightarrow \{\sigma_1, \ldots, \sigma_{|\Sigma|}\}$ and $\delta(S, \sigma_i)$ stands for $\delta(S, \mu^{-1}(\sigma_i))$.)

At the time the process reaches a leaf node, the values of all the variables except the σ_i variables are set by the process at previous recursion levels. Thus the only thing we need to do at a leaf node is test whether $(S, S') = (T, T')$, in which case we do not output anything since it means the witness string at this position is ϵ. If it is not the case that $(S, S') = (T, T')$, then there is exactly one of the σ_i symbols that is *true*, and that will be the the the symbol at this position of the string.

5   Experiments

All the experiments mentioned in this section ran on RIT's Research Computing cluster [28]. Each node is an Intel® Xeon® Gold 6150 CPU @ 2.70GHz. All jobs were limited to 2 GB of RAM memory. The code implementing the ideas in Sects. 3 and 4, as well as the dataset and output from the runs described in Sect. 5.2 are all available in the supplemental materials of this paper[3].

We start this section with a description of the random model used to generate test instances for our QBF encoder and our witness string extraction algorithm.

5.1   Random Model

We used the model proposed by Tabakov and Vardi [31] to generate random NFAs. This model is parameterized by the number of states n, a *transition density* $r = k/n$ where k is the expected total number of transitions in the NFA that are labeled by a fixed symbol $\sigma \in \Sigma$ (hence the total expected number of transitions is $k|\Sigma|$), and a *final state density* $f = m/n$ where m is the expected number of final states. In order to compare our results to previous work by Bonchi and Pous [3], we fix $r = 1.25$ in our NFA generation process. This number was picked in [3] because "Tabakov and Vardi empirically showed that one statistically gets more challenging NFAs with this particular value," though this claim in [31] is with respect to the size of the minimum equivalent DFA problem. The relationship between minimum equivalent DFAs and the equivalence problem of NFAs (which is the problem we address in this paper) is unclear.

We also fix the alphabet $\Sigma = \{a, b\}$. We generated 35 NFAs using different numbers of states and values for f: 5 NFAs with $n = 2$ and $f = 0.05$[4], and 5 NFAs with $n \in \{3, 4\}$ and $f \in \{0.25, 0.50, 0.75\}$. Per the Tabakov and Vardi model, the start state is also always a final state. We encoded the inequivalence problem of every (ordered) pair of (not necessarily distinct) NFAs in the dataset as a QBF for a total of 1225 QBFs.

As one reads the results in Sect. 5.2, it is important to keep in mind that the HKC algorithm by Bonchi and Pous, which is the current state of the art, processed all 1225 equivalence problem instances in our data set in less than 5 s.

5.2   Determining NFA Equivalence via QBFs

We ran the QSAT solvers from the QBF Eval 2020 competition[5] with default flags with some exceptions due to either availability or limitations in our experimental setup. The complete list of solvers used is: CQESTO [15] (in expert mode), QFUN [16] (in expert mode), QuAbS [8,32], and Qute [25].

Table 1 shows results for solving the NFA equivalence instances using solvers that take QCIR-14 as input. In general, most solvers determined the equivalence of about the same fraction of the problems (45%) within the timeout of 15 min. This fraction corresponds to the 550 instances where the number of combined states is at most 6. (In the case of QFun, we had to disable the learning feature to achieve comparable performance, since learning in these formulas seems to require too much memory.) One notable exception is QuAbS which solved 82% of the cases. Taking a close look at the number of instances in which QuAbS timed out, from Table 1 one can deduce that QuAbS timed out exactly on the 15 × 15 instances in which both NFAs had 4 states

---

[3] https://doi.org/10.5281/zenodo.6896217.

[4] We generated only 5 NFAs with two states because r and f do not change a two-state NFA in a meaningful way.

[5] http://www.qbflib.org/solver_view_domain.php?year=2020.

Table 1. Results with a timeout of 15 min per instance. OOM is out-of-memory.

Solver	Flags	Solved	Timeout	OOM	Total	Fraction solved
CQESTO	-es	550	675	0	1225	0.45
GhostQ		550	675	0	1225	0.45
QFUN	-caps -i64 -n4 -b4 -S4	61	135	1029	1225	0.05
QFUN	-caps -i64 -n4 -b4 -S4 -d	110	1113	2	1225	0.09
QuAbS		1000	225	0	1225	0.82
Qute		586	639	0	1225	0.48

(i.e., QuAbS solved every instance in which the combined number of states is at most 7).

The analysis above focuses on the combined number of states and suggests that the hardness of these instances is solely tied to that parameter. However, some of the data we collected suggests otherwise: Qute, although not having a particularly impressive performance over the dataset used for the experiments in this paper, is able to outperform QuAbS consistently on the instance generated from the NFAs depicted in Fig. 3. It is natural to think that we mean the two instances generated from the encoding the inequivalence of (N, N') and (N', N) are easy for Qute, but we do not. In fact, we mean the inequivalence problem of the NFAs in Fig. 3 taken *in the left-to-right order* as they appear in the figure is easy for Qute, which can determine the satisfiability of the instance in under 5 min while QuAbS times out. If we reverse the order of the NFAs in the encoding, both QuAbS and Qute time out on the resulting instance. (Recall that our timeout is 15 min.) This supports a deliberate experiment design of ours: even though NFA equivalence is obviously a symmetric relation, we were interested in finding out if, in the practical setting, there is an order of the input NFAs for which solvers perform better. As it turns out, the 36 cases of $n_{total} = 7$ combined number of states that Qute was able to solve within our timeout are mostly cases (N, N') where (N', N) timed out—in fact, only 2 pairs were such that Qute was able to solve both orderings of the input within the timeout, though it did so in wildly different solving times. This suggest a future research path to identify what properties determine the hardness of an instance and an optimal ordering of the input parameters.

5.3   Extracting Witness Strings

We implemented the process described in Sect. 4 and evaluated this implementation on the satisfiable instances of our dataset. Since QuAbS [8,32] is able to generate certificates and uses QCIR-14 as input format, we used it as the solver for these experiments. Table 2 shows the results.

Table 2. Count of witness length in a range k for top-down solving.

k	$[0, 10)$	$[10, 20)$	$[20, 30)$	$[30, 40)$	$[40, 50)$	$[50, 60)$	Total
Count	319	210	187	92	59	29	896

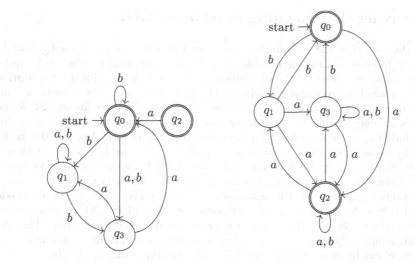

Fig. 3. Two NFAs for which the QBF encoding of their inequivalence problem is easy for Qute and hard for QuAbS.

These results suggest that the majority of the inequivalent instances generated by our random method have witness strings that are much shorter than the upper bound described in Sect. 2. This in turn means that in order to determine satisfiability for these instances we only needed to generate QBFs that are much smaller than the default.

To exploit the idea of solving smaller formulas in the hope that these smaller formulas are found to be satisfiable, we implemented an iterative solving strategy that solves formulas generated from pairs of NFAs but incrementally choosing a level k from which to start the recursion. This experiment revealed that of the 896 instances found to be satisfiable, 736 of them had witness strings of length at most 1, 116 had witness strings of length at most 2 (but not of length at most 1), and 44 had witness strings of length at most 4 (but not of length at most 2) (Table 3). This suggests that in order to make the QSAT approach to NFA equivalence more feasible, it would be good to have a better understanding of witness strings and the relation between their length and the input NFAs.

Table 3. Count of witness strings of length of at most k obtained through an iterative solving strategy. This strategy generates formulas $\varphi_{2^k}(N, N')$ for $k = 0, 1, 2, \ldots$ until a formula is determined to be SAT.

Witness string length	Count
≤ 1	736
≤ 2	116
≤ 4	44
Total	896

6   Verifying the Encoding in Isabelle/HOL

Despite the fact that the encoding we study in this paper is conceptually easy to describe, it involves several small details that become potential pitfalls when implementing such an encoding. In the supplemental materials[6] we provide a formalization of the base case of our recursive formula φ_k (i.e., φ_1) and we prove its correctness. Our formalization is heavily based on the Boolean Expression Checkers library [24] which provides a language to define Boolean constraints. Unfortunately at the time of this writing that language does not extend to quantified Boolean formulas, and this is the reason why we only prove the correctness of the base case (since it is a QBF with just the existential quantifier, so essentially a propositional Boolean formula). As future work, we intend to tackle extending the language of the Boolean Expression Checkers library to support quantifiers. Ultimately, our goal is to replace our current Java code that implements the encoding of the equivalence problem of two NFAs with verified code obtained directly from the code extraction mechanism in Isabelle/HOL. This will most likely require adjustments to the design decisions of our current formalizations, as some parts of our formalization rely on "abstract" datatypes that cannot be extracted as code.

Most notably, our formalization relies heavily on finite sets (the datatype `fset` in Isabelle/HOL) in an attempt to keep the definitions and theorems very close to those available in textbooks that cover NFAs. Unfortunately, the type system in Isabelle/HOL does not make the interaction between finite sets and sets very easy, and key concepts like transitive closures are only defined for (potentially infinite) sets and not for `fsets`. We expect that the completion of our formalization will in turn contribute a number of theorems that are missing in the Isabelle/HOL library regarding `fsets`.

7   Conclusions and Future Work

To summarize, we consider the PSPACE-complete problem of NFA equivalence and evaluate whether tackling this problem through QSAT is feasible, and whether QSAT technology can be used to learn information about the input instance, in particular whether we can use a trace of the solver to learn a witness string that is accepted by one of the input NFAs and rejected by the other. The answer to the first research question is that these formulas still pose serious challenges to non-clausal QSAT solvers, in comparison to the performance of dedicated algorithms. Nevertheless, we show that satisfiability certificates provide a way to extract a witness string from the solving process.

To normalize the results among the different approaches, we will develop and use the same merit function[7] for each approach so that the programming language used (e.g., Bonchi and Pous used OCaml) does not change the results. Related to the HKC algorithm by Bonchi and Pous, we are interested in learning whether there are concepts in specialized algorithms (e.g., "congruence up to") that can be used to develop QSAT algorithms that are specific for the class of QBFs we introduce in this paper.

---

[6] https://doi.org/10.5281/zenodo.6896217.

[7] For example, Bonchi and Pous track the number of processed pairs in their experimental work.

Regarding the formalization, as immediate future work we will extend the Boolean Expression Checkers library [24] to handle quantifiers and provide our formalization as a case study. It would also be interesting to improve the integration of our formalization with other libraries in Isabelle's Archive of Formal Proofs[8], for example the Transition Systems and Automata library [5].

Acknowledgements. We thank the referees for helpful comments and suggestions. We also thank the AAAI-21 Student Abstract referees for their helpful comments on our preliminary work on this topic [22].

References

1. Aho, A.V., Lam, M.S., Sethi, R., Ullman, J.D.: Compilers: Principles, Techniques, and Tools, 2 edn. Addison Wesley (2006)
2. Alpoge, L., Ang, T., Schaeffer, L., Shallit, J.: Decidability and shortest strings in formal languages. In: Holzer, M., Kutrib, M., Pighizzini, G. (eds.) DCFS 2011. LNCS, vol. 6808, pp. 55–67. Springer, Heidelberg (2011). https://doi.org/10.1007/978-3-642-22600-7_5
3. Bonchi, F., Pous, D.: Checking NFA equivalence with bisimulations up to congruence. In: Giacobazzi, R., Cousot, R. (eds.) POPL, pp. 457–468. ACM (2013). https://doi.org/10.1145/2429069.2429124
4. Bonchi, F., Pous, D.: Hacking nondeterminism with induction and coinduction. Commun. ACM **58**(2), 87–95 (2015). https://doi.org/10.1145/2713167
5. Brunner, J.: Transition systems and automata. Archive of Formal Proofs (2017). https://isa-afp.org/entries/Transition_Systems_and_Automata.html. Formal proof development
6. Cruz-Filipe, L., Marques-Silva, J., Schneider-Kamp, P.: Formally verifying the solution to the Boolean Pythagorean triples problem. J. Autom. Reason. **63**(3), 695–722 (2018). https://doi.org/10.1007/s10817-018-9490-4
7. Goldberg, E., Novikov, Y.: Verification of proofs of unsatisfiability for CNF formulas. In: Design, Automation and Test in Europe Conference and Exhibition, pp. 886–891. IEEE (2003). https://doi.org/10.1109/DATE.2003.1253718
8. Hecking-Harbusch, J., Tentrup, L.: Solving QBF by abstraction. In: Proceedings Ninth International Symposium on Games, Automata, Logics, and Formal Verification. Electron. Proc. Theor. Comput. Sci. (EPTCS), vol. 277, pp. 88–102. EPTCS (2018). https://doi.org/10.4204/EPTCS.277.7
9. Heule, M.J.H., Seidl, M., Biere, A.: A unified proof system for QBF preprocessing. In: Demri, S., Kapur, D., Weidenbach, C. (eds.) IJCAR 2014. LNCS (LNAI), vol. 8562, pp. 91–106. Springer, Cham (2014). https://doi.org/10.1007/978-3-319-08587-6_7
10. Heule, M.J., Hunt, W.A., Wetzler, N.: Trimming while checking clausal proofs. In: Formal Methods in Computer-Aided Design, pp. 181–188. IEEE (2013). https://doi.org/10.1109/FMCAD.2013.6679408
11. Holzer, M., Kutrib, M.: Nondeterministic descriptional complexity of regular languages. Int. J. Found. Comput. Sci. **14**(6), 1087–1102 (2003). https://doi.org/10.1142/S0129054103002199
12. Hopcroft, J.: An $n \log n$ algorithm for minimizing states in a finite automaton. Tech. Rep. STAN-CS-71-190, Stanford University (1971)

---

[8] https://www.isa-afp.org/.

13. Hopcroft, J.E., Ullman, J.D.: Introduction to Automata Theory, Languages and Computation. Addison-Wesley (1979)
14. Hoque, K.A., Mohamed, O.A., Abed, S., Boukadoum, M.: An automated sat encoding-verification approach for efficient model checking. In: 2010 International Conference on Microelectronics, pp. 419–422. IEEE (2010). https://doi.org/10.1109/ICM.2010.5696177
15. Janota, M.: Circuit-based search space pruning in QBF. In: Beyersdorff, O., Wintersteiger, C.M. (eds.) SAT 2018. LNCS, vol. 10929, pp. 187–198. Springer, Cham (2018). https://doi.org/10.1007/978-3-319-94144-8_12
16. Janota, M.: Towards generalization in QBF solving via machine learning. In: McIlraith, S.A., Weinberger, K.Q. (eds.) AAAI, pp. 6607–6614. AAAI Press (2018). http://dblp.uni-trier.de/db/conf/aaai/aaai2018.html#Janota18
17. Jirásková, G.: State complexity of some operations on binary regular languages. Theor. Comput. Sci. **330**(2), 287–298 (2005). https://doi.org/10.1016/j.tcs.2004.04.011
18. Jordan, C., Klieber, W., Seidl, M.: Non-CNF QBF solving with QCIR. In: Darwiche, A. (ed.) AAAI Workshop: Beyond NP. AAAI Workshops, vol. WS-16-05. AAAI Press (2016). http://www.aaai.org/ocs/index.php/WS/AAAIW16/paper/view/12601
19. Jussila, T., Biere, A.: Compressing BMC encodings with QBF. Electron. Notes Theor. Comput. Sci. **174**(3), 45–56 (2007). https://doi.org/10.1016/j.entcs.2006.12.022
20. Jussila, T., Biere, A., Sinz, C., Kröning, D., Wintersteiger, C.M.: A first step towards a unified proof checker for QBF. In: Marques-Silva, J., Sakallah, K.A. (eds.) SAT 2007. LNCS, vol. 4501, pp. 201–214. Springer, Heidelberg (2007). https://doi.org/10.1007/978-3-540-72788-0_21
21. Meyer, A.R., Fischer, M.J.: Economy of description by automata, grammars, and formal systems. In: 12th Annual Symposium on Switching and Automata Theory, East Lansing, Michigan, USA, 13–15 October 1971, pp. 188–191. IEEE Computer Society (1971). https://doi.org/10.1109/SWAT.1971.11
22. Miller, H., Narváez, D.E.: Toward determining NFA equivalence via QBFs. In: AAAI-21 Student Abstract (2021). To appear
23. Moore, F.R.: On the bounds for state-set size in the proofs of equivalence between deterministic, nondeterministic, and two-way finite automata. IEEE Trans. Comput. **20**(10), 1211–1214 (1971). https://doi.org/10.1109/T-C.1971.223108
24. Nipkow, T.: Boolean expression checkers. Archive of Formal Proofs (2014). https://isa-afp.org/entries/Boolean_Expression_Checkers.html. Formal proof development
25. Peitl, T., Slivovsky, F., Szeider, S.: Dependency Learning for QBF. In: Gaspers, S., Walsh, T. (eds.) SAT 2017. LNCS, vol. 10491, pp. 298–313. Springer, Cham (2017). https://doi.org/10.1007/978-3-319-66263-3_19
26. Rabin, M.O., Scott, D.S.: Finite automata and their decision problems. IBM J. Res. Dev. **3**(2), 114–125 (1959). https://doi.org/10.1147/rd.32.0114
27. Rintanen, J.: Partial implicit unfolding in the Davis-Putnam procedure for quantified Boolean formulae. In: Nieuwenhuis, R., Voronkov, A. (eds.) LPAR 2001. LNCS (LNAI), vol. 2250, pp. 362–376. Springer, Heidelberg (2001). https://doi.org/10.1007/3-540-45653-8_25
28. Rochester Institute of Technology: Research computing services (2019). https://doi.org/10.34788/0S3G-QD15
29. Savitch, W.J.: Relationships between nondeterministic and deterministic tape complexities. J. Comput. Syst. Sci. **4**(2), 177–192 (1970). https://doi.org/10.1016/S0022-0000(70)80006-X

30. Stockmeyer, L.J., Meyer, A.R.: Word problems requiring exponential time (preliminary report). In: Proceedings of the Fifth Annual ACM Symposium on Theory of Computing, pp. 1–9. STOC 1973, ACM, New York, NY, USA (1973). https://doi.org/10.1145/800125.804029

31. Tabakov, D., Vardi, M.Y.: Experimental evaluation of classical automata constructions. In: Sutcliffe, G., Voronkov, A. (eds.) LPAR 2005. LNCS (LNAI), vol. 3835, pp. 396–411. Springer, Heidelberg (2005). https://doi.org/10.1007/11591191_28

32. Tentrup, L.: Non-prenex QBF solving using abstraction. In: Creignou, N., Le Berre, D. (eds.) SAT 2016. LNCS, vol. 9710, pp. 393–401. Springer, Cham (2016). https://doi.org/10.1007/978-3-319-40970-2_24

33. Wetzler, N., Heule, M.J.H., Hunt, W.A.: DRAT-trim: efficient checking and trimming using expressive clausal proofs. In: Sinz, C., Egly, U. (eds.) SAT 2014. LNCS, vol. 8561, pp. 422–429. Springer, Cham (2014). https://doi.org/10.1007/978-3-319-09284-3_31

34. Yu, S., Zhuang, Q., Salomaa, K.: The state complexities of some basic operations on regular languages. Theor. Comput. Sci. **125**(2), 315–328 (1994). https://doi.org/10.1016/0304-3975(92)00011-F

Targeted Configuration of an SMT Solver

Jan Hůla, Jan Jakubův[✉], Mikoláš Janota, and Lukáš Kubej

Czech Technical University in Prague, Prague, Czech Republic
jakubuv@gmail.com

Abstract. We present a generic method to configure an automated reasoning solver in order to increase its performance on selected target problems. We describe a strategy invention system Grackle that is designed to invent a set of strong and complementary solver strategies. The strategies are then used to train a gradient boosted decision tree model to select the best strategy for a specific input problem. We evaluate our method on the SMT solver Bitwuzla and we obtain a significant increase in the number of solved problems, and a substantial decrease in runtime.

Keywords: Satisfiability Module Theories · Strategy Invention · Strategy Scheduling · Machine Learning

1 Introduction

Automated reasoning solvers, such as automated theorem provers (ATPs), satisfiability modulo theories (SMT) solvers, and SAT solvers, are important tools when dealing with computer mathematics. Formal mathematics and interactive theorem provers (ITPs), such as Mizar [15], Coq [12], and Isabelle [25], are making increasing use of automated reasoning solvers, for example, in the form of dedicated systems called hammers [9]. This involves translating the problems into the language understood by the underlying solver. Problems coming from such translations are quite often different from problems that the solver is optimized for. Hence, optimization of solvers to specific target problems is becoming a topic of increasing importance.

Many solvers allow a user to specify various options to influence the inner workings of the solver. These options are quite often the only way for the user to target the solver to specific problems. By a solver *strategy*, we understand a collection of solver options and their values, which are used to influence the behavior of the solver. In this paper, we consider the important question of an automated configuration of a solver to user-specified problems by deployment of targeted strategies.

This work was supported by the Ministry of Education, Youth and Sports within the dedicated program ERC CZ under the project *POSTMAN* no. LL1902, and by the ERC Starting grant *SMART* no. 714034. This scientific article is part of the *RICAIP* project that has received funding from the European Union's Horizon 2020 research and innovation programme under grant agreement No. 857306.

K. Buzzard and T. Kutsia (Eds.): CICM 2022, LNAI 13467, pp. 256–271, 2022.
https://doi.org/10.1007/978-3-031-16681-5_18

Our method is a combination of *strategy invention* and *strategy selection*. In Sect. 2 and Sect. 3, we present a generic strategy invention system Grackle, designed to invent a set of strong and complementary strategies for an arbitrary parametrized solver. Section 4 then describes a method for selecting the best strategy for a given problem. The strategy selection is implemented by gradient boosted decision trees, and it tries to predict the best strategy based on syntactic features of the problem.

The new system Grackle, a successor of the system called BliStr [31], is a strategy invention system for a generic solver. It is designed to invent a portfolio (collection) of well-performing but complementary strategies targeted to user-specified input problems. In this way, the users can develop their own sets of strategies for specific problems, and does not need to rely on the strategies predefined in the solver. As opposed to Grackle, the ancestor BliStr is hardwired to the first-order logic theorem prover E and does not combine strategy invention with different strategy selection modes.

Grackle is based on an evolutionary algorithm, where strategies are considered "animals", and problems to be solved are considered their "food". Only the animals that consume enough food, that is, solve enough problems, survive to the next generation and are given the chance to conceive an offspring. This algorithm favors animals that consume food that is not consumed by others. This leads to the diversity and complementarity of the invented strategies. The name *Grackle* is motivated by the common name of passerine birds native to North and South America. During evolution, different grackle species developed different bill sizes to be able to feed on different types of nutrients. This decreases competition between different species and increases the chances of their survival. The core of the Grackle algorithm is based on this successful evolutionary strategy.

While the method proposed in this paper is generic, we focus our evaluation on SMT solvers, in particular, on a recent SMT solver Bitwuzla [24]. There are ample connections between SMT and other types of automated reasoning [1,8]. For example, the *bitvector theory* [5] enables reasoning about exact representation of numbers in a computer, i.e., arithmetic modulo 2^n and exact modeling of floating-point numbers. Bitwuzla is the winner of the quantifier-free bitvector (QFBV) category of the SMT competition in 2021.[1] Hence Bitwuzla is a reasonable choice for the experiments. Moreover, we can expect the results with other solvers to be similar since none of the proposed methods are specific to Bitwuzla.

Related Work. Automatic portfolio selection and parameter tuning has long history in automated reasoning. Within the SAT community, the use of machine learning (ML) methods for the selection of a solver from a portfolio was popularized by Leyton-Brown et al. and their system called SATzilla [32]. In the context ATPs, various ML methods for strategy invention were developed [16,18,19,31].

Our work is mostly related to various approaches using optimization and ML methods for solver selection and scheduling, especially in the domain of SMT. Scott et al. developed a system called MachSMT [29] which selects a solver

---

[1] https://smt-comp.github.io/2021/.

from a portfolio of existing solvers based on the feature representation of the given problem. Similarly to us, they use *Bag of Words* features and reduce the dimensionality with PCA. Balunovic et al. [2] use imitation learning techniques to schedule strategies within the Z3 SMT solver; the approach targets a domain specific strategy language of Z3 and therefore strategies are not understood in the same sense as in this paper. Similarly, Ramírez et al. [26] use an evolutionary algorithm to generate strategies for the Z3 solver. For an overview of related use cases of ML methods see one of the survey papers [7,21,30].

Satisfiability Modulo Theories (SMT). SMT solvers are the driving force behind software verification, testing, or synthesis, among others [6,13,14,27]. These applications often require repeated queries to an SMT solver. This means that quick response times of the solver are paramount. An SMT solver receives as input a formula and responds if it is satisfiable or not. Since the problem is generally undecidable, solvers often timeout or give up.

The language and the semantics of the given formula depends on the theories being used. For instance, the formula $(3 < x) \land (x < 4)$ is satisfiable in the theory of real arithmetic but unsatisfiable in the theory of integer arithmetic. The language and possible theories are standardized in the *SMT-LIB standard* [5]. A repository of benchmarks is maintained in the *SMT-LIB* [4]. A combination of theories is called a *logic*. For instance, the logic UFNIA supports uninterpreted functions (theory UF) and nonlinear integer arithmetic (theory NIA). Hence, it is mandatory for each problem file to specify the intended logic in the header.

Contributions. The system Grackle is presented for the first time in this work. While BliStr [31], the ancestor of Grackle, is a strategy invention system for the ATP prover E [28], Grackle supports an arbitrary solver. Moreover, Grackle adds support for an additional external tuner and implements additional features like alternative strategy selection modes and a non-atavistic behavior. Advancements of Grackle over BliStr are detailed in Sect. 3 and they are experimentally evaluated in Sect. 5. The next contribution is a strategy selector designed to select the best problem-specific strategy from the portfolio of strategies invented by Grackle. The selector is described in Sect. 4 and evaluated in Sect. 6.

2   Grackle: Strategy Portfolio Invention System

Grackle[2] is a generalization of the system called BliStr [31], which is based on the same evolutionary algorithm motivated in Sect. 1. BliStr is a strategy portfolio invention system for automated theorem prover E [28]. Its first successor, BliStrTune [19], is an extension of BliStr to handle larger strategy space by a hierarchical strategy invention. The second successor, EmpireTune [18], is an extension that additionally handles another ATP called Vampire [22]. Therefore, Grackle is the third successor of BliStr, and apart of the generalization

---

[2] https://github.com/ai4reason/grackle.

from E/Vampire to a generic solver, it implements other interesting features. This section describes the core of the evolutionary algorithm common both to BliStr and Grackle. Section 3 describes novel features that are specific to Grackle only.

Algorithm 1: GrackleLoop($\mathcal{S}, \mathcal{P}, \beta$)

$\Phi_{strats} \leftarrow \mathcal{S}$ // Initialize the state Φ, and set the initial strategies
loop
  Evaluate($\mathcal{P}, \Phi, \beta$)   // Evaluate all strategies on $\mathcal{P}$ (1)
  $\Phi_{cur} \leftarrow$ Reduce($\mathcal{P}, \Phi, \beta$)   // Select the current generation of strategies (2)
  $s \leftarrow$ Select($\mathcal{P}, \Phi, \beta$)   // Select the strategy to improve (3)
  if s **is** None **then return** Φ // Termination criterion
  $s_0 \leftarrow$ Specialize($s, \mathcal{P}, \Phi, \beta$)   // Improve s on its best problems (4)
  $\Phi_{strats} \leftarrow \Phi_{strats} \cup \{s_0\}$ // Extend the set of strategies

---

Fig. 1. An outline of the Grackle strategy portfolio invention loop.

Figure 1 outlines the core BliStr/Grackle strategy portfolio invention loop. The input of the algorithm is a non-empty initial set of strategies $\mathcal{S}$, and the set of target problems $\mathcal{P}$. The argument β collects additional hyperparameters detailed below. The variable Φ encapsulates the current state, including, for example, the set of all strategies invented so far (Φ_{strats}). The state Φ might be modified during function calls inside the loop, while all the other variables are immutable (read-only). The current state Φ is also the output of the algorithm.

The loop consists of four basic steps. At first, all the known strategies are evaluated on all problems $\mathcal{P}$, and the results are stored in the state Φ. In the second step, a subset of strategies is selected as a current generation (Φ_{cur}), and, in the third step, one of the strategies (s) is selected for specialization. In the fourth step, the selected strategy is specialized for a subset of problems using an external tool for parameter tuning and algorithm configuration. We do not allow the same strategy to be specialized on the same problems more than once. Since we consider only finite sets of problems and possible strategies, the loop must eventually terminate because sooner or later we will run out of strategies to specialize. In practice, however, we allow the user to set the runtime limit by the hyperparameter $\beta_{timeout}$. Detailed description of the individual steps follows.

Step 1: Generation Evaluation (Evaluate). In the first phase, all strategies (Φ_{strats}) are evaluated on all target problems $\mathcal{P}$. This involves running the solver on each problem with a time limit β_{eval} yielding (1) the overall result (solved/unsolved) and (2) a number representing the length of the run (runtime). Both the time limit and the runtime can be specified as a CPU time or as some abstract time (such as the number of instructions) if that is supported by the solver. This information is stored in the state Φ to avoid duplicate evaluations.

Step 2: Generation Reduction (Reduce). In the next step, we select the current generation of strategies Φ_{cur} from all strategies Φ_{strats}. First, we compute for each strategy s its set of best problems $\mathcal{P}_s$, that is, the set of all problems from $\mathcal{P}$ on which s is the best strategy. In the case the best strategy of problem p is not unique, we randomly select one and mark it as the best. Then we select strategies $s \in \Phi_{strats}$ with at least β_{bests} best-performing problems, that is, with $|\mathcal{P}_s| \geq \beta_{bests}$. From the selected strategies, we take only β_{tops} best strategies, where the strategies are compared by the number of best-performing problems ($|\mathcal{P}_s|$). The first restriction keeps only well-performing strategies (strong individuals) and removes redundant strategies (since every solved problem is exactly in one $\mathcal{P}_s$). The second restriction reduces the count of strategies, keeping the size of Φ_{cur} within the selected bound ($|\Phi_{cur}| \leq \beta_{tops}$). This prevents invention of a large number of over-specialized strategies. The function Reduce is depicted in Fig. 4 and further discussed in Sect. 3.

Step 3: Strategy Selection (Select). The next step is to select a single strategy for specialization. As a rule, no strategy can be specialized on the same problems more than once within one execution of the GrackleLoop algorithm. Because the sets of best-performing problems vary in time, the same strategy can be improved more than once, but only on different problems. Our default selection approach is to prefer specialization on diverse problems. Therefore, we prefer to improve strategies whose best-performing problems have not been used for specialization very often. In more detail, for each problem $p \in \mathcal{P}$, we keep a problem specialization counter $\Phi_{spec,p}$ that is increased by $1/|\mathcal{P}_s|$ whenever a strategy s is specialized on p (that is, when $p \in \mathcal{P}_s$). We select the strategy s with (currently) the lowest average $\Phi_{spec,p}$ over its best-performing problems $\mathcal{P}_s$. Ties are broken by higher $|\mathcal{P}_s|$. If no strategy can be selected, the algorithm terminates. The function Select is depicted in Fig. 3 in Sect. 3.

Step 4: Strategy Specialization (Specialize). In this phase, Grackle invents a new strategy by specializing on a subset of problems. The strategy specialization is done by an external parameter tuning software. BliStr uses the ParamILS [17] automated algorithm configuration framework, while Grackle additionally supports the SMAC3 [23] framework. The strategy s is always specialized on its best-performing problems $\mathcal{P}_s \subseteq \mathcal{P}$. Given a strategy s and its set of best-performing problems $\mathcal{P}_s$, the external tuner is launched to find a strategy s_0 with an improved performance on $\mathcal{P}_s$. The idea behind this is that s_0 will become even better than s on $\mathcal{P}_s$, and this shall allow additional problems outside of $\mathcal{P}_s$ to be solved. The external tuner is always launched for a specific wall-clock time limit $\beta_{improve}$. The function Select is depicted in Fig. 2 in Sect. 3.

3   Making the Grackle Fly

This section describes what needs to be done to use the Grackle system and the main differences between Grackle and other members of the BliStr family. Apart from generalization to an arbitrary solver, main Grackle features introduced in this work are alternative methods of strategy selection and generation reduction.

Solver Wrapper. An improvement of Grackle over BliStr is that Grackle is a generalization to an arbitrary solver. To use Grackle with a selected solver, a simple wrapper function must be implemented in Python. This function takes a single problem filename together with a strategy as arguments and launches the specified solver strategy on the specified problem. Then, it must process the solver output and return the result status (solved/unsolved) and performance measurement. The performance measurement can be an arbitrary number in selected units, for example, the CPU time in seconds, or any reasonable abstract performance metric.

Grackle currently implements a solver wrapper for ATP provers E [28], Vampire [22], and Lash [10], and for SMT solvers CVC5 [3] and Bitwuzla [24]. In this paper, we focus on Bitwuzla which is used for evaluation in Sect. 5 and Sect. 6.

Parametrization of the Strategy Space. Apart from the solver wrapper, the space of all considered solver strategies must be described. This strategy space parameterization is passed to one of the supported external parameter tuners that are used to specialize a strategy to specific problems. As noted above, Grackle supports ParamILS [17] and SMAC3 [23] frameworks as external tuners. ParamILS employs an *iterated local search* (ILS) from the initial strategy, occasionally perturbing a strategy to escape from a local optimum. SMAC3 is based on Bayesian optimization in combination with an aggressive racing mechanism to efficiently decide which of the two given strategies performs better. Both frameworks are capable of finding a strategy that performs well on specific target problems.

Furthermore, both ParamILS and SMAC3 use a similar mechanism to describe the space of strategies. Since the mechanism used by SMAC3 is a subset of the one used by ParamILS, we simply use the ParamILS style to accommodate both tuners. The strategy space is described by a finite set of parameters, where each parameter is assigned a finite domain of possible values and the default value. The strategy space can be additionally pruned by specification of *conditional arguments* and *forbidden values*. Conditional arguments specify dependencies among arguments, and forbidden values allow us to specify combinations of parameter values which are banned in a single strategy. Both frameworks additionally require the user to provide a solver wrapper and a performance metric. These are, however, automatically derived from Grackle's solver wrapper. During the specialization of a strategy, the external tuner launches the solver with various strategies. Grackle's hyperparameter β_{cutoff} controls their runtime.

Parallel Tuner Execution. Both ParamILS and SMAC3 provide partial support for parallel execution to speed up the tuning. In the case of ParamILS, multiple independent instances are simply launched in parallel and, thanks to the randomized nature of the algorithm, each instance traverses the space of strategies in a different order. Each instance reports its progress as a triple (s_0, q_0, n_0), where s_0 is the best strategy found so far, and q_0 is the quality of s_0 based on evaluation on n_0 problems. The number n_0 only increases over time as new results are acquired. Finally, we select the strategy with the best quality q_0 among the different parallel runs.

SMAC3 provides similar, but improved, support for parallelization. Again, multiple instances are launched in parallel, but the instances share a common database of results and thus avoid duplicate solver evaluations.

The external tuner, called to specialize strategy s, is always launched with the wall-clock time limit specified by Grackle's hyperparameter $\beta_{improve}$. This time limit is fixed and does not reflect the size of the set of problems used for the specialization ($\mathcal{P}_s$). It can be expected that with smaller problem sets, the tuner can be launched with smaller time limits and still achieve equivalent results. Therefore, Grackle additionally implements the ParamILS extension with restarts and automated termination. We call this extension ResParamILS.

In ResParamILS, multiple ParamILS instances are launched in parallel with the same initial strategy s and the set of target problems $\mathcal{P}_s$. ResParamILS keeps checking the progress of individual instances and waits for the first ParamILS instance to report a strategy s' evaluated on all problems $\mathcal{P}_s$. That is, it waits for some instance to report a triple (s', q', n') where $n' = |\mathcal{P}_s|$. Then ResParamILS enters a stabilization phase and waits for t seconds, where t is the wall-clock time elapsed so far. This stabilization phase allows other instances to evaluate the best strategy on all problems $\mathcal{P}_s$. Then, only the best ParamILS instance is kept running while the other instances are terminated. The terminated instances are then restarted with the best strategy s' found so far as the initial configuration, but keeping the initial set of problems $\mathcal{P}_s$. This process is then iterated and ends when the quality of the best strategy stops improving, that is, when no better strategy can be found. In this way, ResParamILS tries to detect a plateaued state. The hyperparameter $\beta_{improve}$ can be still used as an overall tuning limit.

BliStr does not support parallel execution of the tuner. It was first added in BliStrTune but without ResParamILS. ResParamILS was already partially implemented in EmpireTune, but without any evaluation. Grackle is the first system of the BliStr family to support SMAC3. The number of parallel runs in Grackle is controlled by the hyperparameter β_{cores}.

Strategy Selection Mechanisms. As described in Sect. 2, we prefer specialization to problems that have not been used very often for specialization. This is implemented by keeping a global problem specialization counter $\Phi_{spec,p}$ for every problem p. This counter is initialized to zeros and is updated with every call to the function Specialize, as described in the Fig. 2.

When selecting the strategy to be specialized, the strategies are compared by averaging problem specialization counters over the best-performing problems $\mathcal{P}_s$. For each strategy s, we compute the average problem specialization index $\mathcal{C}_s$ as described in Fig. 3. Since we prefer problems that are not often used for specialization, we prefer smaller values of $\mathcal{C}_s$. Grackle provides several ways to use this index to select the strategy to specialize.

Default Mode. The *default selection mode* is the selection mechanism from BliStr, where the strategies with lower $\mathcal{C}_s$ are preferred. In the case of equal values, which

Function Specialize$(s, \mathcal{P}, \Phi, \beta)$

$\mathcal{P}_s \leftarrow \{p \in \mathcal{P} \mid s$ is the best strategy on problem p among strategies $\Phi_{cur}\}$
$s_0 \leftarrow$ ExternalTuner$(s, \mathcal{P}_s, \beta_{cutoff})$ // *launch the external tuner*
for $p \in \mathcal{P}_s$ **do**   // *update the problem specialization counters* Φ_{spec}
$\quad \lfloor \;\; \Phi_{spec,p} \leftarrow \Phi_{spec,p} + (1/|\mathcal{P}_s|)$
return s_0

Fig. 2. Specialize the strategy s on its best problems $\mathcal{P}_s \subseteq \mathcal{P}$.

Function Select$(\mathcal{P}, \Phi, \beta)$

for $s \in \Phi_{cur}$ **do**
$\quad \mathcal{P}_s \leftarrow \{p \in \mathcal{P} \mid s$ is the best strategy on problem p among strategies $\Phi_{cur}\}$
$\quad \mathcal{C}_s \leftarrow (\sum_{p \in \mathcal{P}_s} \Phi_{spec,p})/|\mathcal{P}_s|$ // *average problem specialization counters*
$\quad \mathcal{Q}_s \leftarrow (\mathcal{C}_s, -|\mathcal{P}_s|)$ // *quality pairs are compared lexicographically*
$\mathcal{S} \leftarrow \{s \in \Phi_{cur} \mid s$ has not been specialized on $\mathcal{P}_s$ yet$\}$ // *skip already done*
return $\arg\min_{s \in \mathcal{S}} \mathcal{Q}_s$

Fig. 3. Select the strategy to be specialized.

happens always in the first iteration as the counters are zeroed, the strategies are compared by their performance, that is, by $|\mathcal{P}_s|$. In the algorithm, this is implemented by constructing the quality pair $\mathcal{Q}_s = (\mathcal{C}_s, -|\mathcal{P}_s|)$. The quality pairs are then compared lexicographically, and the strategy s with the lowest $\mathcal{Q}_s$ is selected for specialization. Since the size of $\mathcal{P}_s$ is reversed in $\mathcal{Q}_s$, we prefer stronger strategies in the case of equal values of $\mathcal{C}_s$. In particular, in the first iteration, the strongest strategy s will be specialized. The default mode is the only strategy selection implemented in BliStr.

Reverse Mode. In practice we often observe that strong strategies are born out of weak ones. Therefore, Grackle additionally provides a way to begin the specification with the weakest strategy. In the *reverse selection mode*, this is implemented by using 'arg max' instead of 'arg min' in the last line of function Select.

Weak Mode. The reverse mode, however, prefers specialization on problems already used for specialization, because it prefers higher values of $\mathcal{C}_s$ in the first element of a quality pair $\mathcal{Q}_s$. Therefore, we implement the *weak selection mode* which uses 'arg min' as in the default mode, but it constructs the quality pair $\mathcal{Q}_s$ as $\mathcal{Q}_s = (\mathcal{C}_s, |\mathcal{P}_s|)$. In this way, we prefer weak strategies, but we still prefer problems not often used for specialization.

Random Mode. Grackle additionally implements a *random selection mode* where the strategy is selected randomly.

Evaluation. All Grackle ancestors support only the default selection mode. The various Grackle selection methods are experimentally evaluated in Sect. 5.

Function Reduce($\mathcal{P}, \Phi, \beta$)

$\mathcal{G} \leftarrow \Phi_{strats}$ // *start with all strategies known so far*
for $s \in \mathcal{G}$ **do**   // *compute the best problems for every strategy*
$\quad \lfloor \; \mathcal{P}_s \leftarrow \{p \in \mathcal{P} \mid s$ is the best strategy on problem p among strategies $\mathcal{G}\}$

$\mathcal{G} \leftarrow \{s \in \mathcal{G} \mid$ if $|\mathcal{P}_s| \geq \beta_{best}\}$ // *keep only strong individuals*
$\mathcal{G} \leftarrow [s \in \mathcal{G} \mid$ sort $\mathcal{G}$ by decreasing $|\mathcal{P}_s|]$ // *list strategies sorted by performance*
$\mathcal{G} \leftarrow \{s \mid s$ is among the first β_{tops} strategies in $\mathcal{G}\}$ // *keep only best strategies*
return $\mathcal{G}$

Fig. 4. Selection of the current generation of strategies.

Atavistic and Non-Atavistic Modes. In genetics, *atavism* is a recurrence of a trait typical for ancestors but not apparent in the current generation. In our context, it can happen that a strategy disappears from the current generation Φ_{cur} but suddenly reappears in one of the following iterations. This happens due to the selection of the current generation in the function Reduce depicted in Fig. 4.

In BliStr and other Grackle ancestors, the current generation is selected out of all known strategies Φ_{strats}. Strategies are filtered by their performance $|\mathcal{P}_s|$ using hyperparameters β_{best} and β_{tops}. Note that $\mathcal{P}_s$ in Reduce is computed with respect to all strategies Φ_{strats}, while in functions Select and Specialize it is computed with respect to the current generation Φ_{cur} only. Since the count $|\mathcal{P}_s|$ can only decrease during the execution of the Grackle loop, a strategy once rejected due to the limit β_{best} can never reappear in future generations. However, a strategy rejected due to the limit β_{tops} can appear in a future generation because the order of the strategy can change. This behavior, called *atavistic*, is the behavior implemented in BliStr.

Grackle additionally supports a *non-atavistic* behavior, where the next generation is selected out of the strategies of the previous generation. This is implemented by setting $\mathcal{G}$ to Φ_{cur} instead of Φ_{strats} in the first line of Reduce. Furthermore, Φ_{cur} is initialized with the initial strategies, and any specialized strategy (any output of Specialize) is always added to the current generation.

4   Strategy Selection with Boosted Decision Trees

Given a large portfolio of strategies obtained by running Grackle, we want to select the best strategy for every problem within the given benchmark set. We achieve this by training a ranking model to rank strategies using features extracted from the problem in question. Then, for an arbitrary input problem, we run the strategy that has the highest rank according to the trained model.

Feature Extraction. We use a simple *Bag of Words* (BOW) as a feature representation of each problem/formula. To compute this representation, we parse the given formula and compute the counts of every unique word. The set of possible words consists of logical operators of the SMT language together with keywords of the given logic. We ignore the concrete names of variables and functions and the concrete values of constants. For example, we do not count how many times a concrete value (such as 1) appears within the formula. Instead, we count the number of occurrences of any constant of a given type (i.e., an integer numeral).

Given a fixed order of the possible words which could appear in an SMT formula, the BOW representation of the formula is a vector in which the n-th element represents how many times the n-th possible word appears within the formula. If we combine multiple logics together, then for a concrete problem (belonging to a specific logic), most of the possible words will not be used and the feature vector will be sparse. Therefore, we reduce the dimension of these vectors by *Principal component analysis* (PCA), and for any new problem, we project its original feature vector to the obtained principal vectors. The simplicity of the feature extraction phase results in a negligible computation time, which is crucial in the setting we are interested in.

Strategy Ranking. To select the strategy on a per-instance basis, we train a ranking model that takes the extracted feature vector and a categorical variable whose possible values correspond to different strategies as input. The model outputs the rank of the strategy for the given problem. For every new problem, we select the strategy with the highest rank. As a ranking model, we use Light-GBM [20] with a default setting and the LambdaRank [11] objective function and train it for 1000 iterations[3].

We follow two different procedures when creating labels for the training dataset, depending on the timeout parameter. When the timeout is small (1 s in our case), we set the rank of a given strategy to 1 or 0 depending on whether the strategy solves the given problem before the timeout or not, respectively. When the timeout is set to higher values (10 s or 60 s in our case), we divide the timeout into five intervals and set the rank of the strategy according to the interval in which it solves the problem. Concretely, if the strategy does not solve the problem before the timeout, we set its rank to 0. If it solves the problem during the last interval (whose endpoint corresponds to the timeout), we set its rank to 1 and so on, until the first interval, which corresponds to a rank equal to 5. In simple words, a shorter solving time corresponds to a higher rank.

For the case of longer timeout (10 s or 60 s), it holds that for any given strategy, the solving times are not distributed uniformly, as can be seen in the lower right part of Fig. 5 that contains a histogram of the solving times under 60 s. The histogram shows that the solving times follow an exponentially decreasing trend. To counteract this nonuniform distribution, we set the endpoints of the intervals according to a power function. That is, we set the endpoint of an interval n to n^p for some $p \in \mathbb{R}$. If we want to have N intervals in total, p is chosen so that

---

[3] We also tried to use a Graph Neural Network which processed formulas represented as directed acyclic graphs but LightGBM ranking with BOW representation provided the best tradeoff in terms of accuracy and speed.

$N^p = timeout$. For example, $N = 5$ and $timeout = 60$ yield $p = \log_5(60) \simeq 2.54$ and interval endpoints: $(1, 5.8, 16.3, 34, 60)$.

The rationale behind the different treatment of the labels for the case of the longer timeout (10 and 60 s) is that it may be easier for the ranking model to distinguish whether the strategy solves the problem in 1 s or 30 s compared to whether it solves it in 200 milliseconds or 600 milliseconds. In our experiments, having multiple ranks for a longer timeout led to better results.

Another option to obtain the ranks of the strategies would be to sort the strategies by solving time and set the rank of each strategy to its order. This would be harder to learn because the prediction would need to be more precise.

5   Evaluation of Grackle Strategy Invention

In this section we experimentally evaluate Grackle strategy invention on an SMT solver Bitwuzla, and on a benchmark of 3000 problems used in the SMT competition in 2021. Bitwuzla's configuration used in the SMT competition is called *smt-comp-mode* and it serves as a baseline in our experiments.

We select the benchmark problems as follows. In the competition, Bitwuzla was launched with 13605 problems and solved 13291 while only 342 problems were not solved. The time limit per problem in the SMT competition was 20 minutes. First, we include all 342 unsolved problems in our benchmark. Next, we remove all problems with runtime smaller than 100 ms to filter out trivial problems. From the remaining problems, we randomly select 2658 problems to obtain the set of 3000 benchmark problems. This set is divided into 2000 training and 1000 testing problems. All experiments in this section are performed with the 2000 training problems, while the testing problems are used for the final evaluation in Sect. 6.

To use Bitwuzla with Grackle we need to parametrize the strategy space. We manually select 39 parameters and their domains based on our intuition. We have 28 boolean parameters, and the overall size of the strategy space is about 10^{15}. To test the parametrization, we launch several instances of SMAC3 without using Grackle. This gives us 20 Bitwuzla strategies that we use as the initial strategies for all Grackle runs described below. During tuning, Bitwuzla is always launched with a 1 second time limit per problem. The best of the initial strategies solves 964 (out of 2000) training problems with 1 second time limit, while Bitwuzla's *smt-comp-mode* solves 906. The union of problems solved by all initial strategies is 1345.

The first experiment tests different external tuners supported by Grackle, that is, SMAC3, ParamILS, and ResParamILS (see Sect. 3). We launch one Grackle instance for each tuner with the runtime limit of 24 hours and with 8 cores per instance. The tuning time for one specialization is set to 5 minutes ($\beta_{improve}$). Individual Bitwuzla runs are limited to 1 second, both during evaluations (β_{eval}) and specializations (β_{cutoff}). We use the non-atavistic behavior, and we restrict the size of the current generation to 30 (β_{tops}). We require every strategy to outperform other strategies in least at 3 problems (β_{bests}).

Table 1. Evaluation of external tuners supported by Grackle.

tuner	solved		specializations		runtime [hh:mm]			termination
	total	new	success	all	eval.	spec.	total	
SMAC3	1449	104	28	128	01:29	10:45	12:15	regular
ParamILS	1459	114	161	186	08:07	15:35	23:56	timeout
ResParamILS	1452	107	194	211	09:58	13:39	23:56	timeout

The results are summarized in Table 1. The column *total* describes the number of problems solved by all invented strategies, and the column *new* shows how many of them are not solved by the initial strategies. The column *success* presents the number of strategies invented by specialization, that is, the count of successful specializations. It can happen, that the outcome of a specialization is some strategy that is already known, that is, Specialize$(s, \mathcal{P}, \Phi, \beta) \in \Phi_{strats}$, in which case we consider the specialization as a failure. The column *all* shows the total number of specializations performed, including failed ones. The columns *runtime* describe Grackle run times in hours and minutes. First two columns describe the time used for evaluations (*eval.*) and the time used for specializations (*spec.*). The *total* runtime additionally covers time used for the reductions and selections of the strategies. The last column describes whether Grackle terminates because no strategy can be specialized or because of the timeout.

First, we observe that the numbers of solved problems are quite similar. We can see that SMAC3 specializations failed much more often. It seems that SMAC3 tends to return the input strategy unless it finds a strictly better one. On the other hand, this happens rarely with ParamILS, where the local search deviates from the initial input strategy quite early. Due to that, ParamILS and ResParamILS managed to invent more strategies, which is a behavior favored by Grackle. Furthermore, the results of ParamILS and ResParamILS can be improved by extending the time limit. We also see that ResParamILS was able to perform more iterations with quite a smaller specialization runtime. This is thanks to its automated termination feature.

While SMAC3 invents 28 strategies, only 20 are needed to cover all solved problems. For ParamILS, this is 21 and for ResParamILS it is 20. From this we can conclude that the invented strategies have a similar strength. Recall that the initial strategies solve 1345 problems. The total number of problems solved by all three runs together is 1481. This means that different tuners invent complementary strategies and that there is no clear winner among the tuners.

In the second experiment, we test different strategy selection and reduction modes from Sect. 3. Here, we restrict our attention to ResParamILS. We test *atavistic* and *non-atavistic* behaviors, combined with three modes of strategy selection (*default, reverse, weak*). This gives us six different Grackle runs. Otherwise, we use the same Grackle hyperparameters as in the first experiment.

The results are presented in Table 2. The first row is the same as the ResParamILS row from the first experiment. We can see that the reverse selection

Table 2. Evaluation of selected strategy selection modes.

atavistic selection		solved total	new	specializations success	all	runtime [hh:mm] eval.	spec.	total	termination
no	default	1452	107	194	211	09:58	13:39	23:56	timeout
no	reverse	1472	127	220	237	12:11	11:22	23:58	timeout
no	weak	1429	84	160	174	09:28	14:17	24:00	timeout
yes	default	1463	118	199	208	09:57	13:42	24:00	timeout
yes	reverse	1425	80	196	212	11:27	12:08	23:55	timeout
yes	weak	1421	76	169	195	09:11	14:32	23:58	timeout

with the non-atavistic behavior solves most problems and invents most strategies. It performs quite differently with the atavistic behavior, where it solves fewer problems even though it invents a large number of strategies. This suggests that the atavistic mode favors the invention of redundant strategies. The weak selection solves the least problems, but it spends the most time by specializations. The total number of problems solved by all six runs is 1500. This again suggests that there is a certain complementarity among the methods.

6   Evaluation of Strategy Selection

In the last Sect. 5 we have described 9 different Grackle runs. We additionally perform several Grackle runs with different hyperparameters, for example, increasing the Bitwuzla time limit to 5 s (β_{eval} and β_{cutoff}). Together, we collect more than 5000 different Bitwuzla strategies, and by iterative greedy cover construction, we select 140 complementary strategies. We evaluate these strategies on the training problems with a 60 s time limit per problem and strategy. This gives us training data for a strategy selector that attempts to select the best strategy for a specific input problem.

While it often makes sense to alternate several strategies within a given time limit, in our context of quantifier-free bit vectors, it is often best to select a single strategy and run it exclusively until the time limit is reached. Therefore, we attempt to construct the selector of a single strategy. We focus on smaller time limits (below 1 min), because we develop methods for a machine-human interaction, keeping in mind limited time resources and impatience of human users. In particular, we test the strategy selection with 3 different time limits (1, 10 and 60 s). The selectors are evaluated on the 1000 testing problems.

To evaluate the selector, we count the number of solved problems for a given timeout and compute the *PAR-2* score (Penalized Average Runtime) commonly used to compare different solvers. The score is the sum of the runtimes of solved problems plus twice the timeout for each unsolved problem. The qualitative and quantitative results can be seen in Fig. 5 and Table 3, respectively. We compare the strategy selector with the default *smt-comp-mode* strategy, the best strategy

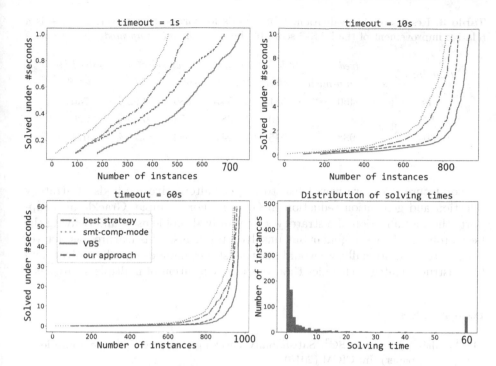

Fig. 5. *Top left, top right, bottom left*: Cactus plots for different timeouts (1 s, 10 s, and 60 s, respectively). The plot shows a comparison of our approach against the virtual best strategy (VBS), the baseline *smt-comp-mode* strategy, and the best invented strategy from the portfolio. *Bottom right*: Histogram of solving times of one selected strategy. The bar at 60 s corresponds to unsolved problems.

from the invented strategies, and the virtual best strategy. The virtual best strategy is a hypothetical selector which would always pick the best performing strategy. The results show that the gain is largest when the time limit is small and there is a substantial gap between the best performing strategy and the virtual best strategy. We observe the clear superiority of our selector.

7   Conclusions and Future Work

We have presented a method for the automated configuration of automated reasoning solvers for specific target problems. We have evaluated the method on the SMT solver called Bitwuzla by targeting the solver to a subset of SMT-LIB benchmarks. We have invented a large amount of Bitwuzla strategies using the Grackle system, that is described for the first time in this work. We have achieved a substantial improvement over the default Bitwuzla mode. The strength of our system is that it both invents new strategies as well as selects the best one to be used on a given problem. To the best of our knowledge, this is the first time such approach was applied to an SMT solver.

Table 3. Results of the evaluation. VBS stands for virtual best solver. *PAR2+* is a relative improvement of the PAR-2 score to the score of *smt-comp-mode* in percent.

time limit	solved by smt-comp-mode	best strategy solved	best strategy PAR2+	our selector solved	our selector PAR2+	solved by VBS
1s	460	541	13.59%	689	39.92%	746
10s	813	840	16.38%	873	50.50%	929
60s	938	945	19.04%	956	46.76%	971

For future work, we would like to inspect alternative methods of strategy selection and generation reduction in Grackle. For example, Grackle also supports the specialization of a strategy s on unsolved problems by interleaving the best problems $\mathcal{P}_s$ with similar but unsolved problems. This feature has not yet been tested. Additionally, we would like to test our methods on other solvers and to construct strategy schedules that alternate execution of multiple strategies.

References

1. Ábrahám, E., et al.: SC²: Satisfiability checking meets symbolic computation - (project paper). In: CICM (2016)
2. Balunovic, M., Bielik, P., Vechev, M.T.: Learning to solve SMT formulas. In: NeurIPS, pp. 10338–10349 (2018)
3. Barbosa, H., et al.: cvc5: a versatile and industrial-strength SMT solver. In: TACAS 2022. LNCS, vol. 13243, pp. 415–442. Springer, Cham (2022). https://doi.org/10.1007/978-3-030-99524-9_24
4. Barrett, C., Fontaine, P., Tinelli, C.: The satisfiability modulo theories library (SMT-LIB). www.SMT-LIB.org (2016)
5. Barrett, C., Fontaine, P., Tinelli, C.: The SMT-LIB Standard: Version 2.6. Technical Report, The University of Iowa (2017). www.SMT-LIB.org
6. Barrett, C.W., Sebastiani, R., Seshia, S.A., Tinelli, C.: Satisfiability modulo theories. In: Biere, A., Heule, M., van Maaren, H., Walsh, T. (eds.) Handbook of Satisfiability, pp. 825–885. IOS Press (2009)
7. Bengio, Y., Lodi, A., Prouvost, A.: Machine learning for combinatorial optimization: a methodological tour d'horizon. Eur. J. Oper. Res. **290**(2), 405–421 (2020)
8. Blanchette, J.C., Böhme, S., Paulson, L.C.: Extending sledgehammer with SMT solvers. J. Autom. Reason. **51**(1), 109–128 (2013). https://doi.org/10.1007/s10817-013-9278-5
9. Blanchette, J.C., Kaliszyk, C., Paulson, L.C., Urban, J.: Hammering towards QED. J. Formalized Reasoning **9**(1), 101–148 (2016)
10. Brown, C.E., Kaliszyk, C.: Lash 1.0 (system description). CoRR abs/2205.06640 (2022)
11. Burges, C., Ragno, R., Le, Q.: Learning to rank with nonsmooth cost functions. Adv. Neural Inf. Process. Syst. **19** (2006)
12. The Coq Proof Assistant. http://coq.inria.fr
13. de Moura, L., Bjørner, N.: Applications and challenges in satisfiability modulo theories. In: Workshop on Invariant Generation (WING) (2012)

14. Godefroid, P., Levin, M.Y., Molnar, D.A.: SAGE: whitebox fuzzing for security testing. Commun. ACM **55**(3), 40–44 (2012)
15. Grabowski, A., Korniłowicz, A., Naumowicz, A.: Mizar in a nutshell. J. Formalized Reasoning **3**(2), 153–245 (2010)
16. Holden, E.K., Korovin, K.: Heterogeneous heuristic optimisation and scheduling for first-order theorem proving. In: Kamareddine, F., Sacerdoti Coen, C. (eds.) CICM 2021. LNCS (LNAI), vol. 12833, pp. 107–123. Springer, Cham (2021). https://doi.org/10.1007/978-3-030-81097-9_8
17. Hutter, F., Hoos, H.H., Leyton-Brown, K., Stützle, T.: ParamILS: an automatic algorithm configuration framework. J. Artif. Intell. Res. **36**, 267–306 (2009)
18. Jakubův, J., Suda, M., Urban, J.: Automated invention of strategies and term orderings for Vampire. In: GCAI (2017)
19. Jakubův, J., Urban, J.: BliStrTune: hierarchical invention of theorem proving strategies. In: CPP (2017). https://doi.org/10.1145/3018610.3018619
20. Ke, G., et al.: LightGBM: a highly efficient gradient boosting decision tree. In: NIPS (2017)
21. Kerschke, P., Hoos, H.H., Neumann, F., Trautmann, H.: Automated algorithm selection: survey and perspectives. Evol. Comput. **27**(1), 3–45 (2019)
22. Kovács, L., Voronkov, A.: First-order theorem proving and VAMPIRE. In: Sharygina, N., Veith, H. (eds.) CAV 2013. LNCS, vol. 8044, pp. 1–35. Springer, Heidelberg (2013). https://doi.org/10.1007/978-3-642-39799-8_1
23. Lindauer, M., et al.: SMAC3: A versatile bayesian optimization package for hyperparameter optimization (2021)
24. Niemetz, A., Preiner, M.: Bitwuzla at the SMT-COMP 2020. CoRR abs/2006.01621 (2020). https://arxiv.org/abs/2006.01621
25. Nipkow, T., Klein, G.: Concrete Semantics - With Isabelle/HOL. Springer, Cham (2014). https://doi.org/10.1007/978-3-319-10542-0
26. Ramírez, N.G., Hamadi, Y., Monfroy, É., Saubion, F.: Evolving SMT strategies. In: ICTAI (2016)
27. Reynolds, A., Kuncak, V., Tinelli, C., Barrett, C., Deters, M.: Refutation-based synthesis in SMT. Formal Methods Syst. Des. **55**(2), 73–102 (2017). https://doi.org/10.1007/s10703-017-0270-2
28. Schulz, S.: System description: E 1.8. In: LPAR, pp. 735–743 (2013)
29. Scott, J., Niemetz, A., Preiner, M., Nejati, S., Ganesh, V.: MachSMT: a machine learning-based algorithm selector for SMT solvers. In: TACAS (2020)
30. Talbi, E.G.: Machine learning into metaheuristics: a survey and taxonomy of data-driven metaheuristics. ACM Comput. Surv. **54**(6), 1–32 (2020)
31. Urban, J.: BliStr: the blind strategymaker. In: Global Conference on Artificial Intelligence, GCAI 2015, Tbilisi, Georgia, 16–19 October 2015 (2015)
32. Xu, L., Hutter, F., Hoos, H.H., Leyton-Brown, K.: SATzilla: portfolio-based algorithm selection for SAT. J. Artif. Intell. Res. **32**, 565–606 (2008)

OuterCount: A First-Level Solution-Counter for Quantified Boolean Formulas

Ankit Shukla[1], Sibylle Möhle[1], Manuel Kauers[2], and Martina Seidl[3](✉)

[1] Institute for Formal Models and Verification, JKU, Linz, Austria
{ankit.shukla,sibylle.moehle-rotondi}@jku.at
[2] Institute for Algebra, JKU, Linz, Austria
manuel.kauers@jku.at
[3] Institute for Symbolic Artificial Intelligence, JKU, Linz, Austria
martina.seidl@jku.at

Abstract. Counting the solutions of symbolic encodings is an intriguing computational problem with many applications. In the field of propositional satisfiability (SAT) solving, for example, many algorithms and tools have emerged to tackle the counting problem. For quantified Boolean formulas (QBFs), an extension of SAT with quantifiers used to compactly encode and solve problems of formal verification, synthesis, planning, etc., practical solution counting has not been considered yet.

We present the first practical counting algorithm for top-level solutions. We prove soundness of our algorithm for true and false formulas and show how to implement it with recent QBF solving technology. Our evaluation of benchmarks from the recent QBF competition gives promising results for this difficult problem.

Keywords: Quantified boolean formulas · Solution counting · #QBF

1 Introduction

QBF solution counting [21], also known as #QBF, is the problem of computing the number of *(counter-)models* of a given *quantified Boolean formula* (QBF). Like #SAT [16], the counting problem for propositional satisfiability (SAT), #QBF is considered to be very hard. It belongs to #PSPACE [21], an intractable problem class. As #SAT is an essential task in many application domains, including probabilistic reasoning [10,33], the analysis of software vulnerability [8,38], and the verification of neural networks [3,29], numerous approaches exist to practically solve #SAT and variations of this problem [11,16]. In contrast, the #QBF problem has only been studied theoretically [17]. Practical counting tools do not exist so far although similar applications as for #SAT can be expected.

This work has been supported by the Austrian Science Fund (FWF) under projects W1255-N23 and P31571-N32, the LIT AI Lab funded by the State of Upper Austria.

With this work, we start to close this gap by presenting the first QBF counter for a large class of QBF problems.

The presence of quantifiers over the Boolean variables poses theoretical as well as engineering challenges to uplifting techniques straightway from propositional logic. The universal and existential quantifiers render the decision problem of QBF PSPACE-complete [35], making QBFs a natural choice for encoding various problems from verification and artificial intelligence [32,34]. Because of the quantification, we observe a duality between true and false problems. This duality does not exist in SAT. If a propositional formula is satisfiable, there exists at least one model, i.e., a truth assignment to the propositional variables under which the formula evaluates to true. If a propositional formula is unsatisfiable, no satisfying assignment exists, i.e., all assignments are counter-models. In contrast, models of a true QBF are functions that tell us how to set the existential variables based on given values of the universal variables. Dually, counter-models of a false QBF are functions that tell us how to set the universal variables based on given values of the existential variables. In both cases, these functions reflect the quantifier dependencies. As existential (resp. universal) variables in the outermost quantifier block do not have any dependencies, their solutions are simply Boolean values in the case of true (resp. false) formulas. Concrete examples of interesting QBF encodings are bounded model checking, planning or formal synthesis. For many of these encodings, the solutions to these problems (e.g., error traces, plans, synthesized programs) are Boolean assignments of the variables in the outermost quantifier block. We are concerned with the question how many such solutions exist for a given QBF.

In this work, we present a concise formulation of the solution counting problem for the variables of the outermost quantifier block and solve it with an enumeration-based approach. The assignments of these variables are of particular interest because their values indicate the solutions to many application problems, e.g., the synthesized implementation in reactive synthesis, the error traces in bounded model checking, or the plans in planning. We implemented our method with recent QBF solving technology and further evaluated our implementation on recent QBF competition benchmarks.

The organization of the paper is as follows. After discussing related work and preliminaries in Sect. 2 and in Sect. 3, we formalize our model-counting approach for true formulas in Sect. 4. On this basis, we present the dual approach for false formulas in Sect. 5. This approach involves some additional transformation steps not necessary for true formulas. Both are implemented with recent QBF solving technology. In Sect. 7, we evaluate our tool on the benchmarks from the last QBF competition. In Sect. 8, we discuss possible future work.

2   Related Work

For many exact counting problems, enumerative approaches are among the first techniques in order to realize counting tools. This applies, for example, to model counting for SAT [9,13], to the counting of minimal inconsistent sets [19,25], to the counting of unsatisfiable sets [2, 6, 36] and satisfiable sets [7], to counting in answer set programming [14], and to the counting of minimal correction sets [27, 28].

An enumeration-based model counter iteratively adds the negation of already found models in the form of so-called *blocking clauses* to the given formula. The addition of blocking clauses excludes the respective models from the solution space until no further models can be found. In the case that a problem has a huge amount of solutions, often such approaches are run only until a certain time limit is reached. In the context of QBF solving, the idea of adding blocking clauses is also known as good (or solution) learning [15, 22, 37], but it has not been used for counting. For SAT, Bayardo and Pehoushek [4] used good learning and the minimization of goods in the context of model counting.

An enumeration-based approach to #SAT was used to enable the formal proof of its correctness [26]. In principle, the proof consists in showing that the detected models are pairwise contradicting and that all models are found. From this, it follows that the corresponding model count is correct. Modern counting tools implement alternative techniques like the detection of connected components, tight integration into the solving process, or hashing solutions (see [11, 16] for more details on propositional model counting).

The enumeration of all models is a very closely related problem. For propositional logic, this problem is known as ALLSAT. In the context of QBF, Ehlers et al. [5] defined the ALLQBF problem as the task to find all assignments of free variables occurring in a given QBF such that the formula evaluates to true. In their work, they realize different learning approaches, including one which iteratively builds a representation of all satisfying assignments in disjunctive normal form (DNF). While they are not interested in calculating the model count, they aim for a DNF representation that is as small as possible while describing all models. Also, they do not consider false formulas.

3 Preliminaries

We consider QBFs of the form $\Pi.\phi$ that consist of a *(quantifier) prefix* $\Pi = Q_1 X_1 \ldots Q_n X_n$ (where $Q_i \in \{\forall, \exists\}$, $Q_i \neq Q_{i+1}$, and $X_1, \ldots, X_n$ are pairwise disjoint and non-empty sets of variables) and the *matrix* ϕ, a propositional formula over variables X_i. We call i the *(quantifier) level* of *quantifier block* $Q_i X_i$, e.g., variables of X_1 are at level 1 of prefix $Q_1 X_1 \ldots Q_n X_n$. If $x \in X_i$ and $y \in X_j$ and $i < j$, then $y < x$ (or If $x_i \in X_i$ and $x_j \in X_j$ and $i < j$, then $x_i < x_j$). For the propositional part of a QBF, we use the standard Boolean connectives $\wedge$ (conjunction), $\vee$ (disjunction), $\rightarrow$ (implication), $\leftrightarrow$ (equivalence), and $\neg$ (negation). A QBF $\Pi.\phi$ is in *prenex conjunctive normal form* (PCNF) if ϕ is a conjunction of clauses. A *clause* is a disjunction of literals and a literal is a variable or a negated variable. For a literal l, $var(l) = x$ if $l = x$ or $l = \neg x$. A *unit clause* contains exactly one literal. An *assignment* σ is a set of literals such that $\sigma \cap \{\neg l \mid l \in \sigma\} = \emptyset$. By $var(\sigma)$ we denote the set $\{var(l) \mid l \in \sigma\}$. We sometimes interpret an assignment as a conjunction of its literals. An X-assignment σ is an assignment with $var(\sigma) \subseteq X$. If $var(\sigma) = X$, then σ is a *full X-assignment*, otherwise it is a *partial X-assignment*. We say that X-assignment τ is an *expansion* of X-assignment σ if $\sigma \subseteq \tau$.

If ϕ is a propositional formula and σ an assignment, then ϕ_σ denotes the formula obtained by setting the variable x to true if $x \in \sigma$, by setting x to false if $\neg x \in \sigma$, and by performing standard simplifications. A QBF $\forall x \Pi.\phi$ (resp. $\exists x \Pi.\phi$) is true if $\Pi.\phi_{\{x\}}$ and $\Pi.\phi_{\{\neg x\}}$ are true (resp. if $\Pi.\phi_{\{x\}}$ or $\Pi.\phi_{\{\neg x\}}$ is true). For example, the QBF $\exists x \forall y.(x \leftrightarrow y)$ is false and the QBF $\forall x \exists y.(x \leftrightarrow y)$ is true. *(Counter-)Models* of QBFs can be described as sets of Boolean functions. Given a true QBF φ, a *model* of φ is a set of functions F such that for each existential variable $x \in var(\varphi)$, there is a function $f_x(y_1, \ldots, y_n) \in F$ with $y_i < x$ and y_i is universal. Further, the propositional formula φ_F which is obtained by replacing all its existential variables x by function $f_x \in F$ is valid iff F is a model. The functions of a model reflect the dependencies of the variables. If the first quantifier block is existential, then the functions of these variables are truth constants, i.e., assignments as introduced above. We are interested in those assignments.

Definition 1. *Let $\varphi = \exists X \Pi.\phi$ be a true QBF and let σ be a partial X-assignment. We call σ a satisfying partial X-assignment if the QBF $\forall X' \Pi.\phi_\sigma$ with $X' = X \setminus var(\sigma)$ is true (note that variables $X \setminus var(\sigma)$ are now universal). Then σ is also called* partial X-model *or* level-1 solution *of φ.*

Based on this definition every expansion of a partial X-model σ to a full X-assignment is an X-model. Hence, all variables of X not mentioned in σ may be set arbitrarily and preserve the satisfiability of the formula.

Example 1. Consider QBF $\exists x_1, x_2 \forall a \exists y.((x_1 \vee \neg x_2 \vee y) \wedge (x_1 \vee \neg x_2 \vee a \vee \neg y))$ with $X = \{x_1, x_2\}$. Then the X-assignment $\{x_1\}$ is a partial X-model, because both full X-assignments $\{x_1, x_2\}$ and $\{x_1, \neg x_2\}$ are X-models. In contrast, $\{\neg x_1\}$ is not a partial X-model, because $\{\neg x_1, x_2\}$ is not an X-model.

A *counter-model* of a false QBF φ is defined dually: a counter-model is a set of functions F such that for each universal variable x there is a function $f_x(y_1, \ldots, y_n) \in F$ with y_i existential and $y_i < x$. Further, φ_F, which is obtained by replacing all universal variables x by f_x, is unsatisfiable. The functions of the outermost universal variables are constants.

Definition 2. *Let $\varphi = \forall X \Pi.\phi$ be a false QBF and let σ be a partial X-assignment. We call σ a falsifying partial X-assignment if QBF $\exists X' \Pi.\phi_\sigma$ with $X' = X \setminus var(\sigma)$ is false. Then σ is also called a* partial X-counter-model *or* level-1 solution *of φ.*

Note that the term *solution* is used for models and counter-models.

4 Counting Models

Given a true QBF $\exists X \Pi.\phi$, we are interested in the number of full X-assignments σ such that $\Pi.\phi_\sigma$ is true. To this end, we enrich the formula with so-called *blocking clauses* until the formula becomes false. From the number of blocking clauses, we can infer the number of satisfying X-assignments, i.e., the number of X-models.

Definition 3 (Blocking Clause). *Let $\exists X \Pi.\phi$ be a true QBF and let σ be a partial X-model of this QBF. Then $\neg\sigma$ is a* blocking clause *of $\exists X \Pi.\phi$.*

By enriching the formula with blocking clauses, we exclude models, i.e., we avoid the same models to be discovered again when evaluating the enriched formula. The following lemma shows that the addition of blocking clauses obtained from a set M of partial X-models eliminates all partial X-models that are expansions of the X-models from M.

Lemma 1. *Let $\varphi = \exists X \Pi.\phi$ be a true QBF and let M be a set of partial X-models of φ. Then there is no partial X-model τ of $\varphi' = \exists X \Pi.(\phi \wedge \bigwedge_{\sigma \in M} \neg\sigma)$ with $\sigma \subseteq \tau$ for all $\sigma \in M$.*

Proof. Let τ be an X-assignment with $\sigma \subseteq \tau$ for some $\sigma \in M$. Then τ cannot be an X-model of φ', because τ falsifies the clause $\neg\sigma$. Hence, τ also falsifies φ'.

Based on blocking clauses, we can realize an enumerative approach to model counting. To this end, we need the following criterion to decide when all X-models have been covered.

Lemma 2. *Let $\varphi = \exists X \Pi.\phi$ be a true QBF and let M be a set of partial X-models. For each full X-model τ of φ, there exists a $\sigma \in M$ with $\sigma \subseteq \tau$ iff the QBF $\varphi' = \exists X \Pi.(\phi \wedge \bigwedge_{\sigma \in M} \neg\sigma)$ is false.*

Proof. We prove both directions by contradictions.
$\Rightarrow$: Assume there is an X-model τ of φ'. Then τ is also an X-model of φ. Since there exists an X-model $\sigma \in M$ with $\sigma \subseteq \tau$, τ cannot be an X-model of φ' by Lemma 1.
$\Leftarrow$: Assume there is an X-model τ of $\Pi.\phi$ such that there is no $\rho \in M$ with $\tau \subseteq \rho$. Then τ is an X-model of $\Pi.(\phi \wedge \bigwedge_{\sigma \in M} \neg\sigma)$ which is false by assumption. $\square$

Corollary 1. *Let $\varphi = \exists X \Pi.\phi$ be a true QBF, let M be a set of full X-models with $\exists X \Pi.(\phi \wedge \bigwedge_{\sigma \in M} \neg\sigma)$ be false. Then φ has $|M|$ X-models.*

Based on this corollary, we could already build an enumeration-based model counter for QBF. A partial X-model describes a set of full X-models and yields a potentially exponentially more compact encoding leading to the following proposition.

Proposition 1. *Let $\varphi = \exists X \Pi.\phi$ be a true QBF and let $M = \{\sigma_1, \ldots, \sigma_n\}$ be a set of partial X-models of φ such that*

1. *$\exists X \Pi.(\phi \wedge \bigwedge_{\sigma \in M} \neg\sigma)$ is false and*
2. *$\sigma_k \in M$ is a partial X-model of $\exists X \Pi.(\phi \wedge \bigwedge_{i=1}^{k-1} \neg\sigma_i)$*

Then the number of X-models of φ is $\Sigma_{\sigma \in M} 2^{|X|-|\sigma|}$.

Proof. Because of the first condition, Lemma 2 applies. Therefore, all full X-models of φ can be obtained by expanding the elements of M. We further need to argue that no full X-model can be obtained by expanding several elements of M. Because of the second condition which imposes an order on the elements of M and the successive application of Lemma 1 it follows that each full X-model is only considered once. As one partial X-model σ of size $|\sigma|$ can be expanded to $2^{|X|-|\sigma|}$ full X-models, we get a total count of full X-models as stated above. □

Example 2. Consider QBF $\exists x_1, x_2 \forall a \exists y.((x_1 \vee \neg x_2 \vee y) \wedge (x_1 \vee \neg x_2 \vee a \vee \neg y))$ with $X = \{x_1, x_2\}$. Furthermore, let $M = \{\{x_1\}, \{\neg x_1, \neg x_2\}\}$, i.e., a set with one partial X-model and a full X-model such that $\{\neg x_1, \neg x_2\}\}$ is an X-model of $\exists x_1, x_2 \forall a \exists y.((x_1 \vee \neg x_2 \vee y) \wedge (x_1 \vee \neg x_2 \vee a \vee \neg y) \wedge \neg x_1)$.

As the QBF $\exists x_1, x_2 \forall a \exists y.((x_1 \vee \neg x_2 \vee y) \wedge (x_1 \vee \neg x_2 \vee a \vee \neg y) \wedge \neg x_1 \wedge (x_1 \vee x_2))$ is false, the formula has *three* $(2+1)$ X-models.

5   Counting Counter-Models

Dually to counting the X-models of a true QBF $\exists X \Pi.\phi$, one can also count the Y-counter-models (falsifying Y-assignments) of a false QBF $\forall Y \Pi.\psi$.

Example 3. The QBF $\forall y_1, y_2 \exists x.(y_1 \vee y_2 \vee x) \wedge (\neg y_1 \vee x) \wedge (\neg x)$ has partial Y-counter-models $\{\neg y_1, \neg y_2\}$ and $\{y_1\}$. The latter can be expanded to full Y-counter-models $\{y_1, y_2\}$ and $\{y_1, \neg y_2\}$.

To count such Y-counter-models we introduce the notion of *blocking cube* as the dual of the notion of blocking clause.

Definition 4 (Blocking Cube). *Let $\forall Y \Pi.\psi$ be a false QBF and let σ be a partial Y-counter-model of this QBF. Then σ is a* blocking cube *of $\forall Y \Pi.\psi$.*

Blocking cubes of the example above are $(\neg y_1 \wedge \neg y_2)$ and (y_1) as well as $(y_1 \wedge y_2)$ and $(y_1 \wedge \neg y_2)$. A blocking cube is a falsifying Y-assignment that we will use to extend a formula such that the same (super-)counter-models are excluded from the set of counter-models of a given formula. Therefore, we have to disjunctively add blocking cubes that have the following properties (as the proofs are similar to the proofs in the previous section, we omit them). In the same way as *partial* models allowed us to get shorter blocking clauses, partial counter-models will help us to get shorter blocking cubes, exponentially decreasing the number of required blocking cubes.

Lemma 3. *Let $\varphi = \forall Y \Pi.\psi$ be a false QBF and let M be a set of partial Y-counter-models of φ. Then there is no partial Y-counter-model ρ of $\varphi' = \forall Y \Pi.(\psi \vee \bigvee_{\sigma \in M} \sigma)$ with $\sigma \subseteq \rho$ for all $\sigma \in M$.*

This lemma ensures that all counter-models of the original formula that can be obtained by expanding a blocking cube are excluded from the set of counter-models of the formula enriched with blocking cubes. The next lemma states that if we disjunctively add all its Y-counter-models to a false QBF, it becomes true.

Lemma 4. *Let $\varphi = \forall Y \Pi.\psi$ be a false QBF and let M be a set of partial Y-counter-models. For each full Y-counter-model ρ of φ, there exists a $\sigma \in M$ with $\sigma \subseteq \rho$ iff the QBF $\varphi' = \forall Y \Pi.(\psi \vee \bigvee_{\sigma \in M} \sigma)$ is true.*

Now we can find the number of Y-counter-models of a false QBF as follows.

Proposition 2. *Let $\varphi = \forall Y \Pi.\psi$ be a false QBF and let $M = \{\sigma_1, \ldots, \sigma_n\}$ be a set of partial Y-counter-models of φ such that*

1. *$\forall Y \Pi.(\psi \vee \bigvee_{\sigma \in M} \sigma)$ is true and*
2. *$\sigma_k \in M$ is a partial Y-counter-model of $X \Pi.(\psi \vee \bigvee_{i=1}^{k-1} \sigma_i)$*

Then the number of Y-counter-models of $\Pi.\psi$ is $\Sigma_{\sigma \in M} 2^{|Y| - |\sigma|}$.

Example 4. Given false QBF $\varphi = \forall y_1, y_2 \exists x.(y_1 \vee y_2 \vee x) \wedge (\neg y_1 \vee x) \wedge (\neg x)$ with partial Y-counter-models $\{\neg y_1, \neg y_2\}$ and $\{y_1\}$. As the QBF

$$\varphi' = \forall y_1, y_2 \exists x.((y_1 \vee y_2 \vee x) \wedge (\neg y_1 \vee x) \wedge (\neg x)) \vee (\neg y_1 \wedge \neg y_2) \vee (y_1)$$

is true, we can conclude that φ has $1 + 2$ counter-models.

Note that in contrast to counting X-models, we lose the PCNF structure when counting Y-counter-models. To obtain a PCNF again, we perform the well-known Plaisted-Greenbaum [31] transformation which introduces additional variables. The following lemma shows that this transformation does not change the number of (counter)-models.

Lemma 5. *Let $\forall Y \Pi.\psi$ be a false QBF and σ be a Y-counter-model of $\Pi.\psi$. Then QBF*
$$\varphi_1 = \Pi \exists t_1, t_2.((t_1 \to \psi) \wedge (t_2 \to \sigma) \wedge (t_1 \vee t_2))$$
has the same number of Y-counter-models as the QBF $\varphi_2 = \Pi.(\psi \vee \sigma)$.

Proof. We show that φ_1 and φ_2 have the same Y-counter-models.

$\Rightarrow$: Let τ be a Y-counter-model of φ_1. The subformula σ which contains only literals from Y has to evaluate to false under τ, because otherwise t_2 would become a pure literal in clause $(t_1 \vee t_2)$, in consequence $\neg t_1$ would become a pure literal as well, and then φ_1 would evaluate to true under τ. Hence, σ has to be false under τ. Then $\neg t_2$ becomes a unit clause, and by propagation t_1 becomes a unit clause as well, simplifying φ_1 under τ to $\Pi.(\psi_\tau)$ which has to be false (because τ is an Y-counter-model). It follows that τ is also a Y-counter-model of φ_2.

$\Leftarrow$: Let τ be a Y-counter-model of φ_2. Then σ is false under τ and so is $\Pi.\psi_\tau$. It follows that τ is a Y-counter-model of φ_1 as well.

It is easy to see that the lemma above directly transfers to extending a QBF with m Y-counter-models by introducing $m + 1$ new variables.

Example 5. Applying the Plaisted-Greenbaum transformation on QBF φ of Example 4 results in $\forall y_1, y_2 \exists x, t_1, t_2, t_3.(t_1 \to (y_1 \vee y_2 \vee x) \wedge (\neg y_1 \vee x) \wedge (\neg x)) \wedge (t_2 \to (\neg y_1 \wedge \neg y_2)) \wedge (t_3 \to (y_1)) \wedge (t_1 \vee t_2 \vee t_3)$ which can efficiently be transformed into PCNF by standard logical rules.

input : QBF $\Phi = QZ\Pi.\phi$
output: Numbers of Z-Solutions of Φ

$c \leftarrow 0; i \leftarrow 0;$
DepQBF.init $(QZ\Pi\exists t_0.(t_0 \rightarrow \phi));$
DepQBF.assume $(t_0);$
$(v, \sigma) \leftarrow$ DepQBF.solve() ;
if $(v = \top \ \& \ Q = \forall) \| (v = \bot \ \& \ Q = \exists)$ **then**
  | return $-1;$
end
if $v = \top$ **then**
  | DepQBF.add $((t_0));$
end
do
  | $c \leftarrow c + 2^{|Z|-|\sigma|};$
  | **if** $v = \top$ **then**
  | | DepQBF.add $(\neg\sigma);$
  | **end**
  | **else**
  | | $i++;$
  | | DepQBF.add $(t_i \rightarrow \sigma);$
  | | DepQBF.assume $((t_0 \vee \ldots \vee t_i));$
  | **end**
  | $(v, \sigma) \leftarrow$ DepQBF.solve() ;
while $(v = \bot \ \& \ Q = \forall) \| (v = \top \ \& \ Q = \exists);$
return $c;$

Algorithm 1: OuterCount (Φ)

6 Implementation

To implement the previously presented approach we use the incremental solving
interface of the search-based QBF solver DepQBF [24] to enrich the formula
with blocking clauses or blocking cubes. In the case of true formulas, in each
solver call, the formula is enriched with additional clauses. In the case of false
formulas, it is necessary to update a clause. For this purpose, we use the push
and pop functions of DepQBF which allows us to add a clause that is only
available for one solver run: we push a clause to DepQBF, run DepQBF and
evaluate the result. If another run is necessary we pop the clause and push the
updated clause to DepQBF. This is necessary to increase the disjunction of the
definitions introduced by the Plaisted-Greenbaum transformation in the case of
counter-model counting. In the following, we use the function assume to indicate
that a clause should only be available temporarily for one solver run and that it
should it be discarded afterwards.

Algorithm 1 takes as input a QBF Φ starting with quantifier block QZ for
which the number of Z-solutions must be determined. To this end, we initialize
the solver with the QBF $QZ\Pi\exists t_0.(t_0 \rightarrow \phi)$ where t_0 is a fresh existential variable
that is added to the innermost scope. During the first solver call, t_0 is assumed

Table 1. Formulas for which the exact count could be found without preprocessing. With preprocessing, more formulas run into timeouts (TO).

	Instance	without preprocessing					with preprocessing				
		Q	# bl	# v	# s	t(s)	Q	# bl	# v	# s	t(s)
True formulas	gttt_1_1...torus_b	∃	17	355	1	53	∀	16	–	–	–
	gttt_2_2...torus_b	∃	9	810	15	103	∃	9	31	15	31
	gttt_2_2....b	∃	9	754	42	1991	∃	9	37	42	192
	k_ph_n-11	∃	5	3	1	0.2	∃	1	110	¿ 1K	TO
	k_ph_n-15	∃	5	3	1	1.3	∃	1	210	¿ 1K	TO
	k_ph_n-18	∃	5	3	1	40	∃	3	454	> 2K	TO
	k_ph_n-19	∃	5	3	1	6.14	∃	3	507	> 2K	TO
	k_ph_n-20	∃	5	3	1	9.72	∃	3	564	> 700	TO
False formulas	arbiter-05-...-depth-8	∀	18	10	1	0.1	∀	14	10	214	24.5
	arbiter-06-...-depth-11	∀	24	12	1	6.5	∀	20	12	342	679
	arbiter-06-...-depth-15	∀	32	12	1	187	∀	28	12	> 40	TO
	arbiter-09-...-depth-15	∀	32	18	1	523	∀	28	12	> 500	TO
	W4...tbm_05..8S...1	∀	20	10	10	60	∀	18	10	10	1023
	W4...tbm_25..7S...3	∀	20	10	10	520	∀	18	10	> 8	TO
	W4...tbm_26..7S...3	∀	20	10	9	689	∀	18	9	> 4	TO

Q ... quantifier type at level 1   #bl ... number of quantifier blocks
#v ... size of first quantifier block
#s ... number of solutions   t(s) ... runtime in seconds

to be true hence it does not have any effect on the solving result. If Q is ∃ (resp., ∀) and Φ is false (resp., true), then -1 is returned as there is nothing to count. If Φ is true, t_0 is permanently added as a unit clause, because in this case it is not required for the normal form transformation. Next, the counting variable c is updated taking into account the size of the found solution. If Φ is true, $\neg\sigma$ is added, otherwise $t_i \to \sigma$ in the form of $|\sigma|$ binary clauses is added. Further, the clause $(t_0 \vee \ldots \vee t_i)$ is assumed for disjunctively excluding the found solutions. The algorithm iterates until the formula becomes false (resp. true) and, in that case, returns c, the number of Z-solutions. Our tool (together with the experimental results of the evaluation described below) is available at https://github.com/marseidl/outer-count.

7   Evaluation

As our tool is the first tool for practical QBF solution counting, there are no other tools to compare with. Further, no benchmarks are available. Therefore, we established two sets of benchmarks. First, we evaluated our tool on the formulas of the PCNF track of QBFEVAL 2020,[1] the most recent QBF competition. In a prerun, we identified 68 true formulas that start with an existential quantifier block, and 34 false formulas that start with a universal quantifier block solvable by plain DepQBF. We also applied the preprocessor bloqqer[2] to obtain a second

---

[1] http://www.qbfeval.org.
[2] http://fmv.jku.at/bloqqer.

set of formulas. Amongst other techniques, bloqqer performs existential variable elimination, universal expansion, blocked clause elimination and equality reasoning. More details on QBF preprocessing and bloqqer can be found in [18]. The preprocessor bloqqer directly solved seven true formulas and eight false formulas, while for nine true formulas and one false formula of the preprocessed benchmarks no solution was found at all. These formulas were excluded from the benchmark set, leading to a set of 52 true formulas and 25 false formulas. All experiments were carried out on a machine with 128 AMD EPYC 7501 processors. For each formula, we limited the time to 3600 s seconds and the memory to 8 GB.

Our tool could determine the exact level-1 solution count for eight true formulas and for seven false formulas. Details on those formulas are shown in the left part of Table 1. The eight true formulas contain three formulas from encodings of a generalization of the two-player-game TIC-TAC-TOE (gttt*) to synthesize a winning strategy for one player [12] and five formulas (k_ph*) of QBF encodings to solve formulas from Modal Logics [30]. While the formulas of the gttt* family have a very deep quantifier structure (up to 17 quantifier blocks) and many variables in the first quantifier block (up to more than 800 variables), the k_ph* formulas have only five quantifier blocks and three variables in the first quantifier block. For the gttt* formulas one, 15, and 42 level-1 solutions were found, the k_ph* formulas have exactly one level-1 solution each. Preprocessing (right part of Table 1) considerably changed the structure of the formulas and also their solution space to some extent. For formula gttt_1_1* the variables of the first existential quantifier block were eliminated, resulting in a formula starting with a universal quantifier block. Hence it does not have any level-1 solutions anymore. For the other two gttt* formulas, the number of variables decreased by preprocessing, but the number of level-1 solutions did not change. This might be an indication that there is potential to optimize the encoding. For the k_ph* formulas the number of quantifier blocks decreased from five to three or even to one. For these formulas, the exact number of level-1 solutions could not be determined within one hour. We found more than 700, 1.200 (twice), 2.000, and 2.500 solutions for the formulas of this family.

The seven false formulas, for which we could determine the exact level-1 solution count, stem from verification. Those formulas have a huge number of quantifier blocks (up to 32) but only few variables in the first quantifier block (up to 18). Also, the number of level-1 solutions is rather small (up to 10). This changes drastically, when preprocessing is enabled. For four formulas the exact level-1 solution count could not be determined, and for formulas with originally one level-1 the solution, several hundred were found (see right part of Table 1).

Figure 1 shows the number of level-1 solutions of all formulas counted within a time-frame of one hour, i.e., we get a lower bound for the level-1 solution count. We observe that in general, preprocessing increases the solution space. This could explain why solvers that do not rely on the formula structure can often solve preprocessed formulas more efficiently. Similarly, as reported for solving #SAT problems, also in the context of QBF counting the number of solutions can get very large.

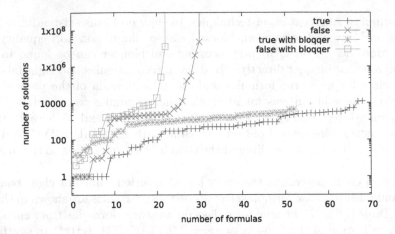

Fig. 1. Solution counts found within one hour and with 8 GB memory limit.

8   Conclusion

In this paper, we presented the first practical QBF model counting approach for the outermost (level-1) quantified variables. To this end, we have lifted the idea of enriching the formula with blocking clauses from SAT to counting the models of true QBFs. We also introduced the dual approach for counting counter-models of false formulas. We applied the incremental solving interface of the state-of-the-art QBF solver DepQBF to enrich the formula with blocking clauses (or blocking cubes), resulting in a very elegant implementation. Our empirical evaluation demonstrates that counting for QBFs is feasible to a certain extent (at least for getting lower bounds). This is particularly interesting to analyze the effects of preprocessing. In the future, we plan to apply solution minimization to exponentially reduce the search space. Short blocking clauses rule out a larger portion of the search space, hence model minimization techniques have been developed for other formalisms [1,20,23]. Further, counting could be directly integrated into a state-of-the-art QBF solver in a similar way it has been done for SAT model counters. The ultimate next step is to generalize our approach for counting functions to perform level-2 counting and beyond.

References

1. Aziz, R.A., Chu, G., Muise, C., Stuckey, P.: #∃SAT: projected model counting. In: Heule, M., Weaver, S. (eds.) SAT 2015. LNCS, vol. 9340, pp. 121–137. Springer, Cham (2015). https://doi.org/10.1007/978-3-319-24318-4_10
2. Bailey, J., Stuckey, P.J.: Discovery of minimal unsatisfiable subsets of constraints using hitting set dualization. In: Hermenegildo, M.V., Cabeza, D. (eds.) PADL 2005. LNCS, vol. 3350, pp. 174–186. Springer, Heidelberg (2005). https://doi.org/10.1007/978-3-540-30557-6_14

3. Baluta, T., Shen, S., Shinde, S., Meel, K.S., Saxena, P.: Quantitative verification of neural networks and its security applications. In: CCS, pp. 1249–1264. ACM (2019)
4. Bayardo Jr., R., Pehoushek, J.D.: Counting models using connected components. In: AAAI/IAAI, pp. 157–162. AAAI Press/The MIT Press (2000)
5. Becker, B., Ehlers, R., Lewis, M., Marin, P.: ALLQBF solving by computational learning. In: Chakraborty, S., Mukund, M. (eds.) ATVA 2012. LNCS, pp. 370–384. Springer, Heidelberg (2012). https://doi.org/10.1007/978-3-642-33386-6_29
6. Bendík, J., Černá, I.: Replication-guided enumeration of minimal unsatisfiable subsets. In: Simonis, H. (ed.) CP 2020. LNCS, vol. 12333, pp. 37–54. Springer, Cham (2020). https://doi.org/10.1007/978-3-030-58475-7_3
7. Bendík, J., Cerna, I.: Rotation based MSS/MCS enumeration. In: LPAR. EPIC, vol. 73, pp. 120–137. EasyChair (2020)
8. Biondi, F., Enescu, M.A., Heuser, A., Legay, A., Meel, K.S., Quilbeuf, J.: Scalable approximation of quantitative information flow in programs. In: VMCAI 2018. LNCS, vol. 10747, pp. 71–93. Springer, Cham (2018). https://doi.org/10.1007/978-3-319-73721-8_4
9. Birnbaum, E., Lozinskii, E.L.: The good old Davis-Putnam procedure helps counting models. J. Artif. Intell. Res. **10**, 457–477 (1999)
10. Chakraborty, S., Meel, K.S., Vardi, M.Y.: Algorithmic improvements in approximate counting for probabilistic inference: From linear to logarithmic SAT calls. In: IJCAI, pp. 3569–3576. IJCAI/AAAI Press (2016)
11. Chakraborty, S., Meel, K.S., Vardi, M.Y.: Chapter 26. Approximate model counting. In: Handbook of satisfiability, frontiers in artificial intelligence and applications, vol. 336, pp. 1015–1045. IOS Press (2021)
12. Diptarama, Yoshinaka, R., Shinohara, A.: QBF encoding of generalized tic-tac-toe. In: QBF@SAT. CEUR Workshop Proceedings, vol. 1719, pp. 14–26. CEUR-WS.org (2016)
13. Dubois, O.: Counting the number of solutions for instances of satisfiability. Theor. Comput. Sci. **81**(1), 49–64 (1991)
14. Gebser, M., Kaufmann, B., Neumann, A., Schaub, T.: Conflict-driven answer set enumeration. In: Baral, C., Brewka, G., Schlipf, J. (eds.) LPNMR 2007. LNCS (LNAI), vol. 4483, pp. 136–148. Springer, Heidelberg (2007). https://doi.org/10.1007/978-3-540-72200-7_13
15. Giunchiglia, E., Narizzano, M., Tacchella, A.: Learning for quantified Boolean logic satisfiability. In: AAAI/IAAI, pp. 649–654. AAAI Press/The MIT Press (2002)
16. Gomes, C.P., Sabharwal, A., Selman, B.: Chapter 25. Model counting. In: Handbook of Satisfiability, Frontiers in Artificial Intelligence and Applications, vol. 336, pp. 993–1014. IOS Press, Amsterdam, Netherlands (2021)
17. Hemaspaandra, L.A., Vollmer, H.: The satanic notations: counting classes beyond #P and other definitional adventures. SIGACT News **26**(1), 2–13 (1995)
18. Heule, M., Järvisalo, M., Lonsing, F., Seidl, M., Biere, A.: Clause elimination for SAT and QSAT. J. Artif. Intell. Res. **53**, 127–168 (2015)
19. Hunter, A., Konieczny, S.: Measuring inconsistency through minimal inconsistent sets. In: KR, pp. 358–366. AAAI Press (2008)
20. Jin, H.S., Han, H.J., Somenzi, F.: Efficient conflict analysis for finding all satisfying assignments of a Boolean circuit. In: Halbwachs, N., Zuck, L.D. (eds.) TACAS 2005. LNCS, vol. 3440, pp. 287–300. Springer, Heidelberg (2005). https://doi.org/10.1007/978-3-540-31980-1_19
21. Ladner, R.E.: Polynomial space counting problems. SIAM J. Comput. **18**(6), 1087–1097 (1989)

22. Letz, R.: Lemma and model caching in decision procedures for quantified Boolean formulas. In: Egly, U., Fermüller, C.G. (eds.) TABLEAUX 2002. LNCS (LNAI), vol. 2381, pp. 160–175. Springer, Heidelberg (2002). https://doi.org/10.1007/3-540-45616-3_12

23. Li, B., Hsiao, M.S., Sheng, S.: A novel SAT all-solutions solver for efficient preimage computation. In: DATE, pp. 272–279. IEEE Computer Society (2004)

24. Lonsing, F., Egly, U.: DepQBF 6.0: a search-based qbf solver beyond traditional QCDCL. In: de Moura, L. (ed.) CADE 2017. LNCS (LNAI), vol. 10395, pp. 371–384. Springer, Cham (2017). https://doi.org/10.1007/978-3-319-63046-5_23

25. McAreavey, K., Liu, W., Miller, P.: Computational approaches to finding and measuring inconsistency in arbitrary knowledge bases. Int. J. Approx. Reason. **55**(8), 1659–1693 (2014)

26. Möhle, S., Biere, A.: Combining conflict-driven clause learning and chronological backtracking for propositional model counting. In: GCAI. EPIC, vol. 65, pp. 113–126. EasyChair (2019)

27. Morgado, A., Liffiton, M., Marques-Silva, J.: MaxSAT-based MCS enumeration. In: Biere, A., Nahir, A., Vos, T. (eds.) HVC 2012. LNCS, vol. 7857, pp. 86–101. Springer, Heidelberg (2013). https://doi.org/10.1007/978-3-642-39611-3_13

28. Narodytska, N., Bjørner, N., Marinescu, M.V., Sagiv, M.: Core-guided minimal correction set and core enumeration. In: IJCAI, pp. 1353–1361. ijcai.org (2018)

29. Narodytska, N., Shrotri, A., Meel, K.S., Ignatiev, A., Marques-Silva, J.: Assessing heuristic machine learning explanations with model counting. In: Janota, M., Lynce, I. (eds.) SAT 2019. LNCS, vol. 11628, pp. 267–278. Springer, Cham (2019). https://doi.org/10.1007/978-3-030-24258-9_19

30. Pan, G., Vardi, M.Y.: Symbolic decision procedures for QBF. In: Wallace, M. (ed.) CP 2004. LNCS, vol. 3258, pp. 453–467. Springer, Heidelberg (2004). https://doi.org/10.1007/978-3-540-30201-8_34

31. Plaisted, D.A., Greenbaum, S.: A structure-preserving clause form translation. J. Symb. Comput. **2**(3), 293–304 (1986)

32. Pulina, L., Seidl, M.: The 2016 and 2017 QBF solvers evaluations (QBFEVAL'16 and QBFEVAL'17). Artif. Intell. **274**, 224–248 (2019)

33. Sang, T., Beame, P., Kautz, H.A.: Performing Bayesian inference by weighted model counting. In: AAAI, pp. 475–482. AAAI Press/The MIT Press (2005)

34. Shukla, A., Biere, A., Pulina, L., Seidl, M.: A survey on applications of quantified Boolean formulas. In: ICTAI, pp. 78–84. IEEE (2019)

35. Stockmeyer, L.J., Meyer, A.R.: Word problems requiring exponential time: preliminary report. In: STOC, pp. 1–9. ACM (1973)

36. Thimm, M., Wallner, J.P.: On the complexity of inconsistency measurement. Artif. Intell. **275**, 411–456 (2019)

37. Zhang, L., Malik, S.: Towards a symmetric treatment of satisfaction and conflicts in quantified Boolean formula evaluation. In: Van Hentenryck, P. (ed.) CP 2002. LNCS, vol. 2470, pp. 200–215. Springer, Heidelberg (2002). https://doi.org/10.1007/3-540-46135-3_14

38. Zhou, Z., Qian, Z., Reiter, M.K., Zhang, Y.: Static evaluation of noninterference using approximate model counting. In: SP, pp. 514–528. IEEE Computer Society (2018)

Computer-Aided Teaching

Experiments with Automated Reasoning in the Class

Isabela Drămnesc[1]([✉]) [ID], Erika Ábrahám[2] [ID], Tudor Jebelean[3] [ID],
Gábor Kusper[4] [ID], and Sorin Stratulat[5] [ID]

[1] West University, Timisoara, Romania
isabela.dramnesc@e-uvt.ro
[2] RWTH Aachen University, Aachen, Germany
abraham@cs.rwth-aachen.de
[3] Johannes Kepler University, Linz, Austria
Tudor.Jebelean@jku.at
[4] Eszterhazy Karoly Catholic University, Eger, Hungary
gkusper@aries.ektf.hu
[5] Université de Lorraine, CNRS, LORIA, 57000 Metz, France
sorin.stratulat@univ-lorraine.fr

Abstract. The European *Erasmus+* project *ARC – Automated Reasoning in the Class* aims at improving the academic education in disciplines related to Computational Logic by using Automated Reasoning tools. We present the technical aspects of the tools as well as our education experiments, which took place mostly in virtual lectures due to the COVID pandemics. Our education goals are: to support the virtual interaction between teacher and students in the absence of the blackboard, to explain the basic Computational Logic algorithms, to study their implementation in certain programming environments, to reveal the main relationships between logic and programming, and to develop the proof skills of the students. For the introductory lectures we use some programs in C and in Mathematica in order to illustrate normal forms, resolution, and DPLL (Davis-Putnam-Logemann-Loveland) with its Chaff version, as well as an implementation of sequent calculus in the Theorema system. Furthermore we developed special tools for SAT (propositional satisfiability), some based on the original methods from the partners, including complex tools for SMT (Satisfiability Modulo Theories) that allow the illustration of various solving approaches. An SMT related approach is natural-style proving in Elementary Analysis, for which we developed and interesting set of practical heuristics. For more advanced lectures on rewrite systems we use the Coq programming and proving environment, in order on one hand to demonstrate programming in functional style and on the other hand to prove properties of programs. Other advanced approaches used in some lectures are the deduction based synthesis of algorithms and the techniques for program transformation.

Keywords: Automated reasoning · Computational logic · Computer aided teaching

ⓒ The Author(s), under exclusive license to Springer Nature Switzerland AG 2022
K. Buzzard and T. Kutsia (Eds.): CICM 2022, LNAI 13467, pp. 287–304, 2022.
https://doi.org/10.1007/978-3-031-16681-5_20

1   Introduction

The international Erasmus+ European Project *ARC - Automated Reasoning in the Class*, running from 2019 to 2022, is a strategic partnership between 5 prestigious universities from Austria, France, Germany, Hungary, and Romania. This consortium of universities represented by the authors is developing the ARC project, which aims at improving the academic education in disciplines related to Computational Logic by using Automated Reasoning tools.

The activities of the project include: the development of the intellectual output as the ARC book that consists in advanced and motivating teaching and learning material dedicated to Computational Logic, five international trainings for academic staff (one at each partner institution), training of more than 5000 students over 8 semesters, an international summer school for students to learn important notions from the ARC book and how to use the developed tools, as well as an international symposium on ARC for academic staff outside of the partnership for disseminating the final results of the project.

The ARC book contains course material that corresponds to the actual and planned lectures at the partner institutions, accompanied and synchronized with interactive exercises and demonstrations using the software tools developed by the partners especially for the purpose of this book, as well as best-practice recommendations to run the lectures based on advanced pedagogical principles as Problem Based Learning [18]. The students addressed are at all levels: bachelor, master, and PhD. In general there are no lecture prerequisites, when necessary the advanced lectures start with an introduction to the needed notions and algorithms.

The purpose of this paper is to present the most important technical aspects of the tools as well as some aspects of our education experiments, which are also somewhat different from the usual ones because they took place in the period of the COVID pandemics, thus mostly in virtual lectures. The tools are used for several education goals:

- to support the interaction between the teacher and the students in a virtual environment, in particular in the absence of the blackboard or other physical devices;
- to illustrate certain computational logic algorithms that are implemented by some of these tools;
- to demonstrate the implementation principles of mathematical logic notions and methods in certain programming environments;
- to help students understand various relationships between logic and programming, in particular the logical basis of computation, program verification, and algorithm synthesis;
- to develop the proof skills of the students and to reveal the relationship between human proving and automated reasoning.

For the theoretical introductory lectures related to the basics of Mathematical Logic we use some programs in C and in Mathematica [50] in order to illustrate basic operations like: simple representation and parsing of formulae, transformation into normal form, resolution, and DPLL (Davis-Putnam-Logemann-Loveland [15]), in particular with an animated graphical interface for explaining the Chaff [35] version of the DPLL algorithm.

For illustrating proof methods in sequent calculus [8] we present a tutorial implementation of this method in the Theorema system [7,45,49], including and original version of sequent calculus that uses unit propagation.

We developed special tools for SAT (propositional SATisfiability [29]), some based on the original methods from some of the partners. The name of the new SAT solver is CSFLOC, which states for Count Subsumed Full-Length Clauses [34]. This solver is an iterative version of the inclusion-exclusion principle, which is used to solve the #SAT problem (that is the problem of counting the solutions of a SAT instance [6]). Thus, unusually, we use idea from the field of #SAT to solve SAT, that is we confront the students with a novel specific approach to this problem.

Furthermore, we address SMT (Satisfiability Modulo Theories [5]) by presenting the principles and the usage of two main tools for SMT solving: *Z3* (a state-of-the-art solver used in numerous applications) and SMT-RAT (the solver developed at the partner institution in Aachen), with the latter having the advantage of being able to present and discuss in detail the most important implementation issues.

An SMT related approach is natural-style proving in Elementary Analysis (the so called ϵ δ proofs for statements with alternating quantifiers), for which we developed and interesting set of heuristics that are very useful in practice, both for training the natural-style human proving, as well as for proof automation.

For more advanced lectures on rewrite systems we used the Coq [44] programming and proving environment, in order on one hand to design programs in functional style and on the other hand to prove their properties. Users interact with Coq via a web interface that mixes lecture content with Coq code that does not require the local installation of the Coq distribution.

Another advanced approach that is used in some of the lectures is the deduction based synthesis of algorithms and special programming transformation techniques that increase the efficiency of algorithms by transforming logical based (pattern matching) programs into functional ones and even into tail recursive and subsequently into loop-based imperative programs.

2 Tools

The tools developed for the project are available online [46] to all partners and to any other interested institutions.

The tools presented in Sects. 2.1 to 2.2 have been used in teaching mainly for the lectures *Automated Theorem Proving*, *Mathematical Logic*, *Automated Reasoning*, *Computational Logic*, *Algorithm Synthesis*, and *Mathematical Theory Exploration*.

2.1   Automated Reasoning in Theorema

The Theorema system is a software system based on Mathematica which offers
a flexible and intuitive user interface, allows the construction of theories, the
development of automated and interactive proofs in natural style (that are similar to human proving), the development of provers specific to several domains,
as well as the direct execution of algorithms that are synthesized by proving.

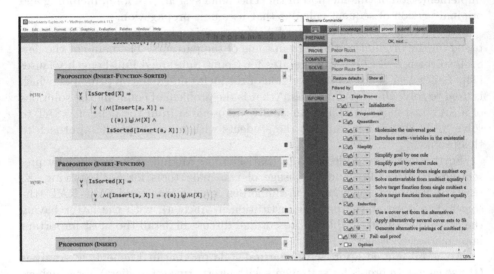

Fig. 1. Theorema notebook and theorema commander

In Fig. 1 we can see two windows: the left hand side represents the Theorema
notebook where the user introduces the knowledge base (formulae as propositions, conjectures, definitions, algorithms), and the right hand side is the Theorema Commander in Prove mode, from where the user can choose the prover to
be applied, can select/deselect inference rules, can change the priority of inference rules, can set the proof depth and time, can choose how to generate the
proof (automatically or interactively), can choose to show the simplified version
of the proof (that includes only the proved branches), etc.

The proofs in Theorema are developed as proof trees of proof situations
(assumptions and one goal). The proof is extended at every transformation of
a proof situation into other proof situation(s) by applying a certain inference
rule. The generated proof is shown in a separate window displaying the proof
information and in the commander it is displayed the generated proof tree as
illustrated in Fig. 2.

The user has the possibility of computing with definitions and algorithms.
Figure 3 illustrates some computations with an algorithm that has been previously synthesized by proving.

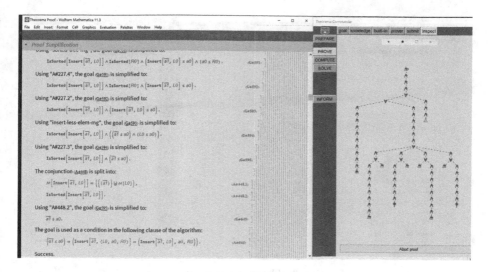

Fig. 2. Theorema proof and the generated proof tree

As a support for teaching proof methods in sequent calculus we developed a Theorema prover that implements sequent calculus, including an original version that uses unit propagation. This prover is used on one hand for presenting automated proofs in natural style that are based on sequent calculus, and on the other hand to study the implementation of such provers in Theorema, namely it shows that, in essence, one only has to specify the inference rules as rewriting steps.

2.2 Synthesis and Transformation of Algorithms

The Algorithm Synthesis Problem consists in: starting from the specification of a problem, given as a pair of input and output conditions, how can we automatically discover an algorithm which satisfies the specification? The concern of algorithm synthesis is to develop methods and tools for mechanizing and automatizing (parts of) the process of finding an algorithm that satisfies a given specification.

As a support for teaching subjects related to algorithm synthesis by proving and mathematical theory exploration we developed a prover [19,20] in the Theorema system that synthesizes a large number of algorithms operating on lists and on binary trees, including sorting algorithms as rewrite programs. A prover in Theorema consists of a collection of inference rules (as rewrite rules). In these related lectures we discuss: the use of multisets in the problem specification (as an easier way to express that two objects have the same elements), we analyse the inference rules, we study the use of several techniques for synthesizing algorithms, the use of different knowledge base (different techniques and different knowledge base can lead to the synthesis of different algorithms), the principles of extracting an algorithm from a proof, as well as the principles of systematic

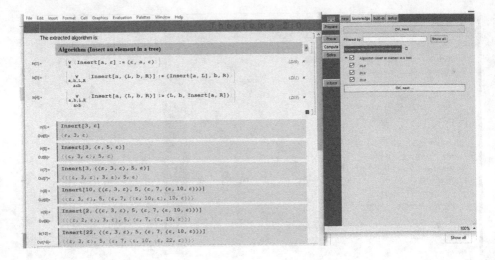

Fig. 3. Theorema compute

exploration of theories. In Fig. 1 we can see on the LHS the specification of the function *Insert* that inserts at the appropriate place an element into a sorted list and on the RHS the list of the prover inference rules. In Fig. 2 we can see a part of the synthesizing proof for this algorithm (LHS) and the structure of the proof tree (RHS). The computation with the synthesized algorithm is presented in Fig. 3.

As a support for teaching subjects related to algorithm transformations, we developed a Theorema tool [21] that transforms recursive algorithms into more efficient ones, which are then transformed into tail recursive, then into functional programs, and then into their corresponding imperative version. In these lectures we expose the systematic principles for transforming an algorithm into the tail recursive version, how to improve the efficiency using a special flag for avoiding the unnecessary recursion, and then how to transform the algorithm into its functional and iterative version.

2.3 Sample Implementations

Computational Logic in C. This teaching experiment presents the design and the implementation in C language of an algorithm that transforms propositional formulae into Negation Normal Form (NNF – only conjunctions and disjunctions, with negation occurring only inside literals). Furthermore the program generates commands for a TeX interface that displays the formula tree and its evolution. This is in fact a starting point for the implementation and visualization of various rewriting and proving methods, as it demonstrates:

- the reading and writing of propositional formulae in reversed Polish notation,
- the tree representation of formulae,
- traversal of the tree in recursive but also in imperative way (for efficiency),

- the efficient transformation into NNF in linear time by destructive update of the formula tree.

The program also generates TeX commands that are used to generate the sequence of pages for the visualization of the algorithm effect.

Mathematical Logic in Mathematica. The Mathematica Computer Algebra system has a programming language that is based on logical rewriting of terms using a powerful pattern matching concept, therefore one can use it for demonstrating the possibility of programming by using essentially logical formulae. Its usage in the class has a double advantage: on one hand the implementation is relatively easy because low level functions like parsing, term matching, term rewriting, etc. are already available, and on the other hand most of the program text is very similar to logical formulae.

For the lectures related to our project we developed various functions implementing: truth table computation (interactive and also automatic), formula rewriting into Negation Normal Form (NNF), Conjunctive Normal Form (CNF), and Disjunctive Normal Form (DNF), the DPLL method (see the definition in Subsect. 2.5), and finally the Chaff implementation of DPLL [35] – see Fig. 4.

On one hand the students have been able to use these demo programs in order to check practically the properties studied in the theoretical part of the lecture, and on the other hand they could understand the implementations and also the close relationship between the logic of term rewriting and programming.

Proof Heuristics in Elementary Analysis. We developed several heuristic techniques for such proofs:

- the S-decomposition method for formulae with alternating quantifiers,
- quantifier elimination by Cylindrical Algebraic Decomposition,
- analysis of terms behavior in zero,
- bounding the ϵ-bounds,
- semantic simplification of expressions involving absolute value,
- polynomial arithmetic and solving,
- usage of equal arguments to arbitrary functions,
- reordering of proof steps in order to insure the admissibility of solutions to meta-variables.

Some of these combine logic with domain specific methods (which is basically equivalent to SMT solving). Moreover, in our teaching context they are very useful for producing proofs that can be easily understood by human readers. These techniques allow to produce natural-style proofs for many interesting examples [27], like convergence of sum and product of sequences, continuity of sum, product and compositions of functions, etc. that have been discussed with the students.

We presented these heuristics in the class, both in lectures related to automated reasoning, but also in lectures dedicated to the training of students in formal mathematical techniques (formalization of mathematical notions, formal proving, etc.).

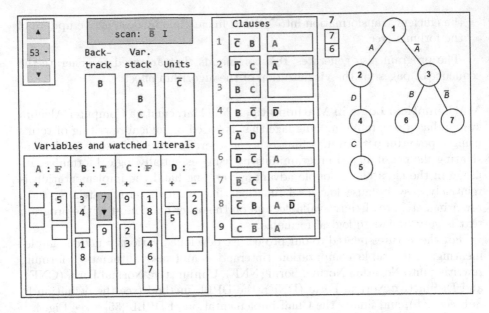

Fig. 4. Visualization of the Chaff implementation of DPLL.

2.4   The Coq Theorem Prover

The Coq proof assistant can be seen as a combination of:

1. a simple but extremely expressive, programming language, and
2. a set of tools for stating logical assertions (including assertions about the behaviour of programs) and giving evidence of their truth.

Started in 1984 by Thierry Coquand and Gérard Huet, it has been extended in 1991 with inductive types by Christine Paulin. Since then, more than 40 people contributed to Coq. It has been used to certify non-trivial theorems, as the 'four colour' and 'odd order' theorems, and applications, as a C compiler. In 2013, it received the prestigious ACM Software System award.

We developed a Coq tutorial based on the material presented at an INRIA summer school [26]. It consists in the following parts, also publicly accessible [11]:

1. Programming with natural numbers and lists (lecture, exercises)
2. Propositions and predicates (lecture, exercises)
3. Making proofs in Coq (lecture, exercises)
4. Proofs about programs (lecture, exercises)
5. Inductive data types (lecture, exercises)
6. Inductive properties I (lecture, exercises)
7. Inductive properties II (lecture, exercises)
8. Recursive functions in Coq (lecture, exercises)

The links above allow to access the material interactively, by the means of *JsCoq* [2], a Jupyter Notebook interface for Coq.

The JsCoq interface is split in two, with its left (Fig. 5) and right (Fig. 6) sides.

Lecture 5: Inductive data types

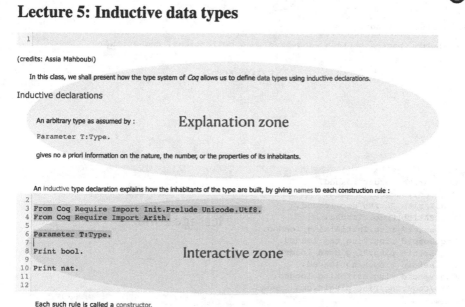

(credits: Assia Mahboubi)

In this class, we shall present how the type system of *Coq* allows us to define data types using inductive declarations.

Inductive declarations

An arbitrary type as assumed by :   **Explanation zone**

```
Parameter T:Type.
```

gives no a priori information on the nature, the number, or the properties of its inhabitants.

An inductive type declaration explains how the inhabitants of the type are built, by giving names to each construction rule :

```
From Coq Require Import Init.Prelude Unicode.Utf8.
From Coq Require Import Arith.

Parameter T:Type.

Print bool.             Interactive zone

Print nat.
```

Each such rule is called a constructor.

Fig. 5. The left side.

One can distinguish the following zones. On the left hand side:

– The *explanation* zone. It contains the explanations given by the lecturer.
– The *interactive zone*. The user can execute the Coq code interactively. Its content can be modified by the user.

On the right hand side:

– The *navigation* zone. The user can navigate through the Coq code using the *up* and *down* arrows. The executed code is grayed in the navigation zone. The user can also execute the code up to the current position of the cursor in the navigation zone by clicking on the 'face-to-face arrows' button. The *cross* button stops the execution of a Coq command, which is useful when it takes too much time. The Coq session can be reset with the rightmost button.
– The *status* zone. It displays the current status of the prover as well as the current goals during the proof sessions.

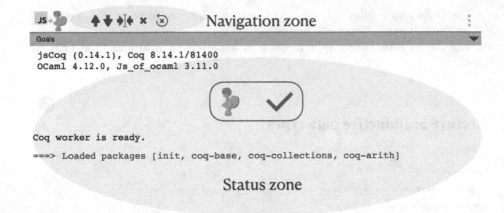

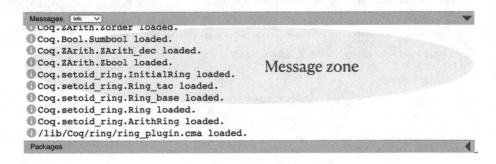

Fig. 6. The right side.

– The *message* zone. It shows the messages given by the Coq prover while
  executing code from the interactive zone.

The tutorial has been taught remotely to Master students in Computer Science at the West University of Timisoara. They appreciated that Coq scripts are mixed with the related explanations, as well as the way to directly change the Coq scripts, execute them and view the results in the same spot. On the other hand, the web interface does not allow yet to save the current state of the scripts in one step; alternatively, the users can manually copy the blocks of Coq script one by one.

2.5   SAT Solvers

In order to keep close contact with the research and with the industrial community, we used in our practical exercises a very popular file format for SAT solvers, namely the DIMACS format [16]. It represents a boolean variable by its

1-based index. A positive number corresponds to a positive literal, a negative number to a negative literal. Each line is a clause, and each line is terminated by a "0".

A DIMACS CNF file must contain a problem line which starts by *p cnf* which tells the number of variables and the number of clauses information:

p cnf #variables #clauses

It can contain also comments, which start by the letter "c". An example for DIMACS CNF file:

```
c This is a SAT problem with 3 variables and 5 clauses.
c It is unsatisfiable.
p cnf 3 5
-1 2 0
-2 3 0
1 -3 0
1 2 3 0
-1 -2 -3 0
```

SAT solvers are based on the Davis-Putnam-Logemann-Loveland (DPLL) [15] algorithm. This algorithm performs Boolean Constraint Propagation (BCP) and backtrack search, i.e., at each node of the search tree it selects a decision variable and assigns a truth value to it, then steps back when a conflict occurs.

Conflict-driven clause learning (CDCL), see Chap. 4 in [6], is based on the idea that conflicts can be exploited to reduce the search space. If the method finds a conflict, then it analyzes this situation, determines a sufficient condition for this conflict to occur, in form of a learned clause, which is then added to the formula, and thus avoids that the same conflict occurs again.

Besides clause learning, lazy data structures are one of the key techniques for the success of SAT solvers, such as "watched literals" as pioneered in 2001, by the CDCL solver Chaff [35].

Since multi-core architectures are common today, the need for parallel SAT solvers using multiple cores has increased considerably.

In essence, there are two approaches to parallel SAT solving. The first group of solvers typically follow a divide-and-conquer approach. They split the search space into several subproblems, sequential workers solve the subproblems. This first group uses relatively intensive communication between the nodes. They do for example load balancing and dynamic sharing of learned clauses.

The second group apply portfolio-based SAT solving. The idea is to run independent sequential SAT solvers with different restart policies, branching heuristics, learning heuristics.

In case of the SAT problem we can speak about the so called threshold phenomenon, or phase transition phenomenon, which is the following observation. If we have a uniform random 3-SAT problem (i.e. each clause contains 3 literals) with n boolean variables and m clauses, then around the threshold $m/n = 4.27$ there are difficult problems for the DPLL algorithm based SAT solvers [23]. Under-constraint problems (where $m/n << 4.27$) are usually very easy for SAT solvers. Again, over-constraint problems (where $m/n >> 4.27$) are in general very easy for them. But there is a subclass of over-constraint problems which is very difficult. This is the class of those problems which are over-constraint and minimal unsatisfiable (UNSAT) at the same time. A SAT problem is minimal UNSAT if and only if all of its clauses are needed in the proof that the problem is UNSAT.

For the lectures related to Computational Logic we have developed a SAT solver, called CSFLOC [34], and some problem generators which can generate minimal UNSAT problems [33].

In the lecture *Applied Logic* we studied several SAT solvers, several problem libraries, and even implemented the well-known algorithms using Python, which is a rapid prototyping programming language.

2.6   SMT Solving

Another class of tools for automated reasoning are *Satisfiability Modulo Theories (SMT) solvers* [5,32]. Also these tools aim at checking the satisfiability of logical formulas in a fully automated manner [6], but in contrast to SAT solvers, they are devoted to formulas from (mostly quantifier-free) first-order logic over different theories. Starting with easier theories (e.g. equality logic with uninterpreted functions) and theories for program verification (e.g. array theory, bit-vector arithmetic), impressive developments have been made also for quantifier-free linear and non-linear arithmetic over the reals and the integers.

SMT solving enjoys an active research community with dedicated conferences [39,42], a SatLive forum [43], and SMT solver competitions [38] between numerous SMT solvers that focus on different theories, algorithms and application areas. One of the main achievements of the community is the SMT-LIB standard input language [3], which is supported by most state-of-the-art SMT solvers. Thus when a problem is encoded in this input language, most SMT solvers (that support the theories used) can be directly fed with it. Using this standard, a large benchmark library [3] has been collected and is publicly available for tool developers and can be used for competitions. Besides command-line SMT-LIB input, some solvers come with APIs for different programming languages (typically C++ and/or Python) that can be used to specify and solve problems.

For educational purposes, besides giving insights into the theoretical backgrounds, in our project we focussed on two tools: Z3 and SMT-RAT. Both support the SMT-LIB standard and offer a C++ API, whereas Z3 has also an API for Python. We use Z3 to train the *usage* of SMT solvers and SMT-RAT to train *tool development.*

Z3 [36] is one of the most popular, most efficient and easy-to-use SMT solvers that support a large number of different theories. Therefore, Z3 is well suited to train the *usage* of SMT solvers. The tool is accessible under [53]. Extensive documentation and numerous examples are provided, which makes it easy to construct a program for the training.

Firstly, in our project we use Z3 to train how to encode and solve problems using the SMT-LIB standard. The rich variety of Z3-supported theories allows to encode and efficiently solve problems from numerous domains [4]. We used these and other examples from [51] to design encoding exercises.

Secondly, we use the Python API [54] that allows an easy employment of Z3 directly from Python programs. Concrete executable examples are provided under [52].

SMT-RAT (SMT Real-Arithmetic Toolbox) [13,14] is an SMT solver with a focus on solving quantifier-free real algebraic problems. Though competitive, the main aim of SMT-RAT is not to be the fastest, but to offer an optimal platform for *implementing* new SMT solving ideas.

To provide optimal support for extensions, SMT-RAT comes with a solid basic support for arithmetic computations and a clear modular structure. SMT-RAT modules implement different decision procedures that can be connected by user-defined strategies: each module tries to solve its input problem and might consult its backends by delegating sub-problems to them. This way we can e.g. use fast but incomplete procedures and assure completeness by providing a complete backend to them.

The latest release [37] offers, besides some non-arithmetic components, the CaRL [9] library for arithmetic datatypes and basic computations with them, and solver modules for e.g. the simplex and the Fourier-Motzkin variable elimination methods for linear problems, for non-linear real arithmetic different adaptions of the cylindrical algebraic decomposition method [10,30], the virtual substitution method [1,12,48], subtropical satisfiability [22], methods based on Gröbner bases [28,47], as well as interval constraint propagation [24,25] and the branch-and-bound technique [17,31] for finding integer solutions.

Besides installation guide, the SMT-RAT documentation [40] also describes built-in support for benchmarking, debugging and testing.

Teaching SMT Solving. Regarding theory, we developed a lecture named *Satisfiability Checking* to convey algorithmic aspects of SMT techniques to students. The lecture covers SAT solving, eager SMT solving for equality logic and uninterpreted functions as well as for bitvector arithmetic, lazy SMT solving as a framework, and dedicated decision procedures for linear and nonlinear real and integer arithmetic theories. We discuss SMT-adaptions of the Gauss and Fourier-Motzkin variable elimination, the simplex algorithm and the branch-and-bound method for linear arithmetic, and the methods of interval constraint propagation, subtropical satisfiability, virtual substitution and cylindrical algebraic decomposition for non-linear real arithmetic.

To support learning, we developed automatic exercise generators for different algorithms, embedded in a GUI for easy usage. The development of such a training tool is challenging, as the generated exercises should satisfy different quality criteria. For example:

- For each trained algorithm, a large number of exercises of comparable difficulty and time effort should be generated.
- A correct answer should be unique and it should be a good indicator that the student understood the trained algorithm. Numerical answers should be syntactically short and easy to type on the keyboard. For multiple choice answers, the probability to correctly guess the answer should be small.
- Offer different difficulty levels, tips, explanations and a full solution.

Regarding practical aspects and implementation, we made extensive experience using SMT-RAT in teaching, in practical courses as well as in more than 50 Bachelor and Master theses [41] for implementing novel ideas for decision procedures.

In general, we observed that good students are very much attracted to SMT solving, as topics in this area are theoretically challenging, but also practical aspects play an important role for efficient implementations. Several of the students who wrote their thesis in SMT solving have been enrolled in PhD studies afterwards.

3   Conclusion

The lectures from the partner universities that are addressed by the project are: *Mathematical Logic, Automated Reasoning, Computational Logic, Automated Theorem Proving, Techniques for Scientific Work, Functional Programming, Logic Programming, Algorithm Synthesis and Mathematical Theory Exploration, Artificial Intelligence, Satisfiability Checking, Applied Logic, Modeling and Verifying Algorithms in Coq, Satisfiability Modulo Theories.* These took place since the Winter Semester 2019 (thus for 7 semesters) with a total of more than 5000 students. Starting with Summer Semester 2020 the lectures have been gradually switched to online mode. On one hand this created the expected difficulties in the interaction with the students, on the other hand this stimulated the creation of specific approaches, reading material, and tools for ensuring the quality of the lectures and for motivating the students. As now we started to switch back to presence lectures, we can notice that the experience and the material created during the pandemics is still very useful. A powerful help for improving the teaching is constituted by the various aspects of cooperation between the project partners: besides the possibility to cross-use some material and tools, we also created specific environments in which the students can access computational resources and licensed software remotely at the other partners.

As a general conclusion we can say that the tools and the practical material developed in this project had a significant positive impact on the teaching process, with visible effects on the quality of the student achievements. In particular, in the context of the pandemics, because of the downgrading of the communication between teachers and students, the lack of these tools would have probably lead to a decrease of student motivation, participation, and learning. Evaluation of the project results has been pursued continuously during the project life, including evaluation from the point of view of the students involved. In Fig. 7

we present the student answers (percentage) to some of the most relevant issues about the lectures.

Issue	Very good	Good	Neutral
Interesting	26	48	19
Well defined goals	35	50	11
Well structured	43	42	11
Helpful materials	38	42	16
Helpful examples	40	45	11
Helps professional development	21	40	25
Teaching methods	22	47	20

Fig. 7. Some evaluation results (percentage of student answers

Further work on the objectives of this project is, among other reasons, strongly motivated by the fact that we come now back to in–presence lectures, and also with a strong uncertainty about possible fall back to virtual mode. Thus, on one hand we can now adapt, test, and evaluate the methods that have been used virtually in order to use them in a classical environment, and on the other hand we have to keep improving the virtual techniques in order to be prepared for future pandemic waves.

Acknowledgements. This work is co-funded by the Erasmus+ Programme of the European Union, project ARC: Automated Reasoning in the Class, 2019-1-RO01-KA203-063943.

References

1. Ábrahám, E., Nalbach, J., Kremer, G.: Embedding the virtual substitution method in the model constructing satisfiability calculus framework. In: proceedings of SC-square'17. CEUR Workshop Proceedings, vol. 1974. CEUR-WS.org (2017). urn:nbn:de:0074–1974-4, http://ceur-ws.org/Vol-1974/EAb.pdf
2. Arias, E.J.G., Pin, B., Jouvelot, P.: jsCoq: Towards hybrid theorem proving interfaces. In: Proceedings of the 12th Workshop on User Interfaces for Theorem Provers, UITP. EPTCS, vol. 239, pp. 15–27 (2016)
3. Barrett, C., Fontaine, P., Tinelli, C.: The satisfiability modulo theories library (SMT-LIB). www.SMT-LIB.org (2016)
4. Barrett, C., Kroening, D., Melham, T.: Problem solving for the 21st century: Efficient solvers for satisfiability modulo theories. Knowledge Transfer Report 3, London Mathematical Society and Smith Institute for Industrial Mathematics and System Engineering (2014)
5. Barrett, C., Sebastiani, R., Seshia, S.A., Tinelli, C.: Satisfiability modulo theories. In: Handbook of Satisfiability, Frontiers in Artificial Intelligence and Applications, chap. 26, vol. 185, pp. 825–885. IOS Press (2009)
6. Biere, A., Heule, M., van Maaren, H., Walsh, T.: Handbook of Satisfiability, Frontiers in Artificial Intelligence and Applications, vol. 185. IOS Press (2009)

7. Buchberger, B., Jebelean, T., Kutsia, T., Maletzky, A., Windsteiger, W.: Theorema 2.0: computer-assisted natural-style mathematics. J. Formalized Reasoning **9**(1), 149–185 (2016). https://doi.org/10.6092/issn.1972-5787/4568
8. Buss, S.R.: An introduction to proof theory. In: Handbook of Proof Theory, pp. 31–35. Elsevier (1998)
9. CArL: Project homepage. https://github.com/smtrat/carl
10. Collins, G.E.: Quantifier elimination for real closed fields by cylindrical algebraic decompostion. In: Brakhage, H. (ed.) GI-Fachtagung 1975. LNCS, vol. 33, pp. 134–183. Springer, Heidelberg (1975). https://doi.org/10.1007/3-540-07407-4_17
11. https://members.loria.fr/SStratulat/files/MVA/index.html
12. Corzilius, F.: Integrating virtual substitution into strategic SMT solving. Ph.D. thesis, RWTH Aachen University, Germany (2016). http://publications.rwth-aachen.de/record/688379
13. Corzilius, F., Kremer, G., Junges, S., Schupp, S., Ábrahám, E.: `SMT-RAT`: an open source `C++` toolbox for strategic and parallel SMT solving. In: Heule, M., Weaver, S. (eds.) SAT 2015. LNCS, vol. 9340, pp. 360–368. Springer, Cham (2015). https://doi.org/10.1007/978-3-319-24318-4_26
14. Corzilius, F., Loup, U., Junges, S., Ábrahám, E.: SMT-RAT: an SMT-compliant nonlinear real arithmetic toolbox. In: Cimatti, A., Sebastiani, R. (eds.) SAT 2012. LNCS, vol. 7317, pp. 442–448. Springer, Heidelberg (2012). https://doi.org/10.1007/978-3-642-31612-8_35
15. Davis, M., Logemann, G., Loveland, D.: A machine program for theorem-proving. Commun. ACM **5**(7), 394–397 (1962). https://doi.org/10.1145/368273.368557
16. http://www.domagoj-babic.com/uploads/ResearchProjects/Spear/dimacs-cnf.pdf
17. Doig, A.G., Land, B.H., Doig, A.G.: An automatic method for solving discrete programming problems. Econometrica **28**, 497–520 (1960)
18. Dolmans, D.H.J.M., De Grave, W., Wolfhagen, I.H.A.P., Van Der Vleuten, C.P.M.: Problem-based learning: future challenges for educational practice and research. Mediac Educ. **39**(7), 732–741 (2005). https://doi.org/10.1111/j.1365-2929.2005.02205.x
19. Dramnesc, I., Jebelean, T.: Synthesis of sorting algorithms using multisets in Theorema. J. Logical Algebraic Methods Program. **119**, 100635 (2020). https://doi.org/10.1016/j.jlamp.2020.100635
20. Dramnesc, I., Jebelean, T., Stratulat, S.: Mechanical synthesis of sorting algorithms for binary trees by logic and combinatorial techniques. J. Symb. Comput. **90**, 3–41 (2019). https://doi.org/10.1016/j.jsc.2018.04.002
21. Dramnesc, I., Jebelean, T.: Implementation of deletion algorithms on lists and binary trees in Theorema. RISC Report Series 20–04, Research Institute for Symbolic Computation, Johannes Kepler University Linz (2020)
22. Fontaine, P., Ogawa, M., Sturm, T., Vu, X.T.: Subtropical satisfiability. In: Dixon, C., Finger, M. (eds.) FroCoS 2017. LNCS (LNAI), vol. 10483, pp. 189–206. Springer, Cham (2017). https://doi.org/10.1007/978-3-319-66167-4_11
23. Friedgut, E.: Appendix by Jean bourgain: sharp thresholds of graph properties, and the k-sat problem. J. Am. Math. Soc. **12**, 1017–1054 (1999)
24. Gao, S., Ganai, M., Ivančić, F., Gupta, A., Sankaranarayanan, S., Clarke, E.M.: Integrating ICP and LRA solvers for deciding nonlinear real arithmetic problems. In: Proceedings of FMCAD'10, pp. 81–90. IEEE (2010)
25. Herbort, S., Ratz, D.: Improving the efficiency of a nonlinear-system-solver using a componentwise Newton method. Technical report 2/1997, Inst. für Angewandte Mathematik, University of Karlsruhe (1997)

26. CEA-EDF-INRIA summer school, INRIA Paris-Rocquencourt, Antenne Parisienne, 2011. https://fzn.fr/teaching/coq/ecole11/
27. Jebelean, T.: A Heuristic Prover for Elementary Analysis in *Theorema*. In: Kamareddine, F., Sacerdoti Coen, C. (eds.) CICM 2021. LNCS (LNAI), vol. 12833, pp. 130–134. Springer, Cham (2021). https://doi.org/10.1007/978-3-030-81097-9_10
28. Junges, S., Loup, U., Corzilius, F., Ábrahám, E.: On Gröbner bases in the context of satisfiability-modulo-theories solving over the real numbers. In: Muntean, T., Poulakis, D., Rolland, R. (eds.) CAI 2013. LNCS, vol. 8080, pp. 186–198. Springer, Heidelberg (2013). https://doi.org/10.1007/978-3-642-40663-8_18
29. Knuth, D.E.: The Art of Computer Programming: Satisfiability, Volume 4, Fascicle 6. Addison-Wesley Professional (2015)
30. Kremer, G., Abraham, E.: Fully incremental cylindrical algebraic decomposition. J. Symb. Comput. **100**, 11–37 (2020). https://doi.org/10.1016/j.jsc.2019.07.018
31. Kremer, G., Corzilius, F., Ábrahám, E.: A generalised branch-and-bound approach and its application in SAT modulo nonlinear integer arithmetic. In: Gerdt, V.P., Koepf, W., Seiler, W.M., Vorozhtsov, E.V. (eds.) CASC 2016. LNCS, vol. 9890, pp. 315–335. Springer, Cham (2016). https://doi.org/10.1007/978-3-319-45641-6_21
32. Kroening, D., Strichman, O.: Decision Procedures: An Algorithmic Point of View. Springer, Heidelberg (2008). https://doi.org/10.1007/978-3-540-74105-3
33. Kusper, G., Balla, T., Biró, C., Tajti, T., Yang, Z.G., Baják, I.: Generating minimal unsatisfiable SAT instances from strong digraphs. In: SYNASC 2020, pp. 84–92. IEEE Computer Society Press (2020). https://doi.org/10.1109/SYNASC51798.2020.00024
34. Kusper, G., Biró, C., Iszály, G.B.: SAT solving by CSFLOC, the next generation of full-length clause counting algorithms. In: 2018 IEEE International Conference on Future IoT Technologies, pp. 1–9. IEEE Computer Society Press (2018). https://doi.org/10.1109/FIOT.2018.8325589
35. Moskewicz, M.W., Madigan, C.F., Zhao, Y., Zhang, L., Malik, S.: Chaff: Engineering an efficient SAT solver. In: Proceedings of the 38th Annual Design Automation Conference, pp. 530–535. Association for Computing Machinery (2001). https://doi.org/10.1145/378239.379017
36. de Moura, L., Bjørner, N.: Z3: an efficient SMT solver. In: Ramakrishnan, C.R., Rehof, J. (eds.) TACAS 2008. LNCS, vol. 4963, pp. 337–340. Springer, Heidelberg (2008). https://doi.org/10.1007/978-3-540-78800-3_24
37. https://github.com/ths-rwth/smtrat/releases
38. https://smt-comp.github.io/
39. https://smt-workshop.cs.uiowa.edu/
40. https://ths-rwth.github.io/smtrat/
41. https://ths.rwth-aachen.de/theses/
42. http://www.satisfiability.org/
43. http://www.satlive.org/smt.html
44. The Coq development team: The Coq Reference Manual. INRIA (2020)
45. https://www.risc.jku.at/research/theorema/software
46. https://arc.info.uvt.ro/?page_id=49
47. Weispfenning, V.: A new approach to quantifier elimination for real algebra. In: Caviness, B.F., Johnson, J.R. (eds.) Quantifier Elimination and Cylindrical Algebraic Decomposition. Texts and Monographs in Symbolic Computation Texts and Monographs in Symbolic Computation, pp. 376–392. Springer, Vienna (1998). https://doi.org/10.1007/978-3-7091-9459-1_20

48. Weispfenning, V.: Quantifier elimination for real algebra - the quadratic case and beyond. Appl. Algebra Eng. Commun. Comput. **8**(2), 85–101 (1997). https://doi.org/10.1007/s002000050055
49. Windsteiger, W.: Theorema 2.0: a system for mathematical theory exploration. In: Hong, H., Yap, C. (eds.) ICMS 2014. LNCS, vol. 8592, pp. 49–52. Springer, Heidelberg (2014). https://doi.org/10.1007/978-3-662-44199-2_9
50. Wolfram Research Inc.: Mathematica, Version 13.0.0. https://www.wolfram.com/mathematica
51. https://ericpony.github.io/z3py-tutorial/guide-examples.htm
52. https://github.com/Z3Prover/doc/tree/master/programmingz3/code
53. https://github.com/z3prover/z3
54. https://theory.stanford.edu/nikolaj/programmingz3.html

Learning to Reason Assisted by Automated Reasoning

Wolfgang Windsteiger[✉] [iD]

Research Institute for Symbolic Computation (RISC), Johannes Kepler University
Linz (JKU), Altenbergerstraße 69, 4040 Linz, Austria
Wolfgang.Windsteiger@risc.jku.at
https://risc.jku.at/m/wolfgang-windsteiger/

Abstract. We report on using logic software in a novel course-format for
an undergraduate logic course for students in computer science or artificial intelligence. Although being designed as the students' basic introduction to the field of logic, the course features a novel structure and it
adds some modern content, such as SAT and SMT solving, to the traditional and established topics, such as propositional logic and first order
predicate logic. The novel course design is characterized by, among others, the integration of existing logic software into the teaching of logic.

In this paper we focus on the module on first-order predicate logic
and the use of the *Theorema* system as a proof-tutor for the students.
We report on statistical evaluation of data collected over two consecutive
years of teaching this course. On the one hand, we asked for feedback of
students on how helpful they felt the software support was. On the other
hand, we evaluated their results in the exams during the course and their
development over the entire teaching period. The performance in exams
is then correlated with students' own perception of the helpfulness of
software.

Keywords: Theorema · Automated theorem proving · Teaching logic

1 Introduction

The new curriculum (2013) for a bachelor study in computer science at JKU Linz
assigned a prominent role to a new modern logic-course [2,7], whose novelties
concentrate on two different aspects:

- we wanted students to see *logic in action* through the use of logic software in
  teaching and
- we wanted to present *modern topics* (such as e.g. automated or interactive
  theorem proving and satisfiability modulo theories) to the students early in
  their studies in addition to classical content such as propositional and first
  order predicate logic.

K. Buzzard and T. Kutsia (Eds.): CICM 2022, LNAI 13467, pp. 305–320, 2022.
https://doi.org/10.1007/978-3-031-16681-5_21

The course is designed consisting of four modules taught by different lecturers, who also use different software in their respective modules. The course starts with the first module on propositional logic (SAT) and finishes with quantifier-free first-order logic with theories in the fourth module (SMT). First-order predicate logic (FO) occupies the two modules (FOA and FOB) in the middle with FOA covering the language itself and its pragmatics, i.e., how first-order logic appears in everyday life of a computer scientist, e.g., when giving an algorithm specification. In this report we concentrate on module FOB on *proving in first-order predicate logic* and discuss how we use the *Theorema* system [6,8] to support the teaching of reasoning in first-order predicate logic. The FOB module consists of *three units* made up from

- a *lecture part*, where theory is presented,
- an *exercise session* based on weekly exercise sheets with examples,
- a *mandatory quiz*, which is a short exam of 15 min duration, and
- an *optional bonus exercise* based on using some logic software.

Moreover, there is one *optional lab exercise*, which is an extended bonus exercise, for the entire module.[1]

The lecture introduces the concept of proof trees for representing proofs in mathematics and computer science, a concrete set of inference rules for first-order predicate logic, and a proof search procedure that applies rules iteratively. The exercise sheets consist of concrete proof problems, starting with abstract inference rules in order to concentrate on the proof search, over formal proving with quantifiers, until finally ending with the informal natural language presentation of formal proofs. In the frame of the bonus exercises students are asked to generate automated proofs for selected exercises from the current exercise sheet, i.e., they should try to let *Theorema* do proofs that they should be able to do manually at the same time. In order to be admitted to submit solutions to a bonus exercise, students must answer a short questionnaire on their own perception of the usefulness of the software and its influence on their personal proving capabilities. In the lab exercise, in contrast, they should submit a hand-written proof of some simple statement plus an automatically generated proof of the same statement by *Theorema*. Both variants of the proof have to be compared in a short oral presentation.

We collect and evaluate statistical data regarding both the students' personal experience as well as their performance. We are interested in how students perceive the interaction with the *Theorema* system from their point of view and whether their exposure to the system shows a measurable influence on their own proving capabilities in the quizzes. In addition, we investigate how the performance of different groups of students develops over time. We present and compare data collected over two consecutive classes taught in winter 2020 and in winter 2021.

---

[1] Note that, in the statistical evaluation presented later, we neglect the final lab exercise at the end of the module, because there is no item following the lab exercise, in which we could measure some influence of doing the lab exercise or not.

	Week 0	Week 1	Week 2	Week 3	Week 4	Week 5
FOB1	FOB1L FOB1E* FOB1B*	FOB1E†	FOB1B† FOB1Q			L
FOB2		FOB2L FOB2E* FOB2B*	FOB2E†	FOB2B† FOB2Q		A
FOB3			FOB3L FOB3E* FOB3B*	FOB3E†	FOB3B† FOB3Q	B

Fig. 1. Nested Module Schedule. * marks publication and † marks deadline.

In Sect. 2 we present the detailed components of the course and their interplay (Sect. 2.1), a short introduction to the *Theorema* system (Sect. 2.2), and the way how logic software is applied in the various parts (Sect. 2.3). The main part is Sect. 3, where we discuss the statistics. The results presented in this work extend preliminary results presented earlier, see [9, 10], by the follow-up statistics of the second year and the respective comparison of the development over the years.

2 The Use of Software in the Teaching of Logic

2.1 Modules, Quizzes, Bonus, Lab

The logic-course is composed of *mandatory* and *elective components* that contribute in different ways to the final grade for the whole course. Lecture and exercise follow the flipped-classroom paradigm, where students need to study the theoretical parts on their own and then come to the class and ask questions, discuss problems, and do examples. What was previously the lecture is now done by the students autonomously with the help of lecture notes, lecture slides, and video recordings done by the lecturers[2]. The former exercise class is now the classroom part where people are physically present.

The FOB module consists of *three units* (FOB1–FOB3) and the grading of the module is based mainly on the mandatory *quizzes* (FOB1Q–FOB3Q) after each unit. Students can enhance their scores through voluntary *bonus exercises* (FOB1B–FOB3B) after each unit and a voluntary *lab exercise* after the entire module. Through publication dates and deadlines for the respective items we enforce a nested sequence of lecture/exercise/quiz/bonus through the module as depicted in Fig. 1. For example, unit FOB2 starts in week 1 with making available the lecture material (FOB2L) for the students and the publication of the exercise sheet (FOB2E) and the bonus exercise (FOB2B), respectively. In week 2 we discuss the exercises, and the respective quiz (FOB2Q) is in week 3

---

[2] Lecture recordings are almost mainstream nowadays, but we switched to flipped-classroom with videos already one year before the pandemic.

giving students one week time to practice and to digest the presentation of the exercises. Figure 1 shows how the units overlap, e.g., in week 2 they finish unit 1 with submitting the bonus FOB1B and doing the quiz FOB1Q, then there is the exercise for unit 2, and after this exercise class they can already start to devote some time to unit 3 due to the publication of FOB3L, FOB3E, and FOB3B.

A bonus exercise spans an entire unit, its publication together with the respective exercise sheet marks the beginning of the unit and its deadline together with the quiz marks the end of the unit. We designed the units like this because the bonus exercise can then serve two purposes: if the bonus is done before the exercise class, the software can be a *tutor*, if it is done after the exercises the software can serve as *checker*, see also Sect. 2.3.

2.2   Interactive Automated Theorem Proving in Theorema

The *Theorema* system aims to be a computer assistant for the working mathematician. Support should be given not only for proving but also for defining and executing algorithms, structuring knowledge, etc. The Theorema project started in the 1990s s with a first prototype implementation in Mathematica 3, a major re-design and a re-implementation based on new features available since Mathematica 7 is now called *Theorema 2.0*. Since the system has been presented to this community several times, we refer to some overview articles [3–6] and only mention some peculiarities that are relevant for the didactical frame, in which the system is used in the logic course.

Working with Theorema 2.0 means to write mathematical content into a *Theorema notebook* and then perform some action on the content using the *Theorema commander*. By mathematical content we mean *informal parts* (basically everything that can be written into a Mathematica notebook, such as text, tables, figures, graphics, etc.) intermixed with *Theorema-specific formal elements* such as definitions or theorems. A Theorema notebook is just a Mathematica notebook using a special stylesheet in order to support special behavior of Theorema-specific items. The Theorema commander is a click-based user interface that supports certain manipulations on formal mathematical content, e.g., proving a formula based on some knowledge base. Every Mathematica user can use the features just described by loading the *Theorema 2.0* add-on package into a Mathematica session. The *Theorema* package is open source and freely available under GNU GPL 3 or higher.

Suppose the task is to prove statement A using knowledge represented by formulas K. The user would first type the statement A into a formula-cell in the notebook, usually inside a named Theorem-, Lemma-, or Proposition-environment. The same would be done for all formulas in K, where knowledge could go into Definition-environments or other theorems, lemmas, or propositions. We want to emphasize that formulas can be used as knowledge even if they have not yet been proven. All environments may carry names as well as each formula inside an environment can carry a separate label as name. Then one would switch to the Theorema commander and choose the PROVE-activity, which guides one through the process of setting up the automated prover. The proof goal is defined by simply selecting the notebook cell containing the goal formula A. Next is the setup of the knowledge base, which is achieved through the *knowledge browser* that shows an outline of each open notebook displaying only formal mathematical content while preserving the sectional structure of the notebooks. Sectional groupings can be collapsed in order to gain an overview over the whole document. For the formal entities, the commander does not display the formulas in detail, it rather shows the formulas' labels only and presents the entire formula as a tool-tip when hovering the mouse over the label. Each formula is accompanied with a checkbox that, when checked, includes the corresponding formula into the knowledge base K.

The next step in the PROVE-activity is the *setup of the prover*. A prover in *Theorema 2.0* consists of a collection of individual inference rules that are applied by a proof search engine in a certain order guided by rule-priorities. The system knows some predefined collections of rules including reasonable priorities that lead to "nice" proofs in many cases. However, the Theorema commander allows one to activate or deactivate every single inference rule, and the pre-assigned priorities can be changed as well. Once all settings are given the prove-task can be submitted. Usually, it is a good strategy to first run the prover with default settings and a limited search depth and search time (both configurable in the prover setup). In case the proof would not succeed, the failing proof can be inspected and checked whether certain settings might be changed in order to prevent the prover from running into an undesired path. Otherwise, search depth and search time can be increased in order to allow the prover to terminate successfully. When the prover stops it writes a link to the automatically generated proof into the notebook just below the environment containing formula A. When clicking the link, a nicely formatted version of the proof explained in natural language is displayed in a separate window, which also offers several options to simplify the proof.

During proof generation and when a proof window is open, the Theorema commander shows a *tree visualization* of the proof search. In a successful proof all nodes belonging to a successful branch are colored green, nodes in failing branches are red, and pending nodes are blue. Pending nodes are proof situations that can still be handled by one of the available rules. Simplification of a successful proof essentially removes all failing branches and pending nodes resulting in an all-green proof tree that in fact corresponds to a formal proof tree as taught in our logic-course. Click-navigation connects the Theorema commander and the proof window, i.e., clicking a cell in the proof window highlights the corresponding node in the tree view, whereas clicking a node in the tree scrolls the proof window to the textual description of the respective proof step. Moreover, when hovering the mouse over a node in the tree, the name of the rule applied in that node is displayed as a tool-tip.

Proving with *Theorema* is an experience of *interactive automated theorem proving*. Although finally a proof is generated in a fully automated fashion, there is a lot of user interaction along the way from stating a theorem until accepting its proof. Sometimes this interaction amounts to adjustments in the prover setup, such as activating or deactivating rules, rearranging rules through refining priorities, or adjusting search depth or search time. It is also common that a proof fails because of missing knowledge. In this case, auxiliary lemmata have to be formulated and passed to the prover and, ideally, also these lemmata are proved then. In any case, all these actions are *inter*actions because they are a reaction to the inspection of the failing proof and an insight regarding the reason for failure. The experienced user can learn a lot from failing proofs even but, admittedly, for beginner students this aspect is probably less pronounced.

2.3 How Software is Embedded into the Course

Our main didactic hypothesis is that for doing (correct and complete) proofs it is beneficial to first get acquainted with the *rules of formal proving* based on the formal language of first-order predicate logic. This contrasts the approach often seen in mathematics, where proving is taught as an "art" that is just demonstrated by many examples without actually telling the rules of the game. Students see many special tricks for particularly interesting special cases but are often insecure how to do the uninteresting routine cases. Therefore, we explain a set of "practical" proof rules for FO, where we neglect aspects of completeness and minimality and concentrate on convenience of the calculus for generating "nice proofs". We then view proofs as proof trees, where each connection from one node to its children must be justified by one of given rules. Finally, we discuss how a proof tree can be presented in natural language intermixed with mathematical formulas in different levels of detail depending on the context with the aim to de-mystify the "art of proving" that they often come across in their mathematics courses.

This view of proving is, in fact, the key philosophy behind the *Theorema* system since its beginnings in the early 1990s.s. It is based on a natural-deduction-like set of inference rules inspired by expert human provers and a pattern-based proof search procedure that aims at finding proofs that are similar to how well-educated human mathematicians would do the proofs. Proof trees are generated in form of proof objects that are on the one hand the basic datastructure guiding the proof search and on the other hand allow nice natural language presentations of proofs. One of the most important guiding principles in *Theorema* is natural style both in input and output. That said, *Theorema* seems a perfect fit for being taken as software support in our logic course. Unlike in other modules, where logic software *employs methods* taught in the course to solve concrete "practical problems", whose size alone would discourage solution attempts without machine support, we employ the *Theorema* software in module FOB to *help students to practice methods* that they should be able to apply even without the help of a computer.

The main idea of using *Theorema* in the frame of the FOB module is that students do the same proofs twice, once by hand and once with *Theorema*. We use the examples from the weekly exercises, which students are required to do by hand, and let them generate Theorema-proofs in the frame of the respective bonus exercises. From the sequence of tasks within one unit shown in Fig. 1,

Sect. 2.1, one sees that students can either do the proofs *manually first* and then *check* whether *Theorema* does them the same way, or they generate an *automated proof first* and then do the manual proof just following the Theorema-model. Both scenarios offer some learning benefit for the students.

– The first approach will typically be chosen by students who are able to do the proofs by hand. Instead of waiting for feedback until the exercise session, they can compare their own proofs to the system-generated versions. They can either feel enforced or they can revise their own versions after comparing to the system proofs.
– The second approach may be a way for students who feel unable to do the proofs by hand. They can let *Theorema* do the proofs, then study the steps the system applied, and finally imitate the system proof in their own formulation (in the best case).

In the frame of the lab exercise, we try to guide the students towards the second approach. They are asked to first do a proof with *Theorema* and then, after understanding the Theorema proof, formulate a proof in their own words. They need to give a short oral presentation comparing the two versions of the proof. It was interesting to observe that many students decided in their hand written versions to deviate from the path that the Theorema proof had marked. In the majority of the cases the argument was that they felt their own version was easier to understand. However, the majority of these cases were logically wrong.

Due to didactical considerations but also organizational reasons, the use of software is on a voluntary basis only. This leads to (at least) two motivations for students to spend time on the software. One group of software users are the curious students that are interested in doing a bit more than required, they want to get as much as possible out of the course. The other group of users are those that need the bonus points in order to pass the course. We have no recordings of students' motives to do the bonus or lab exercises. Hence, we assume the statistics presented in the next section cover both groups equally.

3   Results

Figure 2 shows the distribution of content over the FOB module. Examples to be solved with *Theorema* in the frame of the bonus exercises FOBnB are always examples taken from the corresponding exercise sheet FOBnE. E.g., in FOB3E students have to prove that a function $f: A \to B$ is surjective if its restriction $f|_C: C \to B$ for $C \subseteq A$ is surjective. The usual definition of surjectivity, i.e.,

$$\mathop{\forall}_{y \in B} \mathop{\exists}_{x \in A} f(x) = y$$

and the implicit definition of $f|_C$ to be the unique function $g: C \to B$ with $\mathop{\forall}_{x \in C} g(x) = f(x)$ are given, the task is to apply the definitions correctly and handle the (alternating) quantifiers appropriately depending on whether they appear

	FOBnE/FOBnQ	FOBnB
FOB1	pattern-based proof search procedure with hypothetical inference rules, first-order proofs without quantifiers	first-order proofs without quantifiers from FOB1E
FOB2	first-order proofs with quantifiers	first-order proofs with quantifiers from FOB2E
FOB3	first-order proofs with quantifiers and informal natural language presentation referring to concrete mathematical concepts introduced by definitions; induction proofs	concrete mathematical proofs from FOB3E

Fig. 2. Module FOB unit contents.

in the goal or in the knowledge. Consequently, FOB3B consists of a Theorema notebook containing the theorem and all necessary definitions in Theorema language. In addition, there are some hints concerning prover configuration because with standard settings *Theorema* would explore too many failing branches before it finds the successful proof. The task in FOB3B is then to generate a proof with *Theorema*. Figure 3 shows part of an automated proof as generated by the *Theorema* system and the corresponding proof tree. The screenshot of the proof window does not show the entire proof, the intention is to only give a flavor of the style *how* automated proofs appear for students, namely with natural language explanation of each step and nicely formatted mathematical expressions. Moreover, both the proof as well as the proof tree are hyperlinked, meaning that a mouse-click on a formula in the proof will highlight the corresponding node in the tree, whereas clicking a node in the tree would highlight the corresponding cells in the proof notebook. We claim that this style of proof presentation is feasible for tutoring purposes for undergraduate students.

We try to capture two different aspects of using software in teaching logic: the *personal impression of students* after doing proofs with *Theorema* is collected using a questionnaire, while the *performance of students in the quizzes* is correlated to their participation and their performance in the bonus exercises.

The case study has been done in two consecutive years (winter 2020 and winter 2021) with identical content and only slight organizational changes. Most notably, in 2021 the submission deadline for the bonus exercises was right *before* the corresponding quiz with the consequence that the bonus exercise with *Theorema* had to be completed before doing the quiz. In 2020 the deadline was a little later so that it was possible in principle to do the bonus after the quiz.

3.1 Personal Impression of Students

Before being admitted to submit the solution of a bonus exercise, students were required to answer at most two standard questions about their experiences with

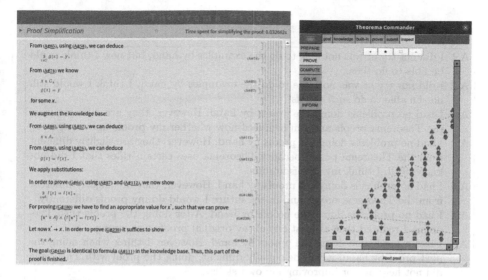

Fig. 3. Final part of a proof generated by *Theorema* (left) and proof tree visualization in the Theorema commander (right).

Theorema depending on whether they were successful in generating an automated proof with the system (Group A) or not (Group B). In each group they had to select the answer that fitted best to their personal situation regarding the relation between automated proving with software support and doing (the same) proofs by hand for the exercises. The possible answers are shown in Fig. 4 and 5, and according to their answers we formed categories A.1–A.9 and B.10–B.16. In both years we had 200–300 self-assessments for each of the bonus exercises and the ratio A:B was a constant $\approx$2:1. With these high numbers we consider the results to be non-random.

First we analyze the relative sizes of the categories in group A and compare the two classes winter 2020 and winter 2021, see Fig. 6. In both instances the top four are A.1, A.4, A.5, and A.8 and they separate quite clearly from the remaining five. Quite encouraging are A.1[3], which ranks first in FOB3B in 2020 and, in 2021, first in average over all three bonus exercises and first overall (groups A and B together). We also highlight A.8[4] particularly in 2021, where this category shows a monotonic increase over time to finally rank clearly first in FOB3B with a huge margin to the second-biggest group. Category A.5[5] displays an interesting development in 2021 from rank 1 with 23% in FOB2B dropping to rank 6 with only 7% in FOB3B.

---

[3] Not able to do the proofs by hand but feel capable after using *Theorema*.

[4] Hard time doing the proofs by hand but feel improvement through using *Theorema*.

[5] No problems doing the proofs by hand but will do proofs differently after having used *Theorema*.

A.1 I did not try or was not able to do the examples by hand, but now I think would be able to do them.

A.2 I did not try or was not able to do the examples by hand. I think I would still not be able to do such proofs.

A.3 I had no problems doing the proofs by hand. However, they are different from the Theorema proofs and I'm confused now whether my proofs are wrong.

A.4 I had no problems doing the proofs by hand. However, they are slightly different from the Theorema proofs because Theorema uses certain rules that I did not know. Still, I think my proofs are fine.

A.5 I had no problems doing the proofs by hand. However, they are slightly different from the Theorema proofs and in the future I would do my proofs differently.

A.6 I had no problems doing the proofs by hand. After doing the proofs with Theorema I realized that at least one of my original proofs was wrong.

A.7 I had a hard time doing the proofs by hand. However, I think when doing the next proof by hand, it will be equally difficult, doing the proof with Theorema did not help me for improving my own skills.

A.8 I had a hard time doing the proofs by hand. After doing the proof with Theorema I understand much better how all of this works. I feel that my own skills improved by using Theorema.

A.9 I don't see any connection between the examples from the exercises and the Bonus Exercise with Theorema

Fig. 4. Possible answers for Group A.

B.10 I did not try or was not able to do these examples by hand. I wanted to see how Theorema does the proofs, but I failed to produce a complete proof.

B.11 I did not try or was not able to do these examples by hand. Theorema is much too complicated for me to use it for such exercises.

B.12 I had no problems doing the proofs by hand. Unfortunately, I failed to produce a complete proof with Theorema. It would have been interesting to compare.

B.13 I had no problems doing the proofs by hand. I'm not interested how an automated proof looks, I have done them by hand anyway.

B.14 I had a hard time doing the proofs by hand. Unfortunately, I failed to produce a complete proof with Theorema. It would have been interesting to compare.

B.15 I had a hard time doing the proofs by hand. I'm not interested how an automated proof looks, I have done them by hand anyway.

B.16 I don't see any connection between the examples from the exercises and the Bonus Exercise with Theorema.

Fig. 5. Possible answers for Group B.

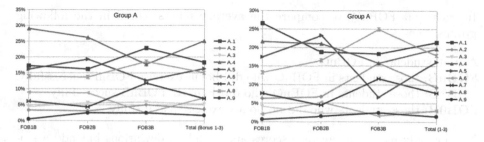

Fig. 6. Development of the group sizes (relative) in self-assessment from FOB1B to FOB3B in group A in winter 2020 (left) and winter 2021 (right).

The situation is less pronounced in group B, which are generally the less interesting cases, since these are the students who failed to generate automated proofs with *Theorema*, see Fig. 7. We are happy that those who consider *Theorema* too complicated (B.11) are a rather small group, and that, at least in 2021, those that were interested in *Theorema* at least in principle (though unsuccessful, categories B.10, B.12, and B.14) dominate those that show no interest at all (B.13 and B.15). Common to both instances is the significant drop of B.12[6] from FOB2B to FOB3B. We think that this is mainly due to fact that less people had no problems doing the proofs by hand, so the fraction of them that also had problems with the software shrunk as a consequence of that.

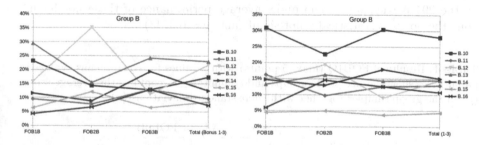

Fig. 7. Development of the group sizes (relative) in self-assessment from FOB1B to FOB3B in group B in winter 2020 (left) and winter 2021 (right).

3.2 Performance in the Quizzes

In addition to the students' *opinion* about the effects of software support on their own proving skills presented in the previous section we now analyze their *performance* in the quizzes. The numbers do not differ much between the two instances of the course, the concrete numbers shown below are from winter 2021.

---

[6] No problems doing the proofs by hand but unable with *Theorema* although keen.

In each quiz FOBnQ we compare the average points scored in the following groups:

All: all students in FOBnQ.
FOBnB: those students in FOBnQ who did bonus exercise FOBnB successfully.
FOB*B: those students in FOBnQ who did FOB1B–FOBnB successfully.
FOB0B: those students who did no bonus exercise successfully.

We not only record the average scores and standard deviations but add a statistical assessment in form of a *(two-sided) Student's T-Test* [1] comparing the sample values of different groups. The Student's T-Test gives a probability (p-value) that the given sample data occur under the hypothesis of equal mean values of underlying distributions. We use a variant of the test that does not assume equal variances of the underlying distributions. When we compare two samples with different averages, then a low p-value means that there is a low probability that the different averages occur although the underlying distribution have equal mean value, i.e., the mean values are equal and the different sample averages just occur by chance. Typically, in statistics, one calls a difference *statistically significant* if $p < 0.05$, so the smaller the numbers in the tables below, the more significant are the different averages in the samples compared. For example, the analysis of quiz FOB1Q (see Table 1) tells us that the average score 4.74 of the 107 students that also did bonus FOB1B is significantly better ($p = 5.65 \times 10^{-6}$) than the 4.36 scored by the 187 students that neglected FOB1B, and still significantly better ($p = 0.0003$) than the 4.50 scored overall by the 294 participants[7]. The under-average performance of those not having done the bonus is statistically not significant ($p = 0.0943$).

Table 1. Results of FOB1Q (max. 5 points) with p-values for equal means. Values in parentheses show the size of the groups and column '$\mu \pm \sigma$' contains the average scores (μ) in the group samples together with their standard deviations (σ).

	$\mu \pm \sigma$	All	FOB0B
All (294)	4.50 ± 0.81	—	—
FOB0B (187)	4.36 ± 0.93	0.0943	—
FOB1B (107)	4.74 ± 0.49	0.0003	5.65×10^{-6}

Table 2 shows the corresponding figures for quiz FOB2Q, which we consider the most difficult quiz because it is the first one with quantifier rules and it is only about quantifier proving whereas in FOB3Q it is quantifier proving as a repetition plus induction proofs. The situation is similar to the one above: those

---

[7] Note that the statistical test gives much more confidence in different mean values than only comparing observed averages and taking into account the standard deviations or variances in the samples.

who do the bonus perform best, followed by the overall average, and the no-bonus-group is the weakest. The different averages are statistically significant, since all p-values are far below 5%. Among the bonus students, we even distinguish between those who did *only FOB2B* and those who did FOB1B–FOB2B. The latter are marginally better scoring 3.87 against 3.79, but this difference cannot be confirmed statistically with $p = 0.6353$, meaning that it can be by chance with a probability of over 60%.

Table 2. Results of FOB2Q (max. 5 points) with p-values for equal means. Values in parentheses show the size of the groups and column '$\mu \pm \sigma$' contains the average scores (μ) in the group samples together with their standard deviations (σ).

	$\mu \pm \sigma$	All	FOB0B	FOB2B
All (290)	3.30 ± 1.29	—	—	—
FOB0B (166)	2.99 ± 1.24	0.0102	—	—
FOB2B (109)	3.79 ± 1.21	0.0006	2.41×10^{-7}	—
FOB*B (91)	3.87 ± 1.20	0.0002	8.43×10^{-8}	0.6353

Finally we discuss our observations in quiz FOB3Q, see Table 3. Again, those who do the bonus perform best, followed by the overall average, and the no-bonus-group is the weakest, but the averages differ much less. The only message that is supported by statistics is that those who did bonus exercises perform better than those who do not. As in FOB2Q, the difference between those who did only the last bonus to those who did all bonus exercises (3.58 vs. 3.68) is not supported by statistics with $p = 0.5620$.

Table 3. Results of FOB3Q (max. 5 points) with p-values for equal means. Values in parentheses show the size of the groups and column '$\mu \pm \sigma$' contains the average scores (μ) in the group samples together with their standard deviations (σ).

	$\mu \pm \sigma$	All	FOB0B	FOB3B
All (282)	3.46 ± 1.05	—	—	—
FOB0B (147)	3.30 ± 1.04	0.1329	—	—
FOB3B (97)	3.58 ± 1.07	0.3560	0.0474	—
FOB*B (64)	3.68 ± 1.10	0.1529	0.0215	0.5620

We also compared the performance of the groups A.1–B.16, see Sect. 3.1, against each other. We do not go into further detail because only a few of the different averages could be confirmed as statistically significant in winter 2020 and, unfortunately, even less in winter 2021.

A final observation we want to share is given in Fig. 8, which shows the development of the performance over time split up into the groups A.1–B.16 from

Sect. 3.1. The striking development of group A.3 is less telling that one might expect because of the small size of that group (4%, 2%, and 3%), hence we neglect A.3 in the further analysis. One can see that subgroups in group A are a bit closer to each other than those in group B. There is a falling tendency from FOB1Q to FOB2Q, regardless of what happened in the bonus, and this is certainly due to the difficulty of FOB2Q, see above. Between FOB2Q and FOB3Q we see more diversity, some groups improve whereas some decline. Among the declining groups, A.2 and B.11 correspond to each other, it contains those that are not able to do the proofs by hand and that see no improvement through software support. Also declining are A.6 and B.13, both having no problems doing the proofs by hand. Those that used *Theorema* successfully at least detected that their hand-proofs are wrong. However, they seem to not have been drawing the right conclusions because still their performance got worse. Among the improving groups, A.7 is a strange phenomenon, because they improve in group A from last position (neglecting A.3) in FOB2Q to second in FOBQ3, although their self-assessment says they had a hard time with hand-proofs and the software did not help them. On the other hand, A.8 claimed to have the feeling their skills had improved through using *Theorema*, but their performance stays constant. Another nice facet is the development of B.14 and B.15 from FOB2Q to FOBQ3. Starting from the same level in FOB2Q, the ones that show interest in the software (although finally failing, B.14) improve in FOB3Q, while those that confess not being interested in the automated proofs show no improvement.

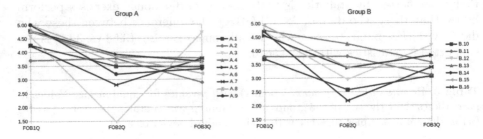

Fig. 8. Development of the performance from FOB1Q to FOB3Q in winter 2021 in group A (left) and group B (right).

4   Conclusion

We report on a big case study using the *Theorema* system as a proof tutor in a big logic course for almost 400 first semester students of computer science or artificial intelligence. The case study consists of a *self-assessment* of students after using the software and a *statistical evaluation* of test results based on groups defined through the answers in the self-assessment. Our didactical hypothesis is that students can improve their own proving skills through working with a natural-style automated theorem prover. Some statistics support this claim, e.g., in general, students that worked with the software performed better than average while the others show results under average. However, there are also surprising results that do not exactly match our expectations, e.g., students, who reported that the automated proofs were of no help, improved, while those, who claimed better understanding through working with the software, did not improve. Our expectation was just the other way round. In this context, it is important to recall that statistical tests can only reveal *correlations* but *no causalities*. This has to be emphasized in particular in our scenario where the groups are not assigned randomly but students actively join a group or not. Now, if one group performs better than another, this can be because students are better *because* of being in that group or because the *better students* (more talented, more interested, or more motivated) *chose* that group. A superior setup would be to divide the entire class into a group that does the bonus exercises and compare them to those that do no bonus exercises, but this is not feasible in our logic course because of the voluntary character of the software-related parts.

In future versions of the case study we have to take measures that answers in the self-assessment become more reliable so that students see no benefit in checking answers that the teachers might like better than others. Moreover, we will also try to investigate the development of individual students during the course like, e.g., follow the paths through which of the categories A.1–B.16 individual students travel from FOB1Q to FOB3Q. In any case, we are happy that the *Theorema* system can be applied in a reasonable way in education with a big group of first-semester students.

References

1. Student's t-test. https://en.wikipedia.org/wiki/Student's_t-test. Accessed 23 May 2022
2. Biere, A., Schreiner, W., Seidl, M., Windsteiger, W.: Logic for computer science (2020). Course in the first year in the bachelor program for computer science at Johannes Kepler University Linz (JKU), taught since 2013
3. Buchberger, B.: Theorema: a proving system based on Mathematica. Math. J. **8**(2), 247–252 (2001)
4. Buchberger, B., et al.: Theorema: towards computer-aided mathematical theory exploration. J. Appl. Log. **4**(4), 470–504 (2006). https://doi.org/10.1016/j.jal.2005.10.006
5. Buchberger, B., et al.: The Theorema project: a progress report. In: Kerber, M., Kohlhase, M. (eds.) Symbolic Computation and Automated Reasoning. Proceedings of CALCULEMUS 2000, Symposium on the Integration of Symbolic Computation and Mechanized Reasoning, pp. 98–113. St. Andrews, Scotland, Copyright: A.K. Peters, Natick, Massachusetts (2000)
6. Buchberger, B., Jebelean, T., Kutsia, T., Maletzky, A., Windsteiger, W.: Theorema 2.0: computer-assisted natural-style mathematics. J. Formaliz. Reason. **9**(1), 149–185 (2016). https://doi.org/10.6092/issn.1972-5787/4568
7. Cerna, D.M., Seidl, M., Schreiner, W., Windsteiger, W., Biere, A.: Computational logic in the first semester of computer science: an experience report. In: CSEDU 2020, pp. 1–8 (2020)

8. Windsteiger, W.: Theorema 2.0: a brief tutorial. In: Jebelean, T., Zaharie, D. (eds.) Proceedings of SYNASC 2017, pp. 1–3. IEEE Explore (2017)
9. Windsteiger, W.: Automated theorem proving in the classroom. In: Janicic, P. (ed.) Proceedings Automated Deduction in Geometry (ADG 2021). Electronic Proceedings in Theoretical Computer Science (EPTCS), vol. 352, pp. 54–63 (2021). https://doi.org/10.4204/EPTCS.352.6
10. Windsteiger, W.: Automated theorem proving in the classroom. RISC Report Series 21–15, Research Institute for Symbolic Computation (RISC), Johannes Kepler University Linz, Altenberger Straße 69, 4040 Linz, Austria (2021). Extended version of keynote talk at ADG 2021 conference

Datasets and System Entries

Making the Census of Cubic Vertex Transitive Graphs Searchable and FAIR

Berčič Katja[1,2] and Koprivec Filip[1(✉)]

[1] Institute of Mathematics, Physics and Mechanics, Ljubljana, Slovenia
`filip.koprivec@imfm.si`
[2] Faculty of Mathematics and Physics, Ljubljana, Slovenia

Abstract. The Census of Cubic Vertex Transitive Graphs, by Potočnik, Spiga, and Verret, has recently been extended. We took this opportunity to make the new dataset as useful as possible according to current best practices and with the current technology. As part of this, we explore the best practices for managing research data in the mathematical community. In this work, we describe the steps we took to make the data as FAIR and as machine actionable as possible. We also describe how we made the data searchable and easy to interact with through a web interface by publishing it on an instance of the platform MathDataHub.

Keywords: Dataset · Graphs · FAIR · MathDataHub

1 Introduction

One of the purposes of compiling mathematical datasets is to make forming conjectures and testing hypotheses easier. An example of such a dataset in the area of group actions on graphs is the Census of Cubic Vertex Transitive Graphs (we will also call it the CVT Census) by Potočnik, Spiga, and Verret. They compiled the list of all 3-regular (cubic) graphs up to order 1280 whose group of automorphisms acts transitively on the vertices, and described the theoretical results and computer calculations needed to determine all such graphs [15]. The dataset was initially published on the authors' website [14] to supplement the paper as downloadable files readable by the computer algebra system Magma [11].

Using the dataset in this format required either running a proprietary computer algebra system or some text processing to explore the data. This has motivated the development of DiscreteZOO [10], the aim of which was to make exploring the dataset easier. DiscreteZOO combines a SageMath package and an online platform for exploring the dataset. In order to support features such as searching and filtering, values for a number of graph invariants were computed

B. Katja—The first author was supported by the AFOSR project no. FA9550-21-1-0024 and by the ARRS research project no. J1-1691.
K. Filip—The second author was partially supported by the ARRS research project no. J1-1691.

K. Buzzard and T. Kutsia (Eds.): CICM 2022, LNAI 13467, pp. 323–328, 2022.
https://doi.org/10.1007/978-3-031-16681-5_22

in SageMath [16] for each of the graphs in the census. The initial part of the census [3] was a test case for MathDataHub [8,9], a platform that generalized and improved on the ideas from the online DiscreteZOO platform.

Since the original publication of the census in 2013, better research data practices have emerged in the broader scientific community. They were published in 2016 as the FAIR guiding principles [17] in a Nature journal, and stress the importance of machine actionability. A set of additional principles called Deep FAIR has been proposed [9] towards making the rich structure of mathematical datasets machine actionable.

In an ongoing project, Potočnik and co-authors are expanding the CVT Census, which has prompted a reevaluation of how the dataset is published online. We hope that this report will seed further discussion of research data in the mathematics community and encourage other dataset authors to consider how to make their datasets more useful: for humans and for computers.

Contribution. In this dataset description, we outline the structure of the dataset and discuss the data representation formats involved. We also describe the steps taken to make the dataset as FAIR as possible and our experience with setting up an instance of the updated MathDataHub platform dedicated to the dataset.

2   A Brief Description of the Contents and Structure

The backbone of the Census of Cubic Vertex-Transitive Graphs is the list of all cubic vertex-transitive graphs on at most 1280 vertices. Each graph carries with it further information: a census-specific index, a canonical labeling, and number of Boolean and integer-valued graph-theoretic properties. The census also contains some of graph names from the literature, such as the "Petersen graph". Almost all formats data types are encoded in standard formats.

The Order of the Vertex Stabilizer of the graph's automorphism group is important in the study of highly symmetric graphs. Some orders are too large to be stored as regular integers (up to 2^{319}), which is why each one is stored as a prime factorization: a list of pairs (p_i, e_i) of primes and exponents such that $\prod_i p_i^{e_i}$ is the order of the vertex stabilizer. This format is more readable and can be used to represent all vertex stabilizer orders in CVT.

Census-Specific Graph Identifiers. During the original dataset generation process, Potočnik, Spiga, and Verret assigned a unique identifier to each graph. It consists of a pair of integers: CVT[n,i] such that n is the order of the graph and i is the consecutive number of that graph among graphs of order n in the census. Here, n is the order of the graph, while i is the consecutive number of that graph among graphs of order n in the census.

Graph Format. The graphs are canonically labeled with nauty [13] (version nauty-27r3) and stored in the widely used sparse6 format [12] for sparse graphs. This is an efficient bit-stream-based representation of a graph as a list of edges. The bit-stream is represented with printable ASCII characters, with

6 bits of information per byte. Canonical labeling combined with the `sparse6` format reduces checking isomorphism between two canonically labeled (with the same algorithm) graphs to string comparison.

Boolean and Integer-Valued Graph-Theoretic Properties. For short descriptions of these properties, we refer the reader to the description of the dataset on Zenodo [7], and just list them here. *Boolean-valued properties*: arc-transitive, bipartite, Cayley, distance regular, distance transitive, edge transitive, Hamiltonian, partial cube, split, strongly regular, Moebius ladder, prism, Split Praeger-Xu graph. *Integer-valued properties*: number of vertices, clique number, diameter, girth, odd girth, triangles count.

Ongoing Work. We are adding two more kinds of information to each graph. The first are the times needed to compute canonical labelings with algorithms such as nauty, Bliss, etc. This is a one-to-many correspondence between a graph and pairs consisting of an algorithm name and time needed to compute the canonical labeling of the graph using that algorithm. The second is information on how to construct graphs as members of graph families. This is a one-to-many correspondence between a graph and constructions consisting of a graph family name and parameters for that family, most of which are integer values.

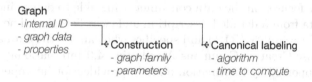

Fig. 1. Intended organization of data in the CVT Census, once the remaining computations are complete.

3   FAIR-ifying the CVT Census on Zenodo

The FAIR principles [17] are guidelines for improving **F**indability, **A**ccessibility, **I**nteroperability, and **R**euse of research data. A good summary is available on the GO FAIR initiative's website [2]. We will use the labels **F**, **A**, **I**, and **R** to refer to the main principles, and labels such as **F4** or **R1.1** to refer to their components. Some FAIR principles are addressed by publishing the dataset in standard format on a repository such as Zenodo [6] under a permissive license: registration in a searchable resource (**F4**), retrievable via a standard protocol, such as HTTP(S) or FTP (**A1**), long term storage of metadata (**A2**, **F4**).

  Findability focuses on finding a dataset (less on finding data within it), with *globally unique and persistent identifiers* playing a key role (**F1-2**, but also **I1-3**). Zenodo provides the identifier for the data, and we included author (ORCID) and reference identifiers (software, papers, etc.), as well as the MSC classification. Better FAIR-ness could be achieved with identifiers from controlled vocabularies (see below). Locally unique identifiers (such as the graph IDs within the dataset and unique column labels), while not required by FAIR, provide a way to refer to objects within the dataset, i.e., the graph (`doi:10.5281/zenodo.6576526`, `CVT[10,3]`). Bundling data and metadata satisfies **F3**.

Accessibility requires that data retrieval should not involve specialised tools and that at least metadata should be stored long term. A dataset available to download via the usual internet protocols (HTTP(S) or FTP) satisfies **A**.

Interoperability relies on standard formats and controlled vocabularies. Zenodo uses a standard metadata format for bibliographic information. CVT information is formatted as easy to parse comma separated values (CSV). The structure description includes format descriptions for `sparse6` [12], the census identifier, graph names, integer factorization, and boolean values, and a list of columns with short descriptions for each table. This provides basic interoperability. Details about the software used to obtain the data (F1-2, I3, R1.2) help determine the usefulness, as well as with actually using the data. We give the version of Nauty (the canonical labeling algorithm). Interoperability (and findability) would be improved by addressing the *lack of controlled vocabularies* in mathematics (F1, I1-2). Such a vocabulary would provide unique identifiers for concepts such as Hamiltonicity and their encodings. In place of controlled vocabularies, we link column identifiers to online resources (such as Wikipedia). We include CVT dataset examples for computer algebra systems in the Zenodo description [7].

Reusability focuses on the data consumer being able to determine the usefulness of the data from a detailed description and clear licensing. The latter is the legal aspect of reusability (R1.1) and specifies who can use the data and how: we chose the license Creative Commons Attribution 4.0 International. The former is partially supported by information already provided for interoperability.

On Zenodo, uploads are organized into communities, which make it easier to find datasets. It did not look like there was an active mathematics research community, so we started one in the hopes that it will get adopted by other dataset authors: the Mathematical Research Data community [5]

4   Making the Dataset Searchable with MathDataHub

MathDataHub is a system and a hosting platform that provides a web interface to datasets. The interface supports easy filtering, searching, and customizable presentation of individual objects. We could have published the updated census on the main MathDataHub instance, where the datasets are organized into versioned snapshots. This was not a great fit, as the census is a part of an ongoing project and the authors want to be able to share recent versions with the research community. To enable faster development and deployment cycle and a tailored presentation of the CVT dataset and objects, we obtained permission from MathDataHub developers to publish the CVT Census on a dedicated instance of MathDataHub [1]. The project repository [4] contains usage and deployment instructions. Our version is forked from the main branch of the project, but we plan to integrate improvements back into the main branch once they are stable.

Setting up a MathDataHub Instance. The biggest challenge was setting up an independent instance in both production and development mode and preparing the configuration for custom deployment. We opted for a Dockerized version for both development and deployment. This requires some familiarity with Docker, but makes work on multiple different operating systems easier. MathDataHub is a Django web application that can work with different database systems. We chose PostgreSQL over the default sqlite3 to enable easier querying and more efficient data insertion for large amounts of data. We updated the default top-level Dockerfile settings to setup the environment variables for PostgreSQL and overrode the environment file (`.env.local`) to point to the correct Docker service. Due to the routing resolution in NextJS and Django, the Docker service name for the frontend server needs to use dashes instead of underscores for word separation (commonly known as *kebab-case*). Removing the dashes and replacing them with underscores resulted in malformed URLs. We prepared a Docker Compose file that covered both the development and production builds for easier debugging. The setup included four different Docker containers.

- Database server: a standard `postgres:14` image with explicit increase of `/dev/shm` size to 2 gigabytes and increased memory limits. This was necessary to ensure that database queries did not run out of `tempfs` storage space during large operations.
- Frontend instance: a modified frontend part of the main Dockerfile which runs a development build of the frontend `mdh` application.
- Development backend instance: a modified backend part of the main Dockerfile. It installs the requirements for the Django app and runs the backend server in development mode using the default Django `runserver` command.
- Full production instance: builds and runs the top-level Dockerfile with production settings and uses `uwsgi` to serve the Django application.

The development frontend and the production instance connect to the same PostgreSQL container and use the same database. This allowed us to share the same imported data both in development version under construction when developing new presenters and sorting orders, while still maintaining high performance on imports through production instance and allowing end-users to access the service. We prepared the compose files and successfully deployed the development and production versions with specific changes.

Setting up a production version of MathDataHub without any code changes should be straightforward for a programmer comfortable with basic bash and Docker commands. Setting up the development version proved to be a bit more involved, as the goal was to tailor the general implementation of MathDataHub to the specific needs of the CVT Census. It should still be within reach for anyone comfortable with Python and TypeScript that can either set up a Docker environment for database or provide the database installation locally.

Changes in the Framework. We added new presenters, custom styling and imports, which did not change the general logic of the framework. The modification process did not pose any specific challenges, at least partly due to the quick feedback from the MathDataHub development team, who also made some small improvements on the main branch. All of this made the development smoother.

Acknowledgements. We would like to thank the MathDataHub developers, especially Tom Wiesing at FAU, for kindly implementing some of the changes we requested: custom templates for presentation of objects, custom presenters for integer factorizations and a custom single collection mode. These features improved the MathDataHub instance for the CVT Census. We would also like to thank Michael Kohlhase for his valuable insights into the FAIR principles.

References

1. Cubic Vertex Transitive Graphs (powered by MathDataHub). http://mdh.graphsym.net/collection/CVT
2. GO Fair initiative: FAIR principles. https://www.go-fair.org/fair-principles/
3. MathDataHub - census of small, connected, cubic, vertex-transitive graphs. https://data.mathhub.info/collection/cvt/
4. Mathdatahub repository. https://github.com/MathHubInfo/mhd
5. Mathematical research data community on Zenodo. https://zenodo.org/communities/math-data/
6. Zenodo: Principles. https://about.zenodo.org/principles/
7. Azarija, J., et al.: Cubic vertex-transitive graphs on up to 1280 vertices. https://doi.org/10.5281/zenodo.6576526
8. Berčič, K., Kohlhase, M., Rabe, F.: Intelligent computer mathematics. In: Intelligent Computer Mathematics (CICM) 2019, pp. 28–43. no. 11617 in LNAI, Springer (2019). https://doi.org/10.1007/978-3-030-23250-4, https://kwarc.info/kohlhase/papers/cicm19-MDH.pdf
9. Berčič, K., Kohlhase, M., Rabe, F.: (Deep) FAIR Mathematics. it - Inf. Technol. **62**(1), 7–17 (2020). https://doi.org/10.1515/itit-2019-0028,https://kwarc.info/kohlhase/submit/it19.pdf
10. Berčič, K., Vidali, J.: DiscreteZOO: a fingerprint database of discrete objects. Math. Comput. Sci. **14**(3), 559–575 (2020). https://doi.org/10.1007/s11786-020-00453-5
11. Bosma, W., Cannon, J., Playoust, C.: The Magma algebra system I. the user language. J. Symbolic Comput. **24**(3—-4), 235–265 (1997). https://doi.org/10.1006/jsco.1996.0125
12. McKay, B.D.: Graph formats. http://users.cecs.anu.edu.au/bdm/data/formats.html
13. McKay, B.D., Piperno, A.: Practical graph isomorphism, II. J. Symb. Comput. **60**, 94–112 (2014). https://doi.org/10.1016/j.jsc.2013.09.003
14. Potočnik, P.: Lists of graphs of a prescribed symmetry type and valence, and some other combinatorial and algebraic structures. https://www.fmf.uni-lj.si/potocnik/work.htm
15. Potočnik, P., Spiga, P., Verret, G.: Cubic vertex-transitive graphs on up to 1280 vertices. J. Symbolic Comput. **50**, 465–477 (2013). https://doi.org/10.1016/j.jsc.2012.09.002
16. The Sage Developers, Stein, W., Joyner, D., Kohel, D., Cremona, J., Eröcal, B.: Sagemath, version 9.0 (2020). http://www.sagemath.org
17. Wilkinson, M.D., et al.: The fair guiding principles for scientific data management and stewardship. Sci. Data **3**, 1–9 (2016). https://doi.org/10.1038/sdata.2016.18

An Evaluation of NLP Methods to Extract Mathematical Token Descriptors

Emma Hamel[(✉)], Hongbo Zheng, and Nickvash Kani

University of Illinois at Urbana-Champaign, Urbana, IL 61801, USA
{ebhamel2,hongboz2,kani}@illinois.edu

Abstract. Mathematical formulae are a foundational component of information in all scientific and mathematical papers. Parsing meaning from these expressions by extracting textual descriptors of their variable tokens is a unique challenge that requires semantic and grammatical knowledge. In this work, we present a new manually-labeled dataset (called the MTDE dataset) of mathematical objects, the contexts in which they are defined, and their textual definitions. With this dataset, we evaluate the accuracy of several modern neural network models on two definition extraction tasks. While this is not a solved task, modern language models such as BERT perform well ($\sim$90%). Both the dataset and neural network models (implemented in PyTorch jupyter notebooks) are available online to help aid future researchers in this space.

Keywords: Mathematical language processing · Dataset · Text summarization · Named entity recognition

1 Introduction

Mathematical formulae are an integral tool to convey information about mathematical concepts and are widely used in academic writing. They can summarize relationships between identifiers in an easily understood and succinct way. In mathematical literature, the text surrounding a formula gives context to the relationship it describes, such as the definitions of the variables and the domain and range of the equation. Although very commonplace, formulae and the texts that describe them are rarely researched in natural language processing (NLP) [4,10].

We define the mathematical token definition extraction (MTDE) task as follows: given a particular block of text containing words and one-or-more mathematical tokens, can we extract the textual descriptor of a particular mathematical token. This problem is very similar to both question-answering and text-summarization tasks but is made more complex with the addition of

---

Supported by University of Illinois at Urbana-Champaign - College of Engineering.

K. Buzzard and T. Kutsia (Eds.): CICM 2022, LNAI 13467, pp. 329–343, 2022.
https://doi.org/10.1007/978-3-031-16681-5_23

mathematical tokens and the different speech patterns found in scientific texts [9,15].

Studying the quirks of mathematical language and ways to automate textual understanding of it can have many applications, such as IR ranking systems or query optimization [1,8]. Token-definition extraction has been a sub-problem for several other papers attempting to embed mathematical equations [6,12,13]. However, the accuracy of the token-definition extraction in these systems is not measured directly.

The contributions of this paper are twofold: First, we introduce a publicly available *human-labeled* dataset containing text segments from STEM papers, a variable token defined in the text segment, and its corresponding definition. The creation of the MTDE dataset is the result of significant manual data-entry by the authors and contains ~10,000 examples of mathematical objects and the contexts in which they are defined. The field of mathematical language processing is hindered by a "fundamental need for datasets and benchmarks, preferably standard ones, to allow researchers to measure the performance of various mathematical language processing techniques and ideas in a objective and statistically significant way, and to measure improvements and comparative progress" [22]. It is the authors hope that this dataset will spur more interest in this field of research.

Secondly, it is important to evaluate the efficacy of modern NLP algorithms on this problem and compare it against past results. The MTDE task can be solved via text generation or estimation of the most likely position of the answer in a given context. Thus, we evaluate the accuracy of two text generation and three question-answering algorithms on the MTDE task using our dataset. We find that NLP models have significantly improved in the past decade with BERT question-answering models achieving an accuracy of 90/85% for short- and long-definition extraction, respectively.

2   Related Work

The field of mathematical language processing has just recently begun to gain traction with several groups seeking to extract definitions for mathematical tokens and equations using surrounding text. In one of the earliest works, Pagel and Schubotz used analytic methods to extract textual descriptions of mathematical tokens in the Wikipedia dataset [16,17]. Using a part-of-speech tagger in conjunction with simple pattern matching, they achieved an accuracy of ~90%.

However, it was soon shown that while these techniques may work well on consistently structured and formatted datasets like Wikipedia, analytic methods are sub-optimal when applied to general scientific and mathematical literature. Kristianto et al. compared the efficacy of several analytic and machine learning methods on a description extraction task [11] using a dataset of mathematical papers (NTCIR-10 [2]). Their research showed that several analytic methods perform relatively poorly (20–50% accurate) while simple, support-vector-machine (SVM) models achieved an accuracy of 60–70% [11].

Our research builds on these past works by evaluating how well modern NLP methods perform using a labeled, STEM-based dataset.

3   Dataset

The MTDE dataset is a collection of text segments from a random sampling of arXMLiv project manuscripts which define one or more variables. The arXMLiv project contains a significant portion of the arXiv preprint manuscripts converted to HTML/MathML from their LaTeX source files [7]. The papers used to generate the dataset cover a wide range of mathematic and scientific disciplines including Physics, Computer Science, and Biology. The dataset identifies a variable contained within the manuscript along with its 'short' (one word) and 'long' (1+ word) definition. In this work, definitions are defined as the noun phrase that best describes a mathematical variable. Each context contains at least one variable and its definition.

While the data was gathered using automated analytic means, each example was manually pruned, cleaned, and labeled by the first two authors of this paper. To ensure consistent labeling, after processing ∼200 data entries, each author would review and label the other author's clean data. Hence, every entry in the dataset has been reviewed twice. Metadata was added to each entry for analysis. The MTDE dataset, along with the code for the models studied in this paper, is available here[1].

3.1   Example Entries

Given that the original manuscript collection was limited to scientific and mathematical research disciplines, the grammatical structures of variable definitions within the context can be organized into a few main categories outlined below:

Simple Definitions. In the most common case, the definition of a mathematical object is the nearest noun phrase to its variable name, as in the following examples:

Here **p** is the electron momentum ...

... direction of the magnetic field **B** ...

Compound Definitions with the Same Definition. In some mathematical texts, variables with the same definition will be defined together. Similar to the previous examples, the nearest noun phrase is typically the correct definition.

We assume that the fluctuations **p**, ξ, **w** are small ...

---

[1] https://github.com/emhamel/Mathematical-Text-Understanding.

Compound Definitions with Different Definitions. Some mathematical objects of interest with different definitions will be defined in close proximity to one another. The variables in these instances are almost always defined in the order in which they are presented in the text.

... where **a** and **b** are about the variance prefactors for free and congested traffic, respectively.

Irregular Definitions. Of course, not all variables are defined as in the three cases above. Sometimes, authors declare variables in nuanced ways for brevity or diversity. The dataset presented in this paper also contains instances of irregular definition constructions, as in the following examples:

... where λn (λ) are the maximum (minimum) eigenvalues of the ...

In the limit of pointlike crosslinks the number N of crosslinks per rod approaches ...

3.2 Dataset Statistics

Figure 1 illustrates the various attributes of the data. The left-most graph reveals that the length of the context text roughly follows a normal distribution, where most contexts are between 5 and 10 words long. The middle-left graph depicts the distribution of long definition length, where most definitions are 2–3 words long. Among the multi-word definitions, only 2,866 are single-word definitions, while the other 7,392 definitions contain multiple words. The next sub-figure shows the distribution of the number of variables defined in the context text and that most contexts only define one mathematical object. The rightmost sub-figure shows an exponential decay in variable to definition distance. It is shown that most definitions can be found within a 3-word radius of the variable name. The sum of these analyses shows that while the token definition is often the nearest noun phrase, compounding factors such as i) multiple variables in context, ii) variable-length definitions, and iii) distant token definitions make the MTDE task difficult.

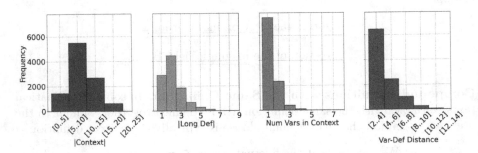

Fig. 1. Statistics of the dataset studied in this work. From left to right, the distributions of context length, the length of the long definition, the number of variables defined in the text, and distance between mathematical object and its definition, are depicted.

4 Models

With the MTDE dataset, we aim to determine how effective modern NLP algorithms are at extracting mathematical descriptors. We chose to test five of the most popular NLP models commonly used for text-summarization and question-answering tasks.[2]:

Vanilla Sequence-to-Sequence (Seq2Seq). The Seq2Seq model generates an output sequence $y = \{y_1, y_2, ..., y_n\}$ given an input sequence $x = \{x_1, x_2, ..., x_n\}$ via the use of an encoder recurrent neural network (RNN) and a decoder RNN. The encoder RNN processes each x_i and computes a hidden state until the sequence is done [19]. Using the final encoder hidden state, the decoder computes a hidden state that is converted to an output token y_i using a probability distribution over all known words in the model's vocabulary. Since there is only one encoder in this model, the variable name and a context text are concatenated together and inputted into the model as one sequence. This method is used for the Transformer Seq2Seq model, Pointer Network, and BERT model. Then the Seq2Seq model will output the sequence of word tokens that represent the definition.

Transformer Seq2Seq Model. Attention-based methods add another layer of information retention to Seq2Seq models via a context vector. This vector maps pertinent information from the input sequence to the output sequence, allowing the model to "focus" on the useful parts of the input. For our model, we used Bahdanau Attention [3]. Similar to the other Seq2Seq model, this model outputs a sequence of word tokens which represent the most likely definition.

Pointer Network. Pointer Networks also use the encoder-decoder architecture. However, instead of finding the conditional probability over the entire vocabulary, Pointer Networks find the probability over the tokens in the input sequence [20] and return a sequence of the most likely positional tokens of the definition. This framework is ideal for definition extraction because the set of possible answers is drastically reduced while still containing the correct definition.

Match Long-Short-Term-Memory (Match-LSTM). Originally proposed for question-answering tasks, the Match-LSTM architecture combines Bahdanau-like attention with Pointer Networks to predict the most likely definition of a variable. The variable name was used in place of the question text and the context was used in place of the answer text. The model studied in this work is a slight modification of the original Match-LSTM model as the mathematical object names in the data are always one-word [21]. First, the hidden

---

[2] Details about each of these models can be found in the Appendix. PyTorch code for each of these models is available here: https://github.com/emhamel/Mathematical-Text-Understanding.

states for the context text and the variable name are calculated separately. Then, the attention weights are calculated for every context-variable hidden state pair, which is then fed into a bi-directional LSTM (referred to as a Match-LSTM). Finally, a Pointer Network layer is deployed to estimate the most likely sequence of positional tokens of the definition in the context from the outputs of the Match-LSTM.

BERT. The BERT technique is implemented in two steps: pre-training and fine-tuning [5]. In the fine-tuning step, a baseline transformer comprised of multiple bi-directional encoders and attention heads is trained on a large corpus of unlabeled text data. The model is then adapted for a specific task in the fine-tuning step. In this step, an output layer specific to a certain NLP task is added to the model and trained on a specialized labeled dataset. For example, in the case of Bert for Question Answering, the BERT model used in this paper, the final output layer is a feed-forward network which indicates the most likely start and end indices of the answer in the input text.

5   Results

The MTDE task can be broken into two sub-tasks. First is the single-word descriptor task which aims to return the most significant single-word description of the mathematical token. However, as mathematical tokens are often described by multiple words ("Planck's constant", "relative dielectric permittivity", etc.) we extend our analysis to extracting multi-word descriptors as well. This section investigates the performance of modern NLP methods on each of these tasks and measures the context attributes which impact model performance.

5.1   Single-Word Descriptor Extraction

The five different model types were first trained to predict the most likely single token definition of a mathematical object. These models either find the word token of the definition (Vanilla Seq2Seq, Transformer Seq2Seq) or the positional index of the definition (Pointer Network, Match LSTM, BERT). Predicting the position of a definition is a much easier task simply because the answer space is much smaller. This is reflected in the results shown in Table 1. All position-based models greatly outperform the word-token-based models in a smaller amount of epochs.

Examining further, the Transformer Seq2Seq model is more accurate than the Vanilla Seq2Seq model. This indicates that, for word-token-based models, attention is needed for optimal accuracy. For the position-based models, accuracy is positively correlated with the complexity of the attention mechanism. The Pointer Network has the simplest of the position-based architectures and the worst accuracy, while the BERT model has the most complex architecture and the best accuracy.

Table 1. Performance metrics for each of the models extracting the **short (single-word) definitions** of mathematical tokens from contexts. Hyper-parameters were manually adjusted to find the optimal performance of each model on the dataset. Note that the adam optimizer was used for every model except for BERT which used the AdamW optimizer [14].

Model Type	Learning Rate	Hidden Size	Batch Size	epochs	Accuracy	F1 Score
Vanilla Seq2Seq	0.0005	1024	128	200	0.708	0.708
Transformer Seq2Seq	0.001	1024	512	100	0.749	0.749
Pointer Network	0.001	516	32	20	0.759	0.878
Match LSTM	0.001	516	128	50	0.809	0.904
BERT	0.00001	-	8	5	0.899	0.893

Table 2. Performance metrics for each of the models extracting the **long (multi-word, variable length) definitions** of mathematical tokens from context. Optimizer details consistent with Table 1.

Model Type	Learning Rate	Hidden Size	Batch Size	epochs	Accuracy	F1 Score
Vanilla Seq2Seq	0.001	1024	256	100	0.298	0.581
Transformer Seq2Seq	0.0005	1024	256	250	0.426	0.646
Pointer Network	0.0005	1024	128	100	0.587	0.362
Match LSTM	0.0005	1024	256	50	0.544	0.359
BERT	0.00005	-	32	5	0.852	0.853

5.2 Multi-word Definition Extraction

Table 2 shows how the models perform when trained to find the most likely multi-word definition for a mathematical object. As compared to Table 1, the accuracies of all the models (besides the BERT model) are significantly worse and generally need larger hidden sizes, batch sizes, and epochs to achieve optimal performance. This is to be expected, as multi-word definition extraction is a much harder problem to learn as long definitions can contain adjectives and prepositional phrases. Additionally, the words of single-word definitions tend to be unique to definitions, meaning that these words typically appear in a context if they are the definition. It follows that the task of single-word extraction can be

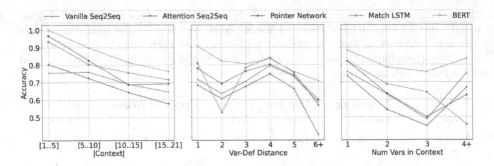

Fig. 2. Effect of different data attributes on model performance on the single-word definition extraction task. From left the right, how the accuracy of all the models studied in this paper is impacted by context length, variable-to-definition distance, and number of variables defined in the context is shown.

learned with grammar, semantics, and probability. This, however, is no longer the case in multi-word definition extraction, which can only be generalized by learning grammar and semantics.

Another noteworthy comparison between Table 2 and Table 1 reflects a shared similar accuracy hierarchy. Table 2 shows that the added attention nearly doubles the accuracy of the Seq2Seq model, which is similar to Table 1. Furthermore, just as in Table 1, Table 2 shows that the position-based models perform better than the word-token-based models. Among the position-based models, the Pointer and the Match LSTM models perform similarly; the BERT model has the best performance out of all the models by a margin of 26%.

In both tasks, the BERT model boasted the best accuracy. This is likely due to a few factors. Firstly, the BERT model is ∼10 times larger than the second largest model studied in these experiments. Secondly, the BERT model find the most probable start and end word positions and not the most probable word position sequence like the other two positional-based models, which is an easier problem to estimate. Finally, the BERT model has the advantage of being pre-trained on a large corpus of text data, which endows it with a comprehensive understanding of English grammar and semantics.

5.3   Data Attributes Impacting Model Accuracy

Several input characteristics affect model performance. In the left sub-figure of Fig. 2, the data in the test dataset was grouped by the length of the context. As the context length increases, the accuracy of the models in general decreases, but some models are more resilient to large contexts than others. In particular, the models which output textual descriptor *positions* perform significantly better than the Seq2Seq-based approaches that attempted to generate the descriptors words.

Next, it is important to analyze the impact of the variable-to-definition distance on model accuracy. The middle sub-figure of Fig. 2 shows that a significant distance between the object and its definition (6+) causes a significant drop in model performance. This sharp performance dip can be attributed to information retention and grammatical conventions. Additionally, it is likely that longer distances appear in longer contexts, which according to Fig. 2 (left) can harm model accuracy.

Finally, the relationship between the number of variables defined in a context and model performance is analyzed in the right sub-figure of Fig. 2. In general, the number of variables defined in the text does not affect model performance. Interestingly, in all these analyses, the models which output positional tokens instead of word tokens perform significantly better than those with word token outputs. This is especially true of the BERT model which performs much better than the other models in both the short- and long-definition tasks while being the least impacted by long-contexts, variable-to-descriptors distances, and multiple variables within the contexts.

6   Conclusion

This work investigated the mathematical token definition extraction task using modern NLP models. Using a custom-created corpus, it was found that models which use many attention mechanisms, as well as large language models such as BERT, perform the best on definition extraction tasks. Further work is needed to fully understand the intricacies of mathematical language processing and gain better accuracy on the dataset presented in this paper. Such improvements could include standard modifications such as adding layers and more complex embeddings. Architectural alterations in the models presented above could also improve the accuracy. Further, models which better account for implicit and domain-specific definitions need to be developed. In the MTDE dataset, all definitions of mathematical objects are present in the contexts. However, some mathematical objects can be defined using the context in which it appears without a definition present. Consider the following example: "The possible values of θ are constrained to be between ...". Here, θ is not defined, but one can reasonably define θ as an angle. This manuscript is the first step in the very complex task of processing scientific and mathematical manuscripts in general.

Appendix

1   Models

Below is a more detailed overview of the models used in this manuscript. Long Short-Term Memory (LSTM) RNNs were used for all of the models (with the exception of BERT).

1.1   Vanilla Sequence-to-Sequence Model

To find the best output sequence, generative models use the probability chain rule, which states that the probability of a sequence is the conditional probability of its tokens. We represent this in the following relation, where T is the training data and θ is the parameters of the RNN:

$$p_\theta(y|T;\theta) = \prod^i p_\theta(y_i|y_{i-1}, y_{i-2}, \ldots, y_1, T; \theta)$$

Seq2Seq-based models only take one sequence as its input. Therefore, a separation token (seen as $< SEP >$ in Fig. 3) was used to concatenate the variable name and it's context. This was done for the Vanilla Seq2Seq model, Transformer Seq2Seq model, Pointer network, and BERT model (Fig. 4).

1.2   Transformer Seq2Seq Model

To create a context vector, an annotation for each input word is generated based on the word itself and its surrounding words. After this, the context vector for the i-th output word is calculated as the weighted sum of the annotations h:

$$c_i = \sum_{j=1}^{k} \alpha_{ij} h_j$$

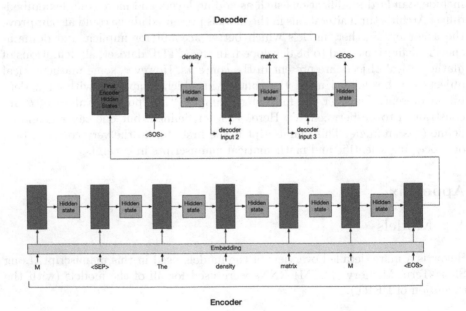

Fig. 3. The Vanilla Seq2Seq architecture

where i is the indicates the length of the output sequence and k is the length of the input sequence. The attention weights are calculated as such:

$$\alpha_{ij} = \frac{\exp\left(a(s_{i-1}, h_j)\right)}{\sum_{k=1}^{T_x} \exp\left(a(s_{i-1}, h_j)\right)}$$

where $a(s_{i-1}, h_j)$ is the alignment function which ranks how likely an input word is related to an output word. The alignment function takes the decoder hidden state of the i-th output word and the annotation of the j-th input word and learns the weights for each input annotation-output word pair via a feed-forward layer.

1.3 Pointer Network

To find the most likely answer out of the input tokens, the Pointer Network computes attention weights of variable length as such:

$$\alpha_j^i = \text{softmax}\left(u_j^i\right)$$

$$u_j^i = v^\top \tanh\left(W_1 e_j + W_2 d_i\right)$$

where e_j is the encoder state at the j-th input word, d_i is the decoder state at the i-th input word, and v, W_1, W_2 are learnable parameters (Fig. 5).

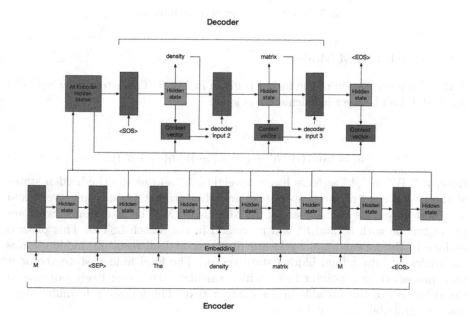

Fig. 4. The Transformer Seq2Seq architecture

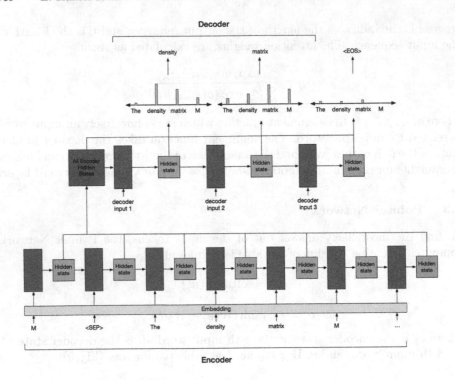

Fig. 5. The Pointer Network architecture

1.4   Match-LSTM Model

A visual representation of the model is given in Fig. 6. The attention weights in the Match-LSTM layer is calculated as such:

$$\alpha_i = w^\top u_i + b$$
$$u_i = \tanh\left(W^v h^v + (W^c h_i^c + W^r h_{i-1}^r + b^c)\right)$$

Here, W^v, W^c, W^r, b^c, w, b are learned weights, h_i^c and h^v are the hidden states of the $i-th$ word of the context text and the variable name respectively, and h_{i-1}^r is the previous hidden state of the Match-LSTM. The attention weights are then combine with h_i^c and h^v and processed in the Match-LSTM. This process is done in the forward and backward direction and all resulting hidden states are coalesced into a final hidden state vector. The final hidden state vector is then processed by a pointer layer, which calculates the most likely position of the definition of the variable in the context text. The following formulas show how the probability β is calculated:

$$\beta_j = \mathrm{softmax}\left(v^\top s_j + b \otimes e_{C+1}\right)$$
$$s_j = \tanh\left(V\bar{H}^r + (W^a h_{k-1}^a + b^a) \otimes e_{C+1}\right)$$

where V, W^a, b^a, v and b are learned weights, $\bar{H}^r$ is the vector of concatenated Match-LSTM hidden states, h^a_{k-1} is the previous hidden state of the pointer LSTM and e_{C+1} is a vector of ones with size C, where C is the length of the context text. β_j is combined with h^c_j and processed by an LSTM.

1.5   Bert-Based Model

In these methods, a pre-trained Hugging Face BertForQuestionAnswering model was fine-tuned to perform the definition-extraction task. The tokenizer used to process the text was a custom Hugging Face tokenizer created by [18]. This tokenizer was trained on a mathematical language corpus comprised of a wide range of education materials used in primary-school, high-school, and college-level courses (Fig. 7).

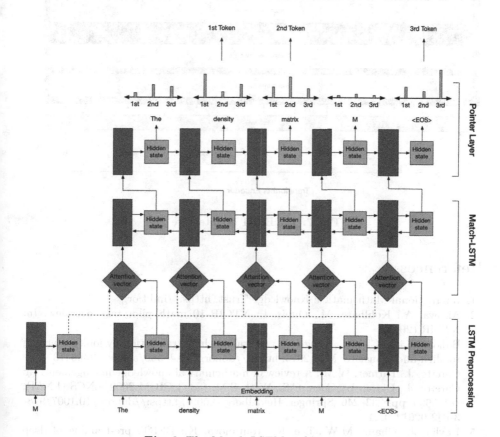

Fig. 6. The Match LSTM architecture

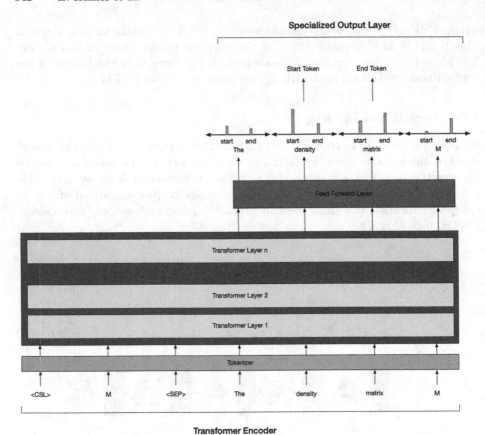

Fig. 7. The BERT architecture

References

1. International Mathematical Knowledge Trust. https://imkt.org/
2. Aizawa, A., Kohlhase, M., Ounis, I.: NTCIR-10 math pilot task overview. In: NTCIR (2013)
3. Bahdanau, D., Cho, K., Bengio, Y.: Neural machine translation by jointly learning to align and translate, May 2016. https://doi.org/10.48550/arXiv.1409.0473
4. Carette, J., Farmer, W.M.: A review of mathematical knowledge management. In: Carette, J., Dixon, L., Coen, C.S., Watt, S.M. (eds.) CICM 2009. LNCS (LNAI), vol. 5625, pp. 233–246. Springer, Heidelberg (2009). https://doi.org/10.1007/978-3-642-02614-0_21
5. Devlin, J., Chang, M.W., Lee, K., Toutanova, K.: BERT: pre-training of deep bidirectional transformers for language understanding. arXiv:1810.04805 [cs], May 2019
6. Gao, L., Jiang, Z., Yin, Y., Yuan, K., Yan, Z., Tang, Z.: Preliminary exploration of formula embedding for mathematical information retrieval: can mathematical formulae be embedded like a natural language? arXiv:1707.05154 [cs], August 2017

7. Ginev, D.: arXMLiv 2020 - an HTML5 dataset for arXiv.org · SIGMathLing (2020)
8. Greiner-Petter, A., et al.: Discovering mathematical objects of interest—a study of mathematical notations. In: Proceedings of The Web Conference 2020, WWW 2020, pp. 1445–1456. Association for Computing Machinery, Taipei, April 2020. https://doi.org/10.1145/3366423.3380218
9. Hirschman, L., Gaizauskas, R.: Natural language question answering: the view from here. Nat. Lang. Eng. **7**(4), 275–300 (2001). https://doi.org/10.1017/S1351324901002807
10. Kohlhase, M. (ed.): MKM 2005. LNCS (LNAI), vol. 3863. Springer, Heidelberg (2006). https://doi.org/10.1007/11618027
11. Kristianto, G.Y., Topić, G., Aizawa, A.: Extracting textual descriptions of mathematical expressions in scientific papers. D-Lib Mag. **20**(11/12) (2014). https://doi.org/10.1045/november14-kristianto
12. Kristianto, G.Y., Topić, G., Aizawa, A.: Utilizing dependency relationships between math expressions in math IR. Inf. Retrieval J. **20**(2), 132–167 (2017). https://doi.org/10.1007/s10791-017-9296-8
13. Kristianto, G.Y., Topic, G., Ho, F.: The MCAT math retrieval system for NTCIR-11 math track, p. 7 (2014)
14. Loshchilov, I., Hutter, F.: Decoupled weight decay regularization, January 2019. https://doi.org/10.48550/arXiv.1711.05101
15. Munot, N., Govilkar, S.: Comparative study of text summarization methods. Int. J. Comput. Appl. **102**, 33–37 (2014). https://doi.org/10.5120/17870-8810
16. Pagael, R., Schubotz, M.: Mathematical language processing project. arXiv:1407.0167 [cs], July 2014
17. Schubotz, M., et al.: Semantification of identifiers in mathematics for better math information retrieval. In: Proceedings of the 39th International ACM SIGIR Conference, SIGIR 2016, pp. 135–144. Association for Computing Machinery, New York, July 2016. https://doi.org/10.1145/2911451.2911503
18. Shen, L., et al.: Backdoor pre-trained models can transfer to all. In: Proceedings of the 2021 ACM SIGSAC Conference on Computer and Communications Security, pp. 3141–3158, November 2021. https://doi.org/10.1145/3460120.3485370
19. Sutskever, I., Vinyals, O., Le, Q.V.: Sequence to sequence learning with neural networks. In: Advances in Neural Information Processing Systems, vol. 27. Curran Associates, Inc. (2014)
20. Vinyals, O., Fortunato, M., Jaitly, N.: Pointer networks. arXiv:1506.03134 [cs, stat], January 2017
21. Wang, S., Jiang, J.: Machine comprehension using match-LSTM and answer pointer, November 2016
22. Youssef, A., Miller, B.R.: Explorations into the use of word embedding in math search and math semantics. In: Kaliszyk, C., Brady, E., Kohlhase, A., Sacerdoti Coen, C. (eds.) CICM 2019. LNCS (LNAI), vol. 11617, pp. 291–305. Springer, Cham (2019). https://doi.org/10.1007/978-3-030-23250-4_20

CICM'22 System Entries

Peter Koepke[1], Anton Lorenzen[1(✉)], and Boris Shminke[2]

[1] University of Bonn, Bonn, Germany
anfelor@posteo.de
[2] Laboratoire J.A. Dieudonné, CNRS and Université Côte d'Azur, Nice, France
boris.shminke@cnrs.fr

Abstract. This consolidated paper gives an overview of new tools and improvements of existing tools in the CICM domain that occurred since the last CICM conference.

Web-Naproche

Anton Lorenzen, Peter Koepke, University of Bonn, Germany

Tool:	Web-Naproche
Version:	1.0
Impl. in:	Haskell & JavaScript
License:	GPL (Naproche) / MIT (web)
Download:	
https://naproche.github.io	

Description. Naproche [3] is a proof assistant for mathematical texts written in a controlled natural language. The input is translated to first order logic and checked using standard reasoning procedures and automated theorem provers. Naproche ships on the Isabelle platform [7] which is recommended for the development of larger and interdependent formalizations.

To lower the entry barrier for a general mathematical audience we have assembled a simple Naproche web-interface that can be opened instantly in all standard web browsers and uses the same checking procedures as the desktop distribution. The user interface consists of an editor window and a feedback panel. The user may write controlled natural language input in ASCII or LATEX into the left editor panel. By clicking the respective button, this input is then either translated into first order logic or proof-checked. The output is shown on the feedback panel to the right.

While our Web-Naproche uses the same reasoning methods as Naproche, it tends to be slower due to the limitations of the web browsers. Nonetheless, many interesting formalizations can be checked and are included as examples, including the maximum principle (of complex analysis), Newman's lemma (on the confluence of rewriting systems), or the Knaster-Tarski fixpoint theorem [2].

Implementation. To build the web-interface we compile Naproche (which is written in Haskell) to JavaScript using the GHCJS compiler [13] and use the

K. Buzzard and T. Kutsia (Eds.): CICM 2022, LNAI 13467, pp. 344–348, 2022.
https://doi.org/10.1007/978-3-031-16681-5_24

generated code as a Web Worker. As workers can run asynchronously, the interface remains responsive while Naproche is active and one can even run several Naproche instances at the same time.

Naproche runs in server-mode and sends all system calls (printing to the console, reading source files and calling a prover) as JSON [4] to the web-interface. The output is displayed in the feedback panel and source files are read from Github. The user can choose via a dropdown menu how calls to the automatic theorem provers are handled: either one can make requests to System on TPTP [14] (which supports all major provers) or requests can be handled in-browser. For the latter, we have compiled SPASS [15] to webassembly. On small examples, SPASS-within-the-browser can be more efficient as it avoids the overhead and delays of creating a connection to System on TPTP.

Considerations. We decided against running Naproche on a server as this would require securing prototypical software against cyber attacks. In our investigations we also compiled Naproche to webassembly using the new Asterius compiler [11], but the generated code turned out to be much slower. Similarly, running eprover [10], which is Naproche's standard ATP, in the browser via webassembly was not efficient.

Applications. The web-interface makes Naproche easily accessible, without software installation and without substantial system requirements. It is a convenient alternative for using Naproche in lectures or conference presentations. The interface works even on smartphones. We are currently designing a tutorial for the web-interface which will teach how to write proofs in Naproche.

Python client for Isabelle server

Boris Shminke, Laboratoire J.A. Dieudonné, CNRS and Université Côte d'Azur, France

Tool:	Python client for Isabelle server
Version:	0.3.5
Impl. in:	Python
License:	Apache-2.0
Download:	
https://pypi.org/project/isabelle-client	

Description. Python client to Isabelle server [16] gives researchers and students using Python as their primary programming language an opportunity to communicate with the Isabelle server through TCP directly from a Python script.

Such an approach helps avoid the complexities of integrating the existing Python script with languages used for Isabelle development (ML and Scala). `isabelle-client` relies on `asyncio`, a Python package providing a high-level interface for TCP communications.

The client supports relaying all the Isabelle server commands described in its manual (see Chap. 4 of [17]) and parses the responses back to Python objects.

The client's distributive package also provides a utility function for starting an instance of the Isabelle server in the background directly from a Python script.

The digital artefact of the current version is available on Zenodo [12]. Interested potential users of the client can also follow an interactive example[1] and the package documentation[2] for more detailed information.

Applications

- A discovery of a proof [5] for an algebraic problem which stood open for two years despite the efforts of specialists in the field.
- The `isabelle-client` running in a Docker container on Binder was used during the practical sessions of the Advanced Logic course taught at the Université Côte d'Azur in the autumn of the 2021–2022 academic year. This use case helped the author to realise the need for ease of the installation procedure on different operating systems and programming environments.
- Also, Fabian Huch used the `isabelle-client` for debugging the "Proving for Fun" backend [6]. Thanks to this use case, the importance of supporting different (and even outdated) versions of Python became known.

Changes from Previous Version. The first public version (0.2.0) of this client [9] worked only for Python 3.7 on GNU/Linux and was supposed to be installed only with the `pip` package manager.

- The current version is available for any Python 3.6+ on GNU/Linux and Windows. Every new build is tested in a continuous integration workflow against each supported Python version.
- The package is hosted now not only on the Python Package Index (PyPI), but also on Conda Forge [1], which enables its installation with both `pip` and `conda` package managers.
- In addition, one can run the client inside a Docker container, for example, in a cloud using Binder [8].
- The current version of the client is tested to work with the latest Isabelle 2021-1 (released in December 2021).
- The last client version returns all server replies as a Python list, not only the last one as it was in the previous version giving more flexibility to the end-user.
- Last but not least, the current version arrives with detailed documentation pages and is nearly 100% covered with unit tests using fixtures for emulating a working Isabelle server behaviour.

---

[1] https://bit.ly/isabelle-client.
[2] https://isabelle-client.rtfd.io.

Acknowledgements. This work has been supported by the French government, through the 3IA Côte d'Azur Investments in the Future project managed by the National Research Agency (ANR) with the reference number ANR-19-P3IA-0002.

References

1. conda-forge Community: The conda-forge Project: Community-based Software Distribution Built on the conda Package Format and Ecosystem. Zenodo (2015). https://doi.org/10.5281/zenodo.4774216
2. De Lon, A., Koepke, P., Lorenzen, A., Marti, A., Schütz, M., Sturzenhecker, E.: Beautiful formalizations in Isabelle/Naproche. In: Kamareddine, F., Sacerdoti Coen, C. (eds.) CICM 2021. LNCS (LNAI), vol. 12833, pp. 19–31. Springer, Cham (2021). https://doi.org/10.1007/978-3-030-81097-9_2
3. De Lon, A., Koepke, P., Lorenzen, A., Marti, A., Schütz, M., Wenzel, M.: The Isabelle/Naproche natural language proof assistant. In: Platzer, A., Sutcliffe, G. (eds.) CADE 2021. LNCS (LNAI), vol. 12699, pp. 614–624. Springer, Cham (2021). https://doi.org/10.1007/978-3-030-79876-5_36
4. Ecma International: Standard ECMA-404: The JSON Data Interchange Syntax (2017). https://www.ecma-international.org/publications-and-standards/standards/ecma-404/
5. Fussner, W., Shminke, B.: Mining counterexamples for wide-signature algebras with an Isabelle server. arXiv:2109.05264 [cs.LO] (2021). https://doi.org/10.48550/ARXIV.2109.05264
6. Haslbeck, M.P.L., Wimmer, S.: Competitive proving for fun. In: Benzmüller, C., Parent, X., Steen, A. (eds.) Selected Student Contributions and Workshop Papers of LuxLogAI 2018. Kalpa Publications in Computing, vol. 10, pp. 9–14. EasyChair (2018). https://doi.org/10.29007/ktx8
7. Isabelle Contributors: The Isabelle 2021–1 release (2021). https://isabelle.in.tum.de/website-Isabelle2021/index.html
8. Jupyter, P., et al.: Binder 2.0 - reproducible, interactive, sharable environments for science at scale. In: Akici, F., Lippa, D., Niederhut, D., Pacer, M. (eds.) Proceedings of the 17th Python in Science Conference, pp. 113–120 (2018). https://doi.org/10.25080/Majora-4af1f417-011
9. Líška, Martin: CICM'21 systems entries. In: Kamareddine, Fairouz, Sacerdoti Coen, Claudio (eds.) CICM 2021. LNCS (LNAI), vol. 12833, pp. 245–248. Springer, Cham (2021). https://doi.org/10.1007/978-3-030-81097-9_20
10. Schulz, S., Cruanes, S., Vukmirović, P.: Faster, higher, stronger: E 2.3. In: Fontaine, P. (ed.) CADE 2019. LNCS (LNAI), vol. 11716, pp. 495–507. Springer, Cham (2019). https://doi.org/10.1007/978-3-030-29436-6_29
11. Shao, C.: Asterius contributors: asterius (2022). https://github.com/tweag/asterius
12. Shminke, B.: Python client for Isabelle server (0.3.5). Zenodo (2022). https://doi.org/10.5281/zenodo.6490275
13. Stegeman, L., GHCJS contributors: GHCJS (2022). https://github.com/ghcjs/ghcjs
14. Sutcliffe, G.: System Description: SystemOnTPTP. In: McAllester, D. (ed.) CADE 2000. LNCS (LNAI), vol. 1831, pp. 406–410. Springer, Heidelberg (2000). https://doi.org/10.1007/10721959_31

15. Weidenbach, C., Dimova, D., Fietzke, A., Kumar, R., Suda, M., Wischnewski, P.: SPASS version 3.5. In: Schmidt, R.A. (ed.) CADE 2009. LNCS (LNAI), vol. 5663, pp. 140–145. Springer, Heidelberg (2009). https://doi.org/10.1007/978-3-642-02959-2_10

16. Wenzel, M.: Isabelle/PIDE after 10 years of development. In: 13th International Workshop on User Interfaces for Theorem Provers (UITP 2018), Oxford, UK (2018). https://sketis.net/wp-content/uploads/2018/08/isabelle-pide-uitp2018.pdf

17. Wenzel, M.: The Isabelle system manual (2021). https://isabelle.in.tum.de/dist/Isabelle2021-1/doc/system.pdf

Author Index

Printed in the United States
by Baker & Taylor Publisher Services